Rare Earth Minerals

Mineralogical Society Series

Series editor
Dr A. P. Jones

The aim of the series is to provide up-to-date reviews through the selected but specialized contributions of leading experts. A particularly attractive feature of the series is the tutorial element, making it suitable for third year undergraduates and researchers of all levels not solely within the Earth Sciences. Each volume is purpose-designed, highly illustrated and serves as an excellent reference tool.

TITLES AVAILABLE

1 **Deformation Processes in Minerals, Ceramics and Rocks**
   Edited by D. J. Barber and P. G. Meredith

2 **High-temperature Metamorphism and Crustal Anatexis**
   Edited by J. R. Ashworth and M. Brown

3 **The Stability of Minerals**
   Edited by Geoffrey D. Price and Nancy L. Ross

4 **Geochemistry of Clay-Pore Fluid Interactions**
   Edited by D. A. C. Manning, P. L. Hall and C. R. Hughes

5 **Mineral Surfaces**
   Edited by D. J. Vaughan and R. A. D. Pattrick

6 **Microprobe Techniques in the Earth Sciences**
   Edited by Philip J. Potts, John F. W. Bowles, Stephen J. B. Reed and Mark R. Cave

7 **Rare Earth Minerals**
   Edited by Adrian P. Jones, Frances Wall and C. Terry Williams

Rare Earth Minerals
Chemistry, origin and ore deposits

Edited by Adrian P. Jones
Department of Geological Sciences, University College London, UK

Frances Wall
Department of Mineralogy, The Natural History Museum, London, UK

and

C. Terry Williams
Department of Mineralogy, The Natural History Museum, London, UK

CHAPMAN & HALL
London · Glasgow · Weinheim · New York · Tokyo · Melbourne · Madras

Published by Chapman & Hall, 2–6 Boundary Row, London SE1 8HN, UK

Chapman & Hall, 2–6 Boundary Row, London SE1 8HN, UK

Blackie Academic & Professional, Wester Cleddens Road, Bishopbriggs, Glasgow G64 2NZ, UK

Chapman & Hall GmbH, Pappelallee 3, 69469 Weinheim, Germany

Chapman & Hall USA, 115 Fifth Avenue, New York, NY 10003, USA

Chapman & Hall Japan, ITP-Japan, Kyowa Building, 3F, 2-2-1 Hirakawacho, Chiyoda-ku, Tokyo 102, Japan

Chapman & Hall Australia, 102 Dodds Street, South Melbourne, Victoria 3205, Australia

Chapman & Hall India, R. Seshadri, 32 Second Main Road, CIT East, Madras 600 035, India

First edition 1996

© 1996 The Mineralogical Society

Typeset in 10/12 pt Times by Best-set Typesetter Ltd., Hong Kong
Printed in Great Britain by Hartnolls Ltd, Bodmin, Cornwall

ISBN 0 412 61030 2

Apart from any fair dealing for the purposes of research or private study, or criticism or review, as permitted under the UK Copyright Designs and Patents Act, 1988, this publication may not be reproduced, stored, or transmitted, in any form or by any means, without the prior permission in writing of the publishers, or in the case of reprographic reproduction only in accordance with the terms of the licences issued by the Copyright Licensing Agency in the UK, or in accordance with the terms of licences issued by the appropriate Reproduction Rights Organization outside the UK. Enquiries concerning reproduction outside the terms stated here should be sent to the publishers at the London address printed on this page.

The publisher makes no representation, express or implied, with regard to the accuracy of the information contained in this book and cannot accept any legal responsibility or liability for any errors or omissions that may be made.

A catalogue record for this book is available from the British Library

Library of Congress Catalog Card Number: 95-71236

∞ Printed on permanent acid-free text paper, manufactured in accordance with ANSI/NISO Z39.48-1992 and ANSI/NISO Z39.48-1984 (Permanence of Paper).

Contents

Colour plates appear between pages 112–113 and pages 176–177

List of contributors	xi
Preface	xiii
Acknowledgements	xv

1	**The rare earth elements: introduction and review** P. Henderson	1
1.1	Introduction	1
1.2	Chemical behaviour	1
1.3	Crustal abundances	6
1.4	Geochemical behaviour: magmatic systems	11
1.5	Geochemical behaviour: aqueous systems	12
1.6	Concluding statement	16
1.7	Review works	16
	Acknowledgements	17
	References	17

2	**Crystal chemical aspects of rare earth minerals** R. Miyawaki and I. Nakai	21
2.1	Introduction	21
2.2	Coordination polyhedra of rare earth atoms	23
2.3	Structural differences between Y-group and Ce-group rare earth minerals	26
2.4	Isomorphous substitutions between rare earth ions and non-rare earth ions, Ca^{2+}, Na^+, Th^{4+}, etc.	32
2.5	Crystal chemistry of gadolinite–datolite group and related minerals	34
2.6	Summary and conclusion	37
	References	37

3	**Perovskites: a revised classification scheme for an important rare earth element host in alkaline rocks** R. H. Mitchell	41
3.1	Introduction	41

3.2	Perovskite structure	42
3.3	Perovskite space	47
3.4	Naturally occurring perovskites	48
3.5	Nomenclature of naturally occurring perovskites	53
3.6	Structural formulae and end-member compositions	57
3.7	Paragenesis	59
3.8	Major element compositional variation of perovskite in alkaline rocks	59
3.9	Rare earth element distribution patterns	69
3.10	Conclusions	70
	Acknowledgements	71
	References	71

4	**Rare earth elements in carbonate-rich melts from mantle to crust**	**77**
	P. J. Wyllie, A. P. Jones and J. Deng	
4.1	Introduction	77
4.2	Formation of carbonate-rich melts in the mantle	78
4.3	Rare earth elements in metasomatized mantle rocks	79
4.4	REE in carbonatites	80
4.5	Bastnäsite synthesis and stability range	81
4.6	Phase equilibrium studies in the system $CaO-La_2O_3-CO_2-H_2O$	84
4.7	Mountain Pass synthetic carbonatite magma	87
4.8	Silicate–carbonate liquid immiscibility	93
4.9	Conclusions and applications	95
	Acknowledgements	97
	References	98

5	**Formation of rare earth minerals in hydrothermal systems**	**105**
	R. Gieré	
5.1	Introduction	105
5.2	RE minerals of hydrothermal origin	106
5.3	Zoning in hydrothermal RE minerals	110
5.4	REE in hydrothermal fluids	113
5.5	Transport of REE	116
5.6	Formation of RE minerals	118
5.7	Alteration of hydrothermal RE minerals	123
5.8	Case study	127
5.9	Conclusions	140
	Acknowledgements	140
	References	141

6	**Rare earth minerals from the syenite pegmatites in the Oslo Region, Norway**	151
	A. O. Larsen	
6.1	Introduction	151
6.2	Geological setting	153
6.3	The rare earth minerals	155
6.4	Concluding remarks	163
	Acknowledgements	164
	References	164
7	**Rare earth element mineralization in peralkaline systems: the T-Zone REE–Y–Be deposit, Thor Lake, Northwest Territories, Canada**	167
	R. P. Taylor and P. J. Pollard	
7.1	Introduction	167
7.2	Geological setting of the Blatchford Lake igneous complex	168
7.3	The Thor Lake rare-metal deposits	170
7.4	The north T-Zone REE–Y–Be deposit	171
7.5	^{40}Ar/^{39}Ar geochronology	179
7.6	Fluid inclusion analysis	181
7.7	Hydrogen and oxygen isotope analysis of fluid inclusion water and host quartz	184
7.8	Carbon and oxygen isotope analysis of carbonates and sulphur isotope analysis of sulphide minerals	187
7.9	Conclusions and summary	188
	Acknowledgements	190
	References	190
8	**Rare earth minerals in carbonatites: a discussion centred on the Kangankunde Carbonatite, Malawi**	193
	F. Wall and A. N. Mariano	
8.1	Introduction	193
8.2	Rare earth elements and their mineral hosts in carbonatites	193
8.3	'Late carbonatites'	194
8.4	Extrusive carbonatites	197
8.5	Formation of rare earth minerals in carbonatites	197
8.6	Rare earth minerals in the Kangankunde Carbonatite Complex	198
8.7	Comparison with the Mountain Pass carbonatite	216
8.8	Other carbonatites with RE mineralization in polycrystalline pseudomorphs	217
8.9	Conclusions	221
	Acknowledgements	222
	References	222

9	**REE distribution and REE carriers in laterites formed on the alkaline complexes of Araxá and Catalão (Brazil)** G. Morteani and C. Preinfalk	227
9.1	Introduction	227
9.2	Laterites and the laterite profile	228
9.3	The behaviour of REE during lateritization	231
9.4	Geology	232
9.5	Geochemistry	234
9.6	REE carriers at Araxá and Catalão	240
9.7	Discussion and conclusions	249
	Acknowledgements	252
	References	252
10	**Authigenic rare earth minerals in karst-bauxites and karstic nickel deposits** Z. J. Maksimović and Gy. Pantó	257
10.1	Introduction	257
10.2	Occurrence	258
10.3	Chemical properties	259
10.4	X-ray powder diffraction study	264
10.5	Rare earth minerals	266
10.6	Genetic considerations	270
10.7	Conclusion	277
	References	278
11	**Rare earth deposits in China** C. Wu, Z. Yuan and G. Bai	281
11.1	Introduction	281
11.2	Types of Chinese rare earth deposits	282
11.3	Spatial distribution and age of formation	288
11.4	Geochemistry	288
11.5	Geological examples	293
11.6	Concluding remarks	305
	Acknowledgements	306
	References	306
12	**Yttrium and rare earth element minerals of the Kola Peninsula, Russia** A. P. Belolipetskii and A. V. Voloshin	311
12.1	Introduction	311
12.2	Late Archaean – Kola stage, 2800–3000 Ma	312
12.3	Early Proterozoic – Karelian stage, 1700–2400 Ma	315
12.4	Palaeozoic – Caledonian–Hercynian stage, 300–400 Ma	316
12.5	Y–REE mineralization in amazonitic rand-pegmatities	317

12.6	Economic REE mineralization	323
	References	324
13	**Analysis of rare earth minerals**	327
	C. T. Williams	
13.1	Introduction	327
13.2	Historical background	327
13.3	Techniques of analysis	328
13.4	Bulk analytical techniques	329
13.5	Microanalytical techniques	335
13.6	Other techniques	345
13.7	Concluding statement	345
	Acknowledgements	346
	References	346
Appendix A	**Glossary of rare earth minerals**	349
A.1	Borates	349
A.2	Carbonates, fluocarbonates and hydroxylcarbonates	349
A.3	Oxides	350
A.4	Halides	351
A.5	Silicates	352
A.6	Phosphates	354
A.7	Arsenates, sulphates and vanadates	354
A.8	Uranyl-carbonates and uranyl-silicates	355
	References	355
Index		357

Contributors

G. Bai	Institute of Mineral Deposits, Chinese Academy of Geological Sciences, Beijing, China
A. P. Belolipetskii	Geological Institute, Kola Science Centre of the Russian Academy of Sciences, 14 Fersman Street, Apatity 184200, Russia
J. Deng	Department of Geology, China University of Geosciences, Xueyuan Road 29, Beijing 100083, China
R. Gieré	Mineralogisch-Petrographisches Institut der Universität, Bernoullianum, CH-4056 Basel, Switzerland, and Geophysical Laboratory, Carnegie Institution of Washington, 5251 Broad Branch Road, NW, Washington DC 20015-1305, USA
P. Henderson	Department of Mineralogy, The Natural History Museum, London SW7 5BD, UK
A. P. Jones	Department of Geology, University College London, Gower Street, London WC1E 6BT, UK
A. O. Larsen	Norsk Hydro a.s., Research Centre Porsgrunn, N-3901, Porsgrunn, Norway
Z. J. Maksimović	Faculty of Mining and Geology, Djušina 7, 11000 Belgrade, Serbia
A. N. Mariano	48 Page Brook Road, Carlisle, MA 01741, USA
R. H. Mitchell	Department of Geology, Lakehead University, Thunder Bay, Ontario P7B 5E1, Canada
R. Miyawaki	Ceramic Technology Department, National Industrial Research Institute of Nagoya, Hirate, Nagoya 462, Japan
G. Morteani,	Lehrstuhl für Angewandte Mineralogie u. Geochemie, Technische Universität München, Lichtenbergstrasse 4, D-85747 Garching, Germany
I. Nakai	Department of Applied Chemistry, Science University of Tokyo, Kagurazaka, Shinjuki, Tokyo 162, Japan
Gy. Pantó	Laboratory for Geochemical Research, Hungarian Academy of Sciences, Budaörsi út 45, H-1112 Budapest, Hungary

LIST OF CONTRIBUTORS

P. J. Pollard	Department of Geology and Key Centre in Economic Geology, James Cook University of North Queensland, Townsville, Queensland 4811, Australia
C. Preinfalk	Lehrstuhl für Angewandte Mineralogie u. Geochemie, Technische Universität München, Lichtenbergstrasse 4, D-85747 Garching, Germany
R. P. Taylor	Department of Earth Sciences, Ottawa-Carleton Geoscience Centre, Carleton University, Ottawa, Ontario K1S 5B6, Canada
A. V. Voloshin	Geological Institute, Kola Science Centre of the Russian Academy of Sciences, 14 Fersman Street, Apatity 184200, Russia
F. Wall	Department of Mineralogy, The Natural History Museum, Cromwell Road, London SW7 5BD, UK
C. T. Williams	Department of Mineralogy, The Natural History Museum, Cromwell Road, London SW7 5BD, UK
C. Wu	Institute of Mineral Deposits, Chinese Academy of Geological Sciences, Beijing, China
P. J. Wyllie	Division of Geological and Planetary Sciences, California Institute of Technology, Pasadena, CA 91125, USA
Z. Yuan	Institute of Mineral Deposits, Chinese Academy of Geological Sciences, Beijing, China

Preface

This book brings together selected papers and some additional contributions based on the conference *Rare Earth Minerals: chemistry, origin and ore deposits* held at the Natural History Museum in April 1993. Interest in the mineralogy and geochemistry of rare earth minerals was well reflected in the 110 conference participants from 26 countries. In compiling this book we have tried to include a tutorial component, to make it accessible and useful to geoscientists who are new to, or have only occasional recourse to, the world of REE. To this we have added a mixture of contributions, some of a review nature, to illustrate the variety of current work on RE minerals. Chapter 1 presents an authoritative review of the REE. The second chapter is a review of the crystal chemical aspects of RE minerals, and a summary of their structure-types. Then follows a review and classification for perovskite which, while not always an RE mineral *sensu stricto*, is a significant host for REE in a number of rock types, especially in alkaline rocks. Chapter 4 shows how an experimental approach has been used to explain the concentration of REE from sources in the Earth's mantle to extraordinary levels in an important crustal carbonatite deposit. Chapter 5 explores the sequential RE minerals developed over lower temperatures from Alpine type metasomatic fluids. Two chapters focus on RE minerals associated with alkaline igneous silicate systems; Chapter 6 is a review of the classic syenite pegmatite localities at Langesundfjord, Norway and Chapter 7 is an in-depth study of multistage fluid-rich processes in RE mineralization in a Canadian intrusion. The importance of carbonatites as hosts for high levels of REE in the crust is also reflected in the detailed example of Kangankunde in Chapter 8. Two chapters present detailed examples of relatively low-temperature RE-mineralization; Chapter 9 examines the RE mineral distribution in economic laterites formed on alkaline complexes in Brazil and Chapter 10 presents unusual RE mineralization associated with karst and bauxite development. Two chapters are devoted to regional reviews with new information in English for the first time: Chapter 11 is of China, which probably contains the world's most important economic REE deposits, and Chapter 12 is of the classic Kola peninsula, which includes the first English publication of an extraordinary rare heavy RE mineral locality. A final chapter details important practical techniques and outlines various procedures in the analysis of RE minerals. We hope that the book will highlight not only what is known and understood about RE minerals, but the large areas that remain to be studied.

We have not included a classical review of RE mineralogy because we believe that much mineral identification will now be achieved by computer methods such as XRD search match and 'MinIdent' but we have included a Glossary with complete list of RE minerals and their formulae to serve as ready reference, and hope that the index will lead the reader to useful examples of interest. Conventions followed here are: formulae as in Clark (1993) *Hey's Mineral Index*. RE minerals are defined and named according to Bayliss and Levinson (1988) as minerals in which the total atomic percentage of the REE and Y are greater than any other element within a single set of crystal-structural sites.

Adrian Jones, Frances Wall and Terry Williams
London, 1995

Acknowledgements

The editors are grateful to Alan Woolley and David Highley, who were members of the organizing committee for the 1993 Rare Earth Minerals conference, and to The Royal Society, the Applied Mineralogy Group of the Mineralogical Society of Great Britain and Ireland, Gesellschaft für Elektrometallurgie mbH, and RTZ Mining and Exploration Limited who provided sponsorship for the conference.

They also thank those who kindly agreed to review chapters. Some reviewers are acknowledged at the end of individual chapters and, in addition, Andrew Clark, Wayne Taylor, Paul Schofield and other anonymous reviewers are thanked for their helpful comments.

CHAPTER ONE
The rare earth elements: introduction and review

P. Henderson

1.1 Introduction

The rare earth elements comprise the following Group IIIA elements of the Periodic Table: scandium (atomic number 21), yttrium (39), lanthanum (57) and the lanthanides, which are the 14 elements from cerium (58) to lutetium (71). In geochemistry and mineralogy usage of the term 'rare earth element' (abbreviated REE) is often restricted to Y, La and the lanthanides. This is adopted in this book. A list of the names and chemical symbols is given in Table 1.1.

The geochemical behaviour of the rare earth elements has been the subject of extensive study over several decades not only because of its intrinsic interest but because it has proved useful in tackling a variety of petrological and mineralogical problems. Their partitioning in igneous systems is particularly suitable for petrogenetic studies; their complexing in aqueous media can be used to gain insights into the proportions of the anionic components, and their redox behaviour can help to delimit the oxygen activity in a system. Their relative immobility during certain types of rock alteration has been used to help establish the provenance or nature of the primary rock. Furthermore, the chemical and physical nature of some of their compounds also give many REE an economic significance.

The aim of this chapter is both to describe the relevant chemical properties and to give a brief overview of how these are of use in tackling mineralogically related problems. There is a substantial body of literature on the mineralogy and geochemistry of these elements, so any review has to be selective, more especially as the field is still developing quite rapidly.

1.2 Chemical behaviour

The REE are a coherent group of elements in that the different elements all show very similar chemical behaviour. The differences are, however, sufficient to be of interest and use especially as, for the most part, they are a smooth and systematic function of atomic number.

Rare Earth Minerals: Chemistry, origin and ore deposits. Edited by Adrian P. Jones, Frances Wall and C. Terry Williams. Published in 1996 by Chapman & Hall. ISBN 0 412 61030 2

Table 1.1 Names and symbols of the rare earth elements

Atomic number	Symbol	Name
39	Y	Yttrium
57	La	Lanthanum
58	Ce	Cerium
59	Pr	Praseodymium
60	Nd	Neodymium
61	(Pm)	(Promethium)[a]
62	Sm	Samarium
63	Eu	Europium
64	Gd	Gadolinium
65	Tb	Terbium
66	Dy	Dysprosium
67	Ho	Holmium
68	Er	Erbium
69	Tm	Thulium
70	Yb	Ytterbium
71	Lu	Lutetium

[a] Promethium has no long-lived nucleii; it has, therefore, no natural abundance.

The rare earth elements are very electropositive so their compounds are generally ionic. From a mineralogical standpoint these compounds are oxides, halides, carbonates, phosphates and silicates, with a few additions such as a borate, arsenate, etc., but not sulphides. Their ionic radii are relatively large and so substitution reactions usually involve the large cations such as calcium or strontium even though some additional charge balancing is often necessary. Their commonest oxidation state is three, with europium existing also in the 2+ state and cerium also in the 4+ state. The proportion of the different oxidation states of europium or of cerium in any system will depend on the temperature, pressure, composition and redox conditions. The systematics of these effects have not yet been well established, which is unfortunate for the mineralogist because of their potential use in studies of mineral and rock genesis.

The arrangement of the electrons around the nucleii of the different REE is a determining factor of the properties of these elements. The electronic configurations (Table 1.2) of the rare earth elements La to Lu involve the regular filling of the inner $4f$ shell (while the outer $5d$ shell remains empty) in proceeding from cerium to ytterbium, but with the exception of gadolinium which has an electron in the $5d$ shell. Lanthanum and lutetium also have one electron in the $5d$ shell. In their oxidized state the elements have no electrons in the outer $5d$ shell, any changes in the number of electrons being reflected in the inner $4f$ level. It is the fact that the change in the electronic configurations of the different rare earths is mostly confined to the inner shells

Table 1.2 Electronic configurations of the rare earth elements

Name	Configuration
Y	[Kr] $4d^15s^2$
La	[Xe] $5d^16s^2$
Ce	[Xe] $4f^26s^2$
Pr	[Xe] $4f^36s^2$
Nd	[Xe] $4f^46s^2$
(Pm)	[Xe] $4f^56s^2$
Sm	[Xe] $4f^66s^2$
Eu	[Xe] $4f^76s^2$
Gd	[Xe] $4f^75d^16s^2$
Tb	[Xe] $4f^96s^2$
Dy	[Xe] $4f^{10}6s^2$
Ho	[Xe] $4f^{11}6s^2$
Er	[Xe] $4f^{12}6s^2$
Tm	[Xe] $4f^{13}6s^2$
Yb	[Xe] $4f^{14}6s^2$
Lu	[Xe] $4f^{14}5d^16s^2$

Configurations of krypton [Kr] and xenon [Xe]:
Kr $1s^22s^22p^63s^23p^63d^{10}4s^24p^6$
Xe $1s^22s^22p^63s^23p^63d^{10}4s^24p^64d^{10}5s^25p^6$

rather than the outer ones that gives these elements their coherence in chemical behaviour. Thus the variation in the ionic radius of the rare earth ions shows a smooth progression with atomic number, for any given oxidation state (Figure 1.1). Yttrium as the 3+ ion has the same electronic configuration as the noble gas krypton, which gives it particular stability, and an ionic radius close in value to that of Ho^{3+}. The radius of any ion is also a function of the size, and hence the coordination number, of the site occupied by that ion in a mineral. In general the greater the coordination number, the greater the ionic radius for a given occupying ion. Ionic radii for the common oxidation states, as compiled by Shannon (1976), are given in Table 1.3, for a variety of coordination numbers. Radius is also a function of ionic charge (Table 1.3).

Consideration of the sizes of trivalent REE in sixfold coordination shows that only a few other ions in the same coordination have sizes between that of the largest, La^{3+} (10.32 nm), and the smallest, Lu^{3+} (8.61). These include Na^+ (10.2), Ca^{2+} (10.00) and Y^{3+} (9.0). Examples of ions somewhat larger than La are Sr^{2+} (11.8) and Ba^{2+} (13.5), and of ions smaller than Lu are Zr^{4+} (7.2) and Hf^{4+} (7.1). Any geochemical behaviour that is signficantly dependent on ionic radius, such as element partitioning between different coexisting minerals, will tend to reflect the variation of ionic radius of the REE. Partitioning is usually a smooth function of atomic number. It follows

INTRODUCTION AND REVIEW

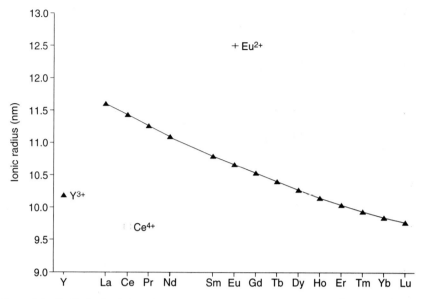

Figure 1.1 Radii (nm) of the trivalent rare earth ions (symbol ▲), Ce^{4+} and Eu^{2+} versus atomic number. (Coordination VIII) Source: Shannon, (1976).

Table 1.3 Ionic radii (nm) of REE ions

Ion	Radii for different coordination numbers					
	VI	VII	VIII	IX	X	XII
Y 3+	9.0	9.6	10.19	10.75		
La 3+	10.32	11.0	11.60	12.16	12.7	13.6
Ce 3+	10.1	10.7	11.43	11.96	12.5	13.4
Pr 3+	9.9		11.26	11.79		
Nd 3+	9.83		11.09	11.63		12.7
Sm 3+	9.58	10.2	10.79	11.32		12.4
Eu 3+	9.47	10.1	10.66	11.20		
Gd 3+	9.38	10.0	10.53	11.07		
Tb 3+	9.23	9.8	10.40	10.95		
Dy 3+	9.12	9.7	10.27	10.83		
Ho 3+	9.01		10.15	10.72	11.2	
Er 3+	8.90	9.45	10.04	10.62		
Tm 3+	8.80		9.94	10.52		
Yb 3+	8.68	9.25	9.85	10.42		
Lu 3+	8.61		9.77	10.32		
Ce 4+	8.7		9.7		10.7	11.4
Eu 2+	11.7	12.0	12.5	13.0	13.5	

Source: Shannon (1976).

that the partition coefficient of a rare earth element can be estimated well if the values for the elements of adjacent atomic numbers are known and provided their oxidation states are the same.

Although the variation in ionic radius with atomic number is smooth, that variation is sufficient in magnitude to be indirectly useful in mineralogy and geochemistry. For example, the partitioning behaviour of the REE will be different for each of two minerals which have different cation coordination sites. This is illustrated in Figure 1.2 for the two minerals zircon and plagioclase feldspar. This kind of behavioural difference can be used in the modelling of rock or mineral genesis.

Also of important use is the variation in the behaviour resulting from differences in oxidation state. Significant evidence for the presence of oxidation states other than 3+ for the REE in natural systems exists only for Eu^{2+} (relative reducing conditions) and Ce^{4+} (relative oxidizing conditions). The sizes of these two ions are sufficiently different from their 3+ counterparts to have a marked effect on their geochemical behaviour. It follows that the REE can be used to assess relative redox conditions in some mineral or rock systems.

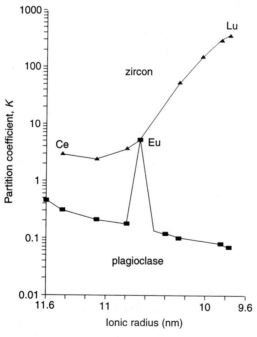

Figure 1.2 Relationships between ionic radii and partition coefficients of rare earth ions between zircon and plagioclase and their coexisting melts. Zircon, symbol ▲, from Nagasawa (1970) and plagioclase, symbol ■, from Nash and Crecraft (1985).

Since the location of a RE ion within a mineral structure is dependent on its ionic charge and size it would clearly be advantageous to be able to determine readily the nature of coordination sites from spectroscopic methods. Unfortunately, REE spectral properties are not affected much by the ion's surroundings because the $4f$ electrons, which determine the behaviour of the ions, are in orbitals that are well shielded by the outer $5s^2$ and $5p^6$ shells. Electronic spectra are therefore of limited use in distinquishing the coordination number of any given rare earth. For the same reason REE compounds are usually the same colour as their aqueous ion. Certain REE-bearing complexes do exhibit phosphorescence, e.g. the red phosphor $Eu^{3+}:Y_2VO_2S$ and the green phosphor $Ce^{3+}:CaS$. Several of these phosphors are of economic importance, for example in the production of televisions and computer monitors.

Mössbauer spectroscopy, which can, for certain elements, provide details of their oxidation states and structural environment, has not been applied much to the REE because of the unsuitability of their nuclei for this technique. It has, however, been used in studies of the oxidation state of Eu (e.g. Aslani-Samin et al., 1987). Extended X-ray absorption fine structure (EXAFS) spectroscopy can be used to establish cation coordinations and has been used with some success in the study of synthetic REE-bearing minerals (e.g. Cressey and Steel, 1988) but, in the case of natural examples, spectral overlaps arising from the presence of several different REE can severely limit the applicability of the technique.

Cathodoluminescence (Marshall, 1988) has proved to be of limited use in studies of rare earth minerals since the emission spectra of the trivalent REE do not give information about the ion's environment because of the shielding of the $4f$ electron shell. The technique has, however, application to those mineralogical problems where the REE are at trace concentrations, for they can serve as activators of cathodoluminescence (CL). It may then be possible to observe stages of crystal growth if rare earth concentrations vary as they may, for example, in the generation of primary and secondary apatite. Care has to be exercised, since several ions besides those of the REE can be CL activators (e.g. Mn^{2+}).

The aqueous chemistry of the REE is dealt with in section 1.5.

1.3 Crustal abundances

The abundances of some of the REE in the Earth's continental crust are very low but not as low as the rarest elements such as gold, mercury and indium. Seven of the REE have abundances comparable with those of other important economic elements (between 1 and $10\,\mu g/g$) which are often not considered to be particularly rare, such as tungsten, tin, arsenic and bromine. Four of the REE have abundances between 15 and $100\,\mu g/g$ along with the elements copper, cobalt, rubidium and zinc. Of all the REE, Ce at an estimated concentration of about 30 μg/g is the most abundant in the crust.

In recent years there have been several estimates made of the composition of the Earth's crust as well as of the upper and lower parts (e.g. Weaver and Tarney, 1984; Taylor and McLennan, 1985; Shaw et al., 1986; and Condie, 1993; see also references cited in Condie, 1993). These authors use different approaches which give some variation to the estimated concentrations of any given element. For example, the abundances of the REE in Condie's map model for the upper crust (in which he used rock proportions based on geological maps) and his restoration model (in which he estimates, and restores, the amount of crust lost by erosion) differ by up to about 6%. However the general nature or pattern of the abundances are similar from one model to another (see below). Abundances as estimated by different authors are given in Table 1.4. They show that if comparison were made with estimates of the average compositions of the core and mantle, the highest concentrations are in the Earth's crust.

A plot of abundances in the upper crust shows a marked oscillation with atomic number (the 'Oddo–Harkins effect', Figure 1.3a). This stems from the production of the REE in nucleosynthetic processes and the greater relative stability of nuclei with even atomic number compared to those with odd atomic number. The plot also shows that there is a trend of decreasing

Table 1.4 Crustal abundances of the REE in ppm

Element	Weaver and Tarney (1984)	Taylor and McLennan (1985)		Shaw et al. (1986)	Condie (1993)	
		Upper crust	Bulk continental crust		Map model	Restoration model
Y	13	22	20	21	29	30
La	27	30	16	32.3	25.6	27.3
Ce	55	64	33	65.6	55.7	59.3
Pr		7.1	3.9			
Nd	23	26	16	25.9	24.6	26.6
Sm	3.9	4.5	3.5	4.51	5.04	5.43
Eu	1.07	0.88	1.1	0.937	1.02	1.01
Gd		3.8	3.3	2.79	4.81	5.11
Tb	0.50	0.64	0.60	0.481	0.76	0.80
Dy		3.5	3.7			
Ho		0.80	0.78	0.623		
Er		2.3	2.2			
Tm	0.23	0.33	0.32			
Yb	1.46	2.2	2.2	1.47	2.33	2.36
Lu		0.32	0.30	0.233	0.43	0.43

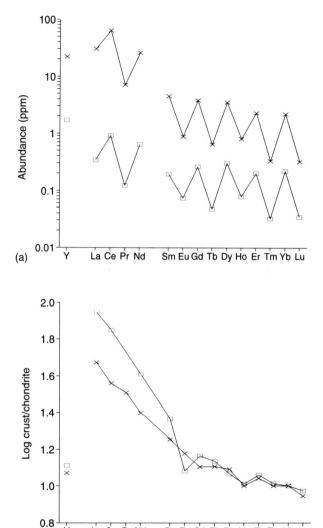

Figure 1.3 (a) Abundances (ppm) of the rare earth elements in the upper crust (from Taylor and McLennan, 1985; symbol ×), and in chondritic meteorites (quoted by Wakita, Rey and Schmitt, 1971; symbol □). (b) Chondrite-normalized abundances of the rare earths in the upper crust (symbol □) and in the bulk crust (symbol ×). (Source: Taylor and McLennan, 1985.)

abundance with atomic number. Note that promethium is virtually absent in the crust since it has no naturally occurring stable or long-lived isotope.

To aid the study of the geochemistry of the REE it has proved useful to normalize the concentration of any given REE in a rock, mineral, fluid, etc. to that in some standard reference material. This is particularly appropriate when using graphical plots of REE data, primarily because it overcomes possible difficulties in comparison etc. brought about by the oscillation in abundances with atomic number. Several reference sets have been proposed and used but one of the best established for a wide range of materials is that given by Wakita, Rey and Schmitt (1971) for a composite of 12 chondritic meteorites (Table 1.5). Figure 1.3a, lower part, shows the variation of the abundances in these meteorites with atomic number. Figure 1.3b gives the 'chondrite-normalized plot' of upper-crustal abundances (i.e. the average concentration of any given element in the crust is divided by the relative concentration in the reference set). Other sets of chondrite normalizing values have been proposed. For example, that by Evensen, Hamilton and O'Nions (1978), while perhaps giving a good average value of a certain type of chondrite, has values which are about 25% lower than those by Wakita, Rey and Schmitt (1971), so the chondrite-normalized plot is proportionally at a higher value. The Wakita, Rey and Schmitt values have been used extensively and, since the purpose of most plots is to make comparisons with other plots, it is recommended that they continue to be used. In the case of sediments and sometimes sedimentary minerals, the practice is to use REE concentrations in an 'average' sediment as normalizing values. A commonly used set of values is that of the North American shales composite (NASC) given in Table 1.5.

Table 1.5 Abundances of the REE in a composite sample of 12 chondrites and in the North American shales composite (NASC)

Element	Abundance (ppm)	
	12 chondrites[a]	NASC[b]
Y	1.7	
La	0.34	32
Ce	0.91	73
Pr	0.121	7.9
Nd	0.64	33
Sm	0.195	5.7
Eu	0.073	1.24
Gd	0.26	5.2
Tb	0.047	0.85
Dy	0.3	–
Ho	0.078	1.04
Er	0.2	3.4
Tm	0.032	0.5
Yb	0.22	3.1
Lu	0.034	0.48

[a] Wakita, Rey and Schmitt (1971), reporting unpublished work by Wakita and Zellmer (1970).
[b] Haskin et al. (1968).

INTRODUCTION AND REVIEW

Chondrite-normalized plots for compositions of the Earth's crust as estimated by other authors are given in Figure 1.4. Studies of these patterns and those for some sediments allow the following conclusions to be drawn about the REE abundances in the crust:

- chondrite-normalized abundances of the REE of relatively lower atomic masses, La to Sm (often referred to as the light REE) are greater than those of higher atomic masses, Gd to Lu (heavy REE);
- typical shales reflect the composition of the exposed crust;
- europium is depleted relative to its adjacent RE elements in chondrite-normalized plots for the upper crust;
- the upper crust is richer in REE than the lower crust.

The rare earths occur only as trace elements in the majority of rock types, whether sedimentary, igneous or metamorphic. In these they will tend to be concentrated in several rock-forming minerals including titanite, apatite, zircon, epidote, garnet and clays. They are often found in abundance in many carbonatites and in some granitic and syenitic pegmatities, and these rock types are host to a range of RE minerals. Many kimberlites can contain several thousand parts per million of REE, as can some lamprophyres,

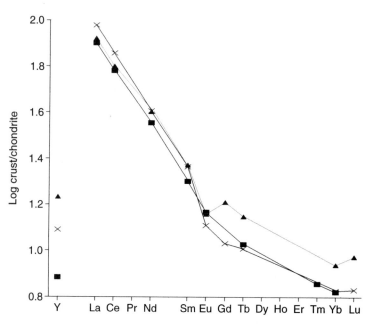

Figure 1.4 Chondrite-normalized crustal abundances of the rare earths by other authors: Weaver and Tarney (1984), symbol ■; Shaw *et al.* (1986), symbol ×; and Condie (1993), symbol ▲.

granites, skarns and other metasomatic deposits. Rare earth minerals sometimes occur as a detrital phase (e.g. the oxide, brannerite, or the phosphates, monazite and xenotime) in sedimentary rocks. Economic deposits include a carbonate-hosted iron and rare earth deposit in Bayan Obo, China, which is one of the world's largest, the famous Mountain Pass carbonatite, California, USA, and monazite placer deposits in Australia. For information on the abundances of REE in different rock types see various chapters in Henderson (1984a).

1.4 Geochemical behaviour: magmatic systems

In silicate magmatic systems the large sizes of the REE ions preclude their entry to any significant extent into many of the common minerals. The rare earth ions show small mineral–melt partition coefficients (where the coefficient, K, is the concentration of the element in the mineral divided by that in the coexisting melt) for minerals with small cation coordination sites such as olivine, pyroxene, magnetite and feldspar. Thus in a cooling basic magmatic system, where these minerals are common, the REE tend to reside in the melt – their concentrations increase significantly in successive melt fractions as crystal fractionation proceeds. Because of this behaviour the REE are referred to as incompatible elements in these systems.

Minerals with larger cation sites, such as zircon, garnet and apatite, readily accept many of the rare earth ions such that they can become a minor component of the composition. In granite pegmatites, for example, the minerals fluorite and titanite can contain significant amounts of the rare earths, occasionally up to a few per cent of REE. Rare earth minerals themselves are commonest in pegmatites, carbonatites and some syenites. Clark (1984) gives a systematic review of rare earth mineralogy.

Rare earth element partitioning between a mineral and a coexisting fluid depends not only on the ionic radius but also on ionic charge, temperature, pressure and composition of the system (for review see Henderson, 1984b).

Europium can exist in both the 2+ and 3+ oxidation states in magmatic systems, depending on the redox potential. Although the ionic radius of Eu^{2+} is larger than that of Eu^{3+} in any given coordination (Table 1.3) its partition coefficient into some minerals is considerably greater than for Eu^{3+}, especially where the exchange involves another 2+ cation (e.g. Ca^{2+}), so avoiding additional charge-balancing exchanges. The partitioning behaviour of europium between a basic magma and plagioclase feldspar is a good example of this phenomenon. This difference in the partitioning of divalent europium relative to the trivalent REE can lead to the occurrence of an europium anomaly, which is a deviation from the trend or line defined by the other REE on a plot of chondrite-normalized abundance versus atomic number (Figures 1.2 and 1.3b). Such an anomaly is referred to as being positive if the point for europium lies above the line and negative if below. Where the existence of Eu^{2+} in a crystallizing closed system causes an europium anomaly in a mineral, a balancing anomaly of opposite sign will occur in the coexisting melt, the size of which will depend on the solid:liquid mass ratio.

Oxidation states other than 3+ are virtually absent in magmatic systems for the other REE. Ce^{4+} exists in many aqueous systems and in these cases leads to a different geochemical behaviour compared to the other REE. Igneous rocks which have interacted with such aqueous systems may then acquire a secondary Ce anomaly. The presence of Ce anomalies in unaltered igneous or metamorphic rocks may be an indicator of source material that included a supracrustal component. Eu^{2+} can also exist in natural aqueous fluids.

Miyawaki discusses the type of substitution reactions of rare earth ions for other ions in Chapter 2. Many authors have reviewed the partitioning data for the REE in both natural and synthetic systems. Henderson (1984b) tabulates average mineral/melt coefficients for different igneous rocks and Jones (1994) reviews the experimental data for REE and other elements, including systems containing carbonate liquids. Some examples are given in Table 1.6 and Figure 1.5.

1.5 Geochemical behaviour: aqueous systems

The rare earth elements are mobile in certain geochemical systems involving aqueous media despite the relative low solubility of their compounds. Direct evidence for this is the analytical data for many natural waters including seawater, hydrothermal fluids, fluid inclusions and some ground waters. Some indirect evidence is from the enhanced concentrations of REE in metasomatized rocks, the changes in REE distribution with weathering, especially in laterite formation, and secondary mineral formation. The mobility is however limited and has led some authors to treat the REE as 'immobile' for practical purposes of modelling particular processes, for example, rock alteration or the mixing of major components in the Earth's crust. Strict immobility, even over small distances, appears to be confined to dry metamorphic systems (e.g. Muecke, Pride and Sarkar, 1979; Merriman, Bevins and Ball, 1986).

The degree of mobility of the REE in any aqueous medium is dependent on a range of factors including pH, Eh, the availability of potential ligands and temperature. The effect of a REE-bearing fluid on a pre-existing rock will in turn depend on the rock–fluid ratio, the constituent minerals and reaction kinetics. Unfortunately the systematics of many of these relationships are not yet well established, although some important advances have recently been made especially in the study of REE complexing in aqueous media.

REE complexing and hydrolysis in aqueous solution is controlled by the

Table 1.6 Mineral/melt partition coefficients for the REE

System:	Mare basalt (experimental) (McKay, 1986)	Komatiitic (MacRae and Russell, 1987)	Diopside–garnet–water (experimental) (Shimizu and Kushiro, 1975)	Silicic (Nash and Crecraft, 1985)	Silicic (Nagasawa, Schreiber and Morris, 1980)	Silicic (Simmons and Hedge, 1978)	Phonolitic (Wörner et al., 1983)	Carbonate melt (experimental) (Brenan and Watson, 1992)
Phases	Olivine	Augite	Garnet	Plagioclase	Perovskite	Titanite	Apatite	Clinopyroxene
La		0.185		0.45	2.62		14.4	
Ce		0.177	0.021	0.31		53.3	24.3	0.16
Pr								
Nd	7E−05	0.312	0.087	0.21		88.3	54.3	
Sm	0.00058	0.595	0.217	0.12	2.7	102	95.2	
Eu		0.447	0.320	5.2	2.34	101	102	
Gd	0.00102	0.678	0.498			102		0.5
Tb				0.12	1.58		41.4	
Dy		0.617	1.06	0.1		80.6		
Ho								
Er		0.724	2.00			58.7		
Tm								
Yb	0.0194	0.871	4.03	0.08	0.488	37.4	8.81	0.3
Lu				0.07	0.411	26.9	3.69	

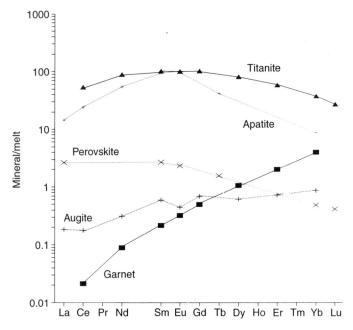

Figure 1.5 Mineral/melt partition coefficients for the REE in various minerals. Sources. Table 1.6.

composition of the solution, its pH and the specific rare earth. The REE bonding to any ligand is dominantly electrostatic rather than covalent in character.

Complexing of the rare earths is important in many natural aqueous systems at neutral and alkaline pH, but much less so at low pH where free rare earth ions tend to be the more stable species. The broad type of complex is also pH dependent as shown for example by carbonate complexes of Eu^{3+}, where $EuCO_3^+$ is the most significant species at pH 6 to 9 but at pH above 9, $Eu(CO_3)_2^-$ becomes more significant. Stability is also a function of temperature for many geochemically relevant complexes. For most rare earths (excepting those of Eu^{3+}) stability increases with temperature over a few hundred °C. For example, Ragnarsdottir, K. V. (pers. com.) shows that the stability of $LaCO_3^+$ increases about twofold from 50 to 350°C. Stability for many complexes also tends to increase with increase in atomic number.

Several authors (e.g. Brookins, 1983, 1989; Wood, 1990a, 1990; Cantrell and Byrne, 1987a, 1987b; Turner, Whitfield and Dickson, 1981; Byrne, Lee and Bingler, 1991; Millero, 1992) have studied and reviewed the solubility and complexing behaviour of rare earths from a geochemical standpoint. They show that in saline waters carbonate complexes of the form $Ln(CO_3)^+$ and $Ln(CO_3)_2^-$ are important, with a tendency for the former to be common when Ln = light REE and the latter with heavy REE. In sea water, hydroxide, sulphate and halide complexes exist, as well as free 3+ ions, but these are subordinate to the carbonate complexes.

Hydroxide complexes (e.g. $Ln(OH)_3^\circ$) are likely to be important at high pH values although hydrolysis reactions of the form:

$$Ln^{3+}{}_{aq} + H_2O_l = Ln(OH)^{2+}{}_{aq} + H^+$$

are not important for trivalent ions but start to be significant in the case of Ce^{4+}.

The halides form some of the more important complexing groups in natural aqueous systems. Evidence for this is not only direct (Brookins, 1989) but also indirect in that associations of fluoride-bearing minerals with rare earth enrichment, as in some metasomatic deposits, are quite common. Trivalent rare earths form strong complexes with the fluoride ion and the strength increases from La to Lu, but only relatively weak complexes with the chloride ion. These show little change in stability with atomic number at 25°C. However, Ragnarsdottir, K. V. (pers. com.) shows that for saline granitic fluids at 500°C and 2 kbar, the most important species are $LnCl^{2+}$ and LnF^{2+} for the light REE but only LnF^{2+} for the heavy REE.

Redox equilibria of Eu and Ce ions in aqueous solution have been studied by both theoretical and empirical approaches. Sverjensky (1984) has shown theoretically that the Eu^{2+}/Eu^{3+} ratio is strongly temperature dependent but only weakly pressure dependent. Above 250°C the dominant form is likely to be Eu^{2+} (single ion or complex) whereas at 25°C the trivalent state will be almost exclusively present. At intermediate temperatures both valency states will occur. These generalized conclusions are for the typical range of oxygen fugacity and pH values in many natural aqueous systems at the Earth's surface. Clearly such findings have significant implications for interpreting europium anomalies in a hydrothermal system because the partitioning of Eu between fluid and solid will be dependent on its valency state which, in turn, is strongly dependent on temperature at any given oxygen fugacity, and on pH. Brookins (1983) using a more empirical approach constructed Eh–pH diagrams of REE species in a simple aqueous system. The behaviour of Ce is in contrast to the other REE, with a much reduced Ce^{3+} stability field (at low pH) and the presence of a significant CeO_2 stability field (at neutral and high pH values). The results from these two approaches help to illustrate the complex interactions of the primary variables in any natural aqueous solution and that it is not possible simply to relate quantitative differences in the behaviour of Ce or Eu to redox potential alone.

In many natural systems the solid-phase chemistry of the REE may be the determining factor in establishing REE concentrations in the aqueous medium, without attainment of equilibrium. REE can be strongly adsorbed onto clay minerals and iron-bearing colloids. In ocean systems baryte, phosphorites and manganese-rich encrustations can be REE scavengers but with significant variability in the REE concentrations and patterns of the solid phases. Under acid pH conditions, such as in surface weathering, REE may be leached from clays and other minerals, and readsorbed or redeposited elsewhere, sometimes to form RE minerals.

A good example of this last phenomenon is to be found in laterite formation. Here, the rare earths will be leached from the upper horizons of the altered rock sequence, fractionated and redeposited lower down. Cerium, through the oxidizing conditions, will often show a behaviour different from the other REE and may remain in the residual horizon. Under some conditions, especially if the REE concentrations are high enough, individual RE minerals such as lanthanite or florencite, may form. A study of laterite formation from an initial syenite has been made by Braun *et al.* (1993), and the mineralogical hosts of REE in Brazilian laterites formed from alkaline rocks are presented by Morteani and Preinfalk (Chapter 9).

1.6 Concluding statement

The REE have proved to be extremely useful in helping to understand and quantify numerous Earth processes. Their geochemical coherence coupled with subtle behavioural differences resulting from the smooth change in ionic radii with atomic number, and from the presence of more than one common valency state in the cases of Eu and Ce, are the prime reasons for this usefulness. Thus, observed REE abundances and fractionations have been used to interpret many petrogenetic processes, especially in igneous systems involving crystal fractionation, partial melting or magma mixing. As our understanding of their transport properties becomes better, so too will our ability to apply these to unravelling the complex problems of ore formation, rock alteration and the genesis of crustal fluids.

Closely related to these studies is the use of the rare earth isotopes in petrogenesis and geochronology. There is now a substantial body of literature on the systematics and use of such isotopes. Although the majority of studies have been concerned with problems in petrology or in the study of large-scale systems, such as mantle domains, it is clear that they are of increasing relevance to mineralogical issues, especially those relating to the sources of mineralizing fluids. There is still much to be learned from the application of the REE to a wide range of investigation.

1.7 Review works

The following are useful review books and chapters on the analysis, mineralogy and geochemistry of the rare earth elements:

REFERENCES

Bunzli, J.-C. G. and Choppin, G. R. (eds) (1989) *Lanthanide Probes in Life, Chemical and Earth Sciences. Theory and Practice*, Elsevier, 432 pp. (Selected chapters: 1, 3, 9, 10 and 11.)

Henderson, P. (ed.) (1984) *Rare Earth Element Geochemistry*, Elsevier, 510 pp.

Lipin, B. R. and McKay, G. A. (eds) (1989) *Geochemistry and Mineralogy of Rare Earth Elements. Reviews in Mineralogy*, **21**, Mineralogical Society of America, 348 pp.

Möller, P. (1983) Lanthanoids as a geochemical probe and problems in lanthanoid geochemistry. Distribution and behaviour of lanthanoids in non-magmatic-phases, in *Systematics and the Properties of the Lanthanides* (ed. S. P. Sinha), Reidel Publishing Co., NATO ASI Series, pp 561–610.

Acknowledgements

My thanks to Dr Linda Campbell for help in the preparation of this chapter.

References

Aslani-Samin, S., Binczydka, H., Hafner, S. S. *et al.* (1987) Crystal chemistry of europium in feldspars. *Acta Crystallogr*, **A43** (Suppl.) C155.

Braun, J.-J., Pagel, M., Herbillon, A. and Rosin, C. (1993) Mobilization and redistribution of REEs and thorium in a syenitic lateritic profile: a mass balance study. *Geochim. Cosmochim. Acta*, **57**, 4419–34.

Brenan, J. M. and Watson, E. B. (1992) Partitioning of trace elements between carbonate melt and clinopyroxene and olivine at mantle P–T conditions. *Geochim. Cosmochim. Acta*, **55**, 2203–14.

Brookins, D. G. (1983) Eh–pH diagrams for the REE at 25°C and 1 bar pressure. *Geochem. J.*, **17**, 223–9.

Brookins, D. G. (1989) Aqueous geochemistry of rare earth elements. In *Geochemistry and Mineralogy of Rare Earth Elements*, (eds B. R. Lipin and G. A. McKay), *Reviews in Mineralogy*, **21**, Mineralogical Society of America, pp. 201–25.

Byrne, R. H., Lee, J. H. and Bingler, L. S. (1991) Rare earth element complexation by PO_4^{3-} ions in aqueous solution. *Geochim. Cosmochim. Acta*, **55**, 2729–35.

Clark, A. M. (1984) Mineralogy of the rare earth elements. In *Rare Earth Element Geochemistry*, (ed. P. Henderson), *Developments in Geochemistry*, **2**, Elsevier, pp. 33–61.

Condie, K. C. (1993) Chemical composition and evolution of the upper continental crust; contrasting results from surface samples and shales. *Chem. Geol.*, **104**, 1–37.

Cressey, G. and Steel, A. T. (1988) An EXAFS study of Gd, Er and Lu site location in the epidote structure. *Phys. Chem. Minerals*, **15**, 304–12.

Evensen, N. M., Hamilton, P. J. and O'Nions, R. K. (1978) Rare earth abundances in chondritic meteorites. *Geochim. Cosmochim. Acta*, **42**, 1199–212.

Haskin, L. A., Haskin, M. A., Frey, F. A. and Wildeman, T. R. (1968) Relative and absolute terrestrial abundances of the rare earths. In *Origin and Distribution of the Elements*, **2**, (ed. L. H. Ahrens), Pergamon Press, pp. 889–911.

Henderson, P. (ed.) (1984a) *Rare Earth Element Geochemistry*, (advisory ed. W. S. Fyfe), *Developments in Geochemistry*, **2**, Elsevier, 510 pp.

Henderson, P. (1984b) General geochemical properties and abundances of the rare earth elements. In *Rare Earth Element Geochemistry*, (ed. P. Henderson (advisory ed. W. S. Fyfe), *Developments in Geochemistry*, **2**, Elsevier, pp. 1–32.

Jones, J. H. (1994) Experimental trace element partitioning. In *AGU Handbook of Geophysical Constants*, **3**, (ed. T. J. Ahrens), in press.

MacRae, N. D. and Russell, M. R. (1987) Quantitative REE SIMS analyses of komatiite pyroxenes, Munro Township, Ontario, Canada. *Chem. Geol.*, **64**, 307–17.
Marshall, D. J. (1988) *Cathodoluminescence of Geological Materials*, Unwin Hyman, 146 pp.
McKay, G. A. (1986) Crystal/liquid partitioning of REE in basaltic systemis; extreme fractionation of REE in olivine. *Geochim. Cosmochim. Acta*, **50**, 69–79.
Merriman, R. J., Bevins, R. E. and Ball, T. K. (1986) Petrological and geochemical variations within the Tal y Fan intrusion: a study of element mobility during low-grade metamorphism with implications for petrotectonic modelling. *J. Petrol.*, **27**, 1409–36.
Millero, F. J. (1992) Stability constants for the formation of rare earth inorganic complexes as a function of ionic strength. *Geochim. Cosmochim. Acta*, **56**, 3123–32.
Muecke, G. K., Pride, C. and Sarkar, P. (1979) Rare earth element geochemistry of regional metamorphic rocks, in *Origin and Distribution of the Elements*, **2**, (ed. L. H. Ahrens), Pergamon Press, pp. 449–64.

Nagasawa, H., Schreiber, H. D. and Morris, R. V. (1980) Experimental mineral/liquid partition coefficients of the rare earth elements (REE), Sc and Sr for perovskite, spinel and melilite. *Earth Planet. Sci. Lett.*, **46**, 431–7.
Nash, W. P. and Crecraft, H. R. (1985) Partition coefficients from trace elements in silicic magmas. *Geochim. Cosmochim. Acta*, **49**, 2309–22.

Shannon, R. D. (1976) Revised effective ionic radii and systematic studies of interatomic distances in halides and chalcogenides. *Acta Crystallogr.*, Sect. A, **32**, 751–67.
Shaw, D. M., Cramer, J. J., Higgines, M. D. and Truscott, M. G. (1986) Composition of the Canadian Precambrian shield and the continental crust of the earth. *Geol. Soc. Spec. Publ.*, **24**, Blackwell, pp. 275–82.
Shimizu, N. and Kushiro, I. (1975) The partitioning of rare earth elements between garnet and liquid at high pressures: preliminary experiments. *Geophys. Res. Lett.*, **2**, 413–16.
Simmons, E. C. and Hedge, C. E. (1978) Minor element and Sr-isotope geochemistry of Tertiary stocks, Colorado mineral belt. *Contrib. Mineral. Petrol.*, **67**, 379–96.
Sverjensky, D. A. (1984) Europium redox equilibria in aqueous solution. *Earth Planet. Sci. Lett.*, **67**, 70–8.

Taylor, S. R. and McLennan, S. M. (1985) *The continental crust: its composition and evolution*. Blackwell, 312 pp.
Turner, D. R., Whitfield, M. and Dickson, A. G. (1981) The equilibrium speciation of dissolved components in freshwater and seawater at 25°C and 1 atm pressure. *Geochim. Cosmochim. Acta*, **45**, 855–81.

Wakita, H., Rey, P. and Schmitt, R. A. (1971) Abundances of the 14 rare-earth elements and 12 other trace elements in Apollo 12 samples: five igneous and one breccia rocks and four soils. *Proc. Second Lunar Sci. Conf.*, **2**, The M.I.T. Press, pp. 1319–29.
Weaver, B. L. and Tarney, J. (1984) Major and trace element composition of the continental lithosphere. In *Structure and Evolution of the Continental Lithosphere*, (eds H. N. Pollack and V. R. Murthy), *Phys. Chem. Earth*, **15**, Pergamon Press, pp. 39–68.
Wood, S. A. (1990a) The aqueous geochemistry of the rare-earth elements and yttrium 1.

REFERENCES

Review of available low-temperature data for inorganic complexes and the inorganic REE speciation of natural waters. *Chem. Geol.*, **82**, 159–86.

Wood, S. A. (1990b) The aqueous geochemistry of the rare-earth elements and yttrium 2. Theoretical prediction of speciation in hydrothermal solutions to 350°C at saturated water vapour pressure. *Chem. Geol.*, **88**, 99–125.

Wörner, G., Beusen, J.-M., Duchateau, N. *et al.* (1983) Trace element abundances and mineral/melt distribution coefficients in phonolites from the Laacher See Volcano (Germany). *Contrib. Mineral. Petrol.*, **84**, 152–73.

CHAPTER TWO
Crystal chemical aspects of rare earth minerals

R. Miyawaki and I. Nakai

2.1 Introduction

More than 150 independent species of rare earth minerals have been described. Levinson (1966) defined rare earth minerals as those minerals containing rare earth elements (REE) as essential constituents. Following this Levinson criteria, the International Mineralogical Association (IMA) decided that the correct name of a rare earth mineral should be written with the group name followed by the chemical symbol of the dominant REE in parentheses: e.g. monazite-(La) [$LaPO_4$] and monazite-(Ce) [$CePO_4$]. RE minerals are therefore defined and named, according to Bayliss and Levinson (1988), as minerals in which the total atomic percentage of the REE and Y are greater than any other element within a single set of crystal-structural sites. For this classification a knowledge of both the crystal structure and chemical composition is required. In this study we have tentatively treated minerals containing significant amounts of REE as rare earth minerals.

Trivalent (REE^{3+}) is the most frequently observed valency for the REE in minerals. The ionic radii of trivalent rare earth ions are about 1 Å, which are comparable to those of Ca^{2+}, Na^+, Th^{4+}, etc., and are significantly larger than those of other trivalent ions such as B^{3+}, Al^{3+}, Fe^{3+} and so on (Figure 2.1). The high oxidation state together with large ionic radii accounts for the characteristic behaviour of rare earths in crystal structures, as well as in geological processes.

The REE are conveniently grouped into two groups for crystal chemical purposes: the 'Ce-group' composed of lanthanides from La to Eu and the 'Y-group' composed of Y and lanthanides from Gd to Lu. The former are also referred to as the light REE, and the latter, except yttrium, as the heavy REE. According to the distribution of rare earths, rare earth minerals can be classified into three types: one with a predominance of the smaller Y-group rare earths; one rich in the larger Ce-group rare earths; and one in which the crystal structures accept both Y-group and Ce-group rare earths: e.g. gadolinite-(Y) [$Y_2FeBe_2Si_2O_{10}$] and gadolinite-Ce [$Ce_2FeBe_2Si_2O_{10}$] (Miyawaki and Nakai, 1993a).

Rare Earth Minerals: Chemistry, origin and ore deposits. Edited by Adrian P. Jones, Frances Wall and C. Terry Williams. Published in 1996 by Chapman & Hall. ISBN 0 412 61030 2

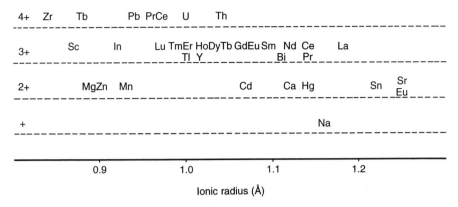

Figure 2.1 Ionic radii and valencies of rare earth ions and other ions with sizes compatible to rare earths. (Data from Shannon and Prewitt, 1969, 1970.)

Crystal structures have been analysed and reported for about half of the known rare earth minerals. Structure analyses have revealed that many rare earth minerals represent compositions within solid solutions involving rare earth ions and other heterovalent ions such as Na^+, Ca^{2+} and Th^{4+} where ionic radii considerations are met. It is known that similarity of both ionic radii and electronegativities are important factors for isomorphous substitutions, whereas the similarity of chemical properties is sometimes not so significant. On the other hand, compensation of charge mismatch is necessary for the substitutions between trivalent rare earth ions and other ions with different valences. Studies of the isomorphous substitution between rare earth ions and other heterovalent ions observed in rare earth minerals are important for understanding the behaviour of rare earth elements in solids. In view of the importance of the rare earth minerals as natural resources as well as to earth sciences, we have started to review their crystal structures (Miyawaki and Nakai, 1987, 1988, 1989, 1990, 1991, 1992, 1993a, 1993b). In this series of publications we have attempted to provide complete data necessary for understanding the crystal structures of the rare earth minerals. In this chapter we summarize our work to provide a basis for the understanding of the structures of rare earth minerals.

Rare earth minerals can be classified into the following six classes based on the types of anionic groups in their crystal structures (Miyawaki and Nakai, 1987, 1993a):

1. minerals with crystal structures containing isolated triangular anionic groups – carbonates: e.g. lanthanite-(La) [$La_2(CO_3)_3 \cdot 8H_2O$] and bastnäsite-(Ce) [$Ce(CO_3)F$];
2. minerals with crystal structures containing only isolated tetrahedral

anionic groups – phosphates: e.g. xenotime-(Y) [YPO$_4$] and monazite-(Ce) [CePO$_4$];
3. minerals with crystal structures containing anionic groups of tetrahedral ions – silicates: e.g. thalenite-(Y) [Y$_3$Si$_3$O$_{10}$(OH)] and cerite-(Ce) [Ce$_9$($\square$,Ca)(Fe,Mg)(SiO$_4$)$_6$(SiO$_3$OH)(OH)$_3$];
4. minerals with crystal structures containing anionic groups of tetrahedral and octahedral ions – aluminosilicates and titanosilicates: e.g. allanite-(Ce) [Ca(Ce,Ca)Al(Al,Fe)(Fe,Al)(Si$_2$O$_7$)(SiO$_4$)O(OH)] and mosandrite [Na(Na,Ca)$_2$(Ca,Ce)$_4$(Ti,Nb,Zr)(Si$_2$O$_7$)$_2$(O,F)$_2$F$_2$];
5. minerals with crystal structures containing anionic groups of octahedral ions – titanates and niobates: e.g. aeschynite-(Ce) [(Ce,Ca,Fe,Th)(Ti,Nb)$_2$(O,OH)$_6$] and fergusonite-(Y) [YNbO$_4$];
6. minerals with crystal structures without anionic groups – fluorides and simple oxides: e.g. fluocerite-(Ce) [CeF$_3$] and cerianite-(Ce) [(Ce,Th)O$_2$].

2.2 Coordination polyhedra of rare earth atoms

The rare earths exhibit seven kinds of coordination number between six and 12. In almost all cases the coordination polyhedra of rare earth atoms are not regular polyhedra, but rather distorted polyhedra. The chemical bonds between rare earths and anions possess largely ionic character and bond angles are not defined by the shape of the simple molecular orbital as in the case of covalent bonds. This is one of the causes of the distortion of the coordination polyhedra. Using these considerations, the coordination polyhedra composed of rare earths (RE) and surrounding anions (X) can be classified into the groups shown in Figure 2.2, including some imaginary forms (Miyawaki and Nakai, 1993a).

The REX$_6$ octahedron is a typical coordination polyhedron for 6-coordinated rare earths. Examples of this coordination are yttrium-group RE sites in yttrialite-(Y) [(Y,Th)$_2$Si$_2$O$_7$] (Batalieva and Pyatenko, 1972) and monteregianite-(Y) [Na$_4$K$_2$Y$_2$Si$_{16}$O$_{38}$·10H$_2$O] (Ghose, Sen Gupta and Campana, 1987). Although octahedral calcium sites partially substituted by Ce-group rare earths are found in agrellite [Na(Ca,Ce,Na)$_2$Si$_4$O$_{10}$F] (Ghose and Wan, 1979) and dollaseite-(Ce) [(Ca,Ce)(Ce,Ca)Mg$_2$Al(Si$_2$O$_7$)(SiO$_4$)(OH)F] (Peacor and Dunn, 1988), no octahedral site dominantly occupied by Ce-group rare earths has been reported. An octahedral site would be too small for the larger Ce-group rare earths. No trigonal prism has been reported as the coordination polyhedron of a rare earth.

Two kinds of polyhedra have been reported for 7-coordinated rare earth atoms, the REX$_7$-monocapped octahedron and REX$_7$-monocapped trigonal prism. These coordination polyhedra can be derived from an octahedron and a trigonal prism, respectively, by the addition of one vertex on a face. However, distorted versions of these polyhedra resemble each other and are often difficult to distinguish clearly. The coordination of the Ce atom in

CRYSTAL CHEMICAL ASPECTS

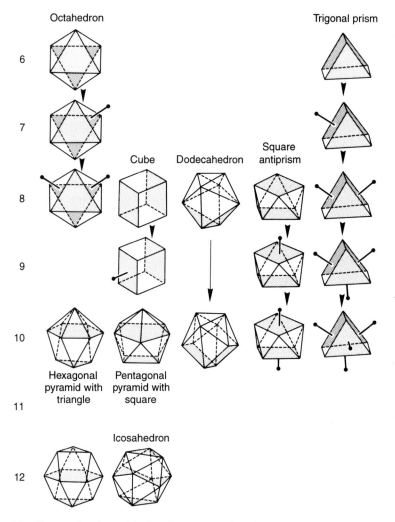

Figure 2.2 The coordination polyhedra about rare earth cations present in rare earth minerals. The numbers on the left side are coordination numbers.

sazhinite-(Ce) [$Na_2Ce(Si_6O_{14}OH) \cdot nH_2O$] was described as monocapped octahedra (Shumyatskaya *et al.*, 1980), and that of yttrotungstite-(Y) [$YW_2O_7OH \cdot H_2O$] was described as a monocapped trigonal prism (Davis and Smith, 1971).

The 8-coordination polyhedra, the most common polyhedra for rare earths, are the REX_8-bicapped octahedron, cube, dodecahedron, square antiprism and bicapped trigonal prism. The bicapped octahedron is observed in the structure of vitusite-(Ce) [$Na_3Ce(PO_4)_2$] (Karpov et al., 1980). Fergusonite-β-(Y) [$YNbO_4$] (Kinzhibalo et al., 1982) and the zirconolite-group minerals have distorted cubic coordinations of the RE sites in their structures. The D2d (or Δ)-dodecahedron is one of the most frequently observed polyhedra for 8-coordinated rare earth sites, e.g. the Y site in xenotime-(Y) [YPO_4] (Krstanović, 1965) and the Ce site in cerite-(Ce) [$Ce_9(\square,Ca)(Fe,Mg)(SiO_4)_6$ $(SiO_3OH)(OH)_3$] (Moore and Shen, 1983). The coordination polyhedra of the (Ca,Ce,Na) sites in agrellite [$Na(Ca,Ce,Na)_2Si_4O_{10}F$] (Ghose and Wan, 1979) and one of the two Y sites in ashcroftine-(Y) [$K_5Na_5(Y,Ca)_{12}Si_{28}O_{70}$ $(OH)_2(CO_3)_8 \cdot 8H_2O$] (Moore et al., 1987) can be regarded as the bicapped trigonal prism. The bicapped trigonal prism can be converted to a square antiprism by flattening the pair of triangles of the two caps to a square. The square antiprisms are often found in rare earth minerals such as gadolinite-(Y) [$Y_2FeBe_2Si_2O_{10}$] (Miyawaki et al., 1984) and aeschynite-(Y) [(Y,Ca,Th) $(Ti,Nb)_2(O,OH)_6$] (Aleksandrov, 1962). Coordination polyhedra of D2d-dodecahedron, square antiprism and cube are found in the sites for both Y-and Ce-group rare earths.

The REX_9-tricapped trigonal prism is frequently found in the 9-coordinated RE sites. A typical example is the symmetrical $YO_6(OH)_3$ polyhedron in agardite-(Ce) [$CeCu_6(AsO_4)_3(OH)_6 \cdot 3H_2O$] (Hess, 1983). This type of polyhedron is also found in the structures of belovite-(Ce) [$Sr_3NaCe(PO_4)_3$ (OH)] (Nadezhina, Pushcharovskii and Khomyakov, 1987) and gagarinite-(Y) [$(Y,Ca)_2(Na,\square)F_6$] (Voronkov, Shumyatskaya and Pyatenko, 1962). The crystal structure of tveitite-(Y) [$Ca_{14}Y_5F_{43}$] (Bevan, Strähle and Greis, 1982) contains the REX_9-monocapped square antiprism, and that of sahamalite-(Ce) [$(Mg,Fe)Ce_2(CO_3)_4$] (Pertlik and Preisinger, 1983) contains the REX_9-monocapped cube.

The REX_{10}-tetracapped trigonal prism is observed in the crystal structure of carbocernaite [$(Sr,Ce,Ba)(Ca,Na)(CO_3)_2$] (Shi, Ma and Peng, 1982). The RE polyhedra in lanthanite-(La) [$La_2(CO_3)_3 \cdot 8H_2O$] (Dal Negro, Rossi and Tazzoli, 1977) can be described as an REX_{10}-bicapped square antiprism (Miyawaki and Nakai, 1993a) or an REX_{10}-bicapped dodecahedron (Shinn and Eick, 1968). The RE polyhedron in huanghoite-(Ce) [$BaCe(CO_3)_2F$] (Qian et al., 1982) is an REX_{10}-hexagonal pyramid with a triangle. The B site in burbankite [$(Na,Ca,\square)_3(Sr,Ba,Ca,Ce)_3(CO_3)_5$] (Effenberger et al., 1985) occupied with Ca and RE atoms forms a combination of a pentagonal pyramid with a square. No example of these larger polyhedra occupied by smaller Y-group rare earths has been reported.

The REX_{12}-tetracapped cube, which consists of a hexagon and two triangles, was reported for the La site in davidite-(La) [$(La,Ca,Th)(Y,U,Fe)$ $(Ti,Fe)_{20}(O,OH)_{38}$] (Gatehouse, Grey and Kelly, 1979). This polyhedron is closely related to the icosahedron. The coordination numbers of rare earth atoms tend to increase in the order of silicate, phosphate, carbonate. This is explained by a decrease in the bond strengths of the oxygen atoms in

SiO_4^{4-}, PO_4^{3-} and CO_3^{2-} ions, which have average negative charges of -1, $-3/4$ and $-2/3$ per oxygen atom, respectively.

2.3 Structural differences between Y-group and Ce-group rare earth minerals

The most frequently observed coordination number of the rare earths is eight, and the coordination polyhedra of rare earths are as large as those of sodium, calcium, thorium ions and so on. Y-group rare earths and Ce-group rare earths show differences in their coordination numbers (Figure 2.3). The coordination numbers of the crystallographic sites dominantly occupied by Y-group rare earths range from six to nine, while those of the Ce-group rare earths range from seven to 12. The coordination numbers of calcium resemble those of the Ce-group, accounting for the isomorphous substitution often observed between calcium and Ce-group rare earths.

The differences in coordination numbers and sizes of coordination polyhedra between Y- and Ce-group rare earths are reflected in the crystal structures of minerals containing isolated triangular anionic groups (carbonates) or isolated tetrahedral anionic groups (phosphates). These minerals do not have infinite framework structures, and the anionic groups in the structures are isolated from each other and are connected by rare earth atoms to form three-dimensional structures. Therefore, the size of the RE-polyhedron affects the whole structure.

Although both tengerite-(Y) and lanthanite-(La) are sesqui-type hydrous rare earth carbonates [$RE_2(CO_3)_3 \cdot nH_2O$], their crystal structures are quite different. The crystal structure of tengerite-(Y) (Miyawaki, Kuriyama and Nakai, 1993) is built up of corrugated sheets of 9-coordinated Y polyhedra and CO_3 triangles. The sheets are connected directly by other CO_3 triangles to form a three-dimensional structure (Figure 2.4). On the other hand, the crystal structure of lanthanite-(La) (Dal Negro, Rossi and Tazzoli, 1977) has two rare earth sites which are 10-coordinated by the oxygen atoms of the carbonate ions and water molecules. The La polyhedra share corners and edges to form platy sheets (Figure 2.5). Lanthanite-(La) is a highly hydrated mineral ($8H_2O$) and the sheets are connected only by hydrogen bonds.

The dependence of structure type of sesqui-type hydrous rare earth carbonates on the ionic radii of rare earths was confirmed by experimental synthesis (Nagashima, Wakita and Mochizuki, 1973). Rare earth carbonates with lanthanite-type structure can only be obtained with larger Ce-group rare earth ions (La, Ce, Nd and Sm), but those with tengerite-type structure can be synthesized with smaller rare earth ions (Nd, Sm, Gd, Dy, Er and Y).

The effect of the difference in ionic radii of rare earths on the crystal structure can be demonstrated for some rare earth carbonates (Miyawaki and Nakai, 1993a). The lattice parameters of kimuraite-(Y) [CaY_2

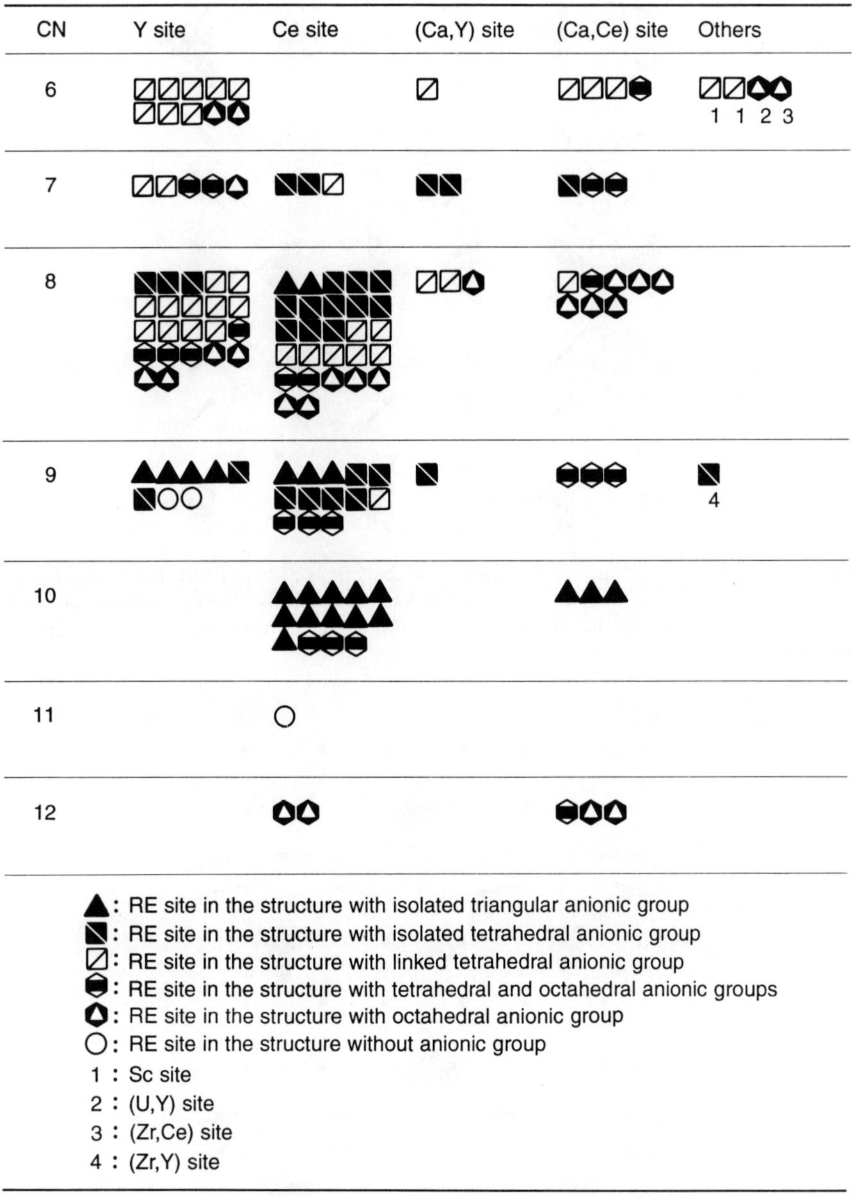

STRUCTURAL DIFFERENCES BETWEEN Y- AND CE-GROUPS

CN	Y site	Ce site	(Ca,Y) site	(Ca,Ce) site	Others
6	▨▨▨▨▨ ▨▨▨◉◓	▨		▨▨▨◉	▨▨◓◓ 1 1 2 3
7	▨▨◉◉◓	◣◣▨	◣◣	◣◉◉	
8	◣◣◣▨ ▨▨▨▨ ▨▨▨◉ ◉◉◓◓ ∞	▲▲◣◣◣ ◣◣▨▨ ▨▨▨▨ ◉◓◓◓ ∞	▨▨◓	▨◉◓◓◓ ◉◉◉	
9	▲▲▲▲◣ ◣◯◯	▲▲▲◣◣ ◣◣◣▨ ◉◉◉	◣	◉◉◉	◣ 4
10		▲▲▲▲▲ ▲▲▲▲ ▲◉◉◉		▲▲▲	
11		◯			
12		◉◓◓		◉◓◓	

▲ : RE site in the structure with isolated triangular anionic group
◣ : RE site in the structure with isolated tetrahedral anionic group
▨ : RE site in the structure with linked tetrahedral anionic group
◉ : RE site in the structure with tetrahedral and octahedral anionic groups
◓ : RE site in the structure with octahedral anionic group
◯ : RE site in the structure without anionic group
1 : Sc site
2 : (U,Y) site
3 : (Zr,Ce) site
4 : (Zr,Y) site

Figure 2.3 The coordination number of rare earths in the crystal structures of rare earth minerals.

CRYSTAL CHEMICAL ASPECTS

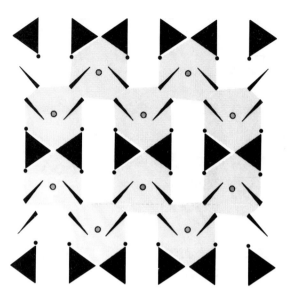

Figure 2.4 Projection of the crystal structure of tengerite-(Y) onto (010). Black triangles and open circles represent carbonate and rare earth ions, respectively. The coordination polyhedra forming corrugated sheets are indicated by shadowing. (Data from Miyawaki, Kuriyama and Nakai, 1993).

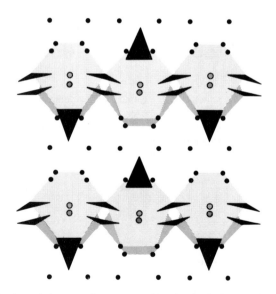

Figure 2.5 Projection of the crystal structure of lanthanite-(La) onto (001). Black triangles and open circles represent carbonate and rare earth ions, respectively. The coordination polyhedra forming platy sheets are indicated by shadowing. (Data from Dal Negro, Rossi and Tazzoli, 1977.)

$(CO_3)_4 \cdot 6H_2O]$ and lokkaite-(Y) $[CaY_4(CO_3)_7 \cdot 9H_2O]$ suggest that these Y-group dominant rare earth minerals are structurally related to tengerite-(Y) $[Y_2(CO_3)_3 \cdot 2-3H_2O]$ with corrugated sheets. The similarity of the lattice parameters between calkinsite-(Ce) $[Ce_2(CO_3)_3 \cdot 4H_2O]$ (Pecora and Kerr, 1953) and lanthanite-(La) $[La_2(CO_3)_3 \cdot 8H_2O]$ indicates the presence of the lanthanite-type platy sheets in the crystal structure of calkinsite-(Ce).

Another pair of rare earth minerals showing the effect of ionic radius is xenotime-(Y) $[YPO_4]$ and monazite-(Ce) $[CePO_4]$. The crystal structure of xenotime-(Y) (Krstanović, 1965) is isostructural with that of zircon $[ZrSiO_4]$. The yttrium atoms in xenotime-(Y) form 8-coordinated polyhedra which connect isolated PO_4 tetrahedra (Figure 2.6). Although the crystal structure of monazite-(Ce) (Ueda, 1967) shows close relationships to that of xenotime-(Y), the difference in the predominant rare earth reflects on the two crystal structures. While the coordination number of the cerium atoms in the crystal structure of monazite-(Ce) (Figure 2.7) is regarded as 8-coordinated, which is the same as that of yttrium in xenotime-(Y), the mean Ce–O distance in monazite-(Ce), 2.45 Å, is slightly larger than the mean Y–O distance in xenotime, 2.42 Å. The arrangements of the PO_4 tetrahedra, which are connected by rare earth atoms in both structures, are distinct due to the differences in volumes and shapes of the rare earth polyhedra. Wakefieldite-(Ce) $[CeVO_4]$ (Baudracco-Gritti et al., 1987) is a Ce–V-analogue of

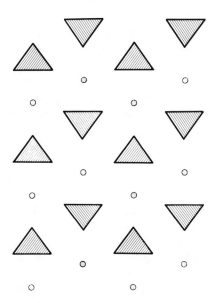

Figure 2.6 Projection of the crystal structure of xenotime-(Y) onto (010). Triangles and open circles represent phosphate and rare earth ions, respectively. Only the atoms at levels of $0 < y < 0.5$ are drawn. (Data from Krstanović, 1965.)

Figure 2.7 Projection of the crystal structure of monazite-(Ce) onto (100). Triangles and open circles represent phosphate and rare earth ions, respectively. Only the atoms at levels of $0 < x < 0.5$ are drawn. (Data from Ueda, 1967.)

xenotime-(Y). The replacement of PO_4 tetrahedra with larger VO_4 enables this mineral to accept larger cerium ions in the xenotime-type structure.

In contrast to the above, structures having a tight framework, e.g. gadolinite-(Y) and aeschynite-(Y), can accept both the Ce- and Y-group rare earths without rearrangement of the structure. Some minerals exhibit both Y-group and Ce-group rare earths as the dominant species, e.g. gadolinite-(Y) and gadolinite-(Ce) (Segalstad and Larsen, 1978). The crystal structure of gadolinite-(Y) (Miyawaki, Nakai and Nagashima, 1984) consists of sheets which are formed by the corner-sharing linkages of SiO_4 and BeO_4 tetrahedra. In these minerals the rare earth atoms are situated between the sheets together with iron atoms (Figure 2.8).

In the crystal structure of aeschynite-(Y) (Aleksandrov, 1962) (Nb,Ti) octahedra form a three-dimensional framework by corner sharing and edge sharing linkages. The rare earth atoms in aeschynite-(Y) are situated in the tunnel-like spaces of the framework, and form distorted 8-coordinated square antiprisms (Figure 2.9).

Bastnäsite-(Ce) $[Ce(CO_3)F]$ is one of the most common rare earth fluocarbonate minerals. In contrast with tengerite-(Y) and lanthanite-(La), the bastnäsite structure can accept both the Ce- and Y-group rare earths, and bastnäsite-(Y) (Mineev *et al.*, 1970) is known as a rare earth mineral in

STRUCTURAL DIFFERENCES BETWEEN Y- AND CE-GROUPS

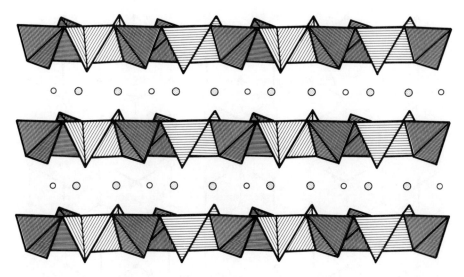

Figure 2.8 Projection of the crystal structure of gadolinite-(Y) onto (010) showing the rare earth (larger shadowed circle) and iron (smaller open circle) between the sheets of SiO_4 and $BeO_3(OH)$ tetrahedra. (Data from Miyawaki, Nakai and Nagashima, 1984).

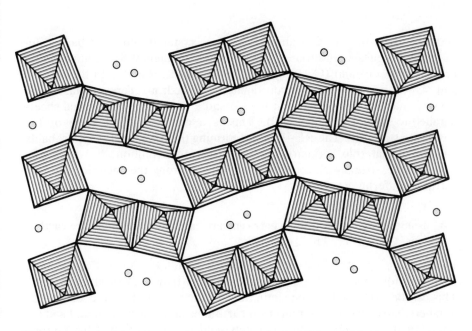

Figure 2.9 Projection of the crystal structure of aeschynite-(Y) onto (001). The rare earth ions (shadowed circles) are located in the tunnel of $(Ti,Nb) [O,(OH)]_6$ octahedra parallel to the a axis. (Data from Aleksandrov, 1962.)

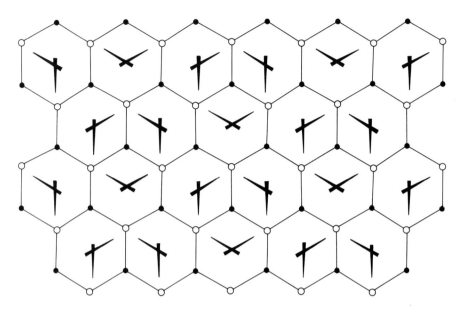

Figure 2.10 Projection of the crystal structure of bastnäsite-(Ce) onto (001). The rare earth (open circle) and fluoride (solid circle) ions form a hexagonal network, and carbonate ions (black triangles) are located between the networks. (Data from Ni, Hughes and Mariano, 1993.)

addition to bastnäsite-(Ce) and -(La). The crystal structure of bastnäsite-(Ce) (Ni, Hughes and Mariano, 1993; Terada, Nakai and Kawashima, 1993) consists of (CeF) and (CO_3) layers. The Ce atom is bonded to the three F and six O atoms to form 9-coordinated polyhedron. Terada, Nakai and Kawashima (1993) reported that the Ce and F atoms form hexagonal sheets parallel to (0001) (Figure 2.10). This is a characteristic feature of the bastnäsite structure. The RE–F bonds forming the hexagonal sheet probably have a special role, primarily in the structure definition, but also help stabilize the structure, accounting for effects from the inclusion of rare earths of differing ionic radii.

2.4 Isomorphous substitutions between rare earth ions and non-rare earth ions, Ca^{2+}, Na^+, Th^{4+}, etc.

Isomorphous substitutions between RE^{3+} and Ca^{2+} or Na^+ are more frequently observed in Ce-group rich minerals than in Y-group rich ones. This can be explained by the similarities of the ionic radii and coordination numbers between the Ce-group rare earth ions and the calcium and sodium ions. On the other hand, thorium ions tend to be substituted by Y-group rare earth ions rather than by Ce-group ones in the octahedral sites, as observed in brannerite [(U,Th,Ca,Y)(Ti,Fe)$_2$O$_6$] (Szymanski and Scott,

1982) and davidite-(La) $[(La,Ca,Th,\square)(U,Y)(Fe,Mg,\square)_2(Fe,Ti,\square)_6Ti_{12}O_{38}]$ (Gatehouse, Grey and Kelly, 1979).

The mutual substitution between heterovalent ions must be accompanied by compensation of their charge differences. The charge compensation mechanisms found in the structures of rare earth minerals were classified into the following four types (Miyawaki and Nakai, 1993a):

1. substitution accompanying vacancies;
2. coupled substitution within a rare earth site;
3. coupled substitution at two independent sites;
4. substitution accompanying valence variation.

In almost all rare earth minerals, complex solid solutions result from these isomorphous substitutions, as illustrated in the following examples.

2.4.1 Type 1: substitution accompanying vacancies ($\square$)

The substitution $3Ca^{2+} = 2RE^{3+} + \square$ can be observed in hellandite $[(Ca,Y,\square)_4(Y,\square)_2(Al,Fe)B_4Si_4O_{18}(OH,O)_4(OH)_2]$ (Mellini and Merlino, 1977) and semenovite-(Ce) $[(Ce,Na,\square)_2(Na,\square,Ce)_2(Fe,Mn,\square)(\square,Fe,Mn)(Ca,Na)_8Si_8(Si,Be)_6O_{40}(OH,O)_2(OH,F)_6]$ (Mazzi et al., 1979) combined with the substitution of type 3. Although the range of solid solution is not narrow, a mineral with the end member compositions has not yet been observed. Gagarinite-(Y) $[(Y,Ca)_2(Na,\square)F_6]$ (Voronkov, Shumyatskaya and Pyatenko, 1962) shows the substitution $RE^{3+} + \square = Ca^{2+} + Na^+$, which is unique to this mineral. The substitution accompanying a vacancy site tends to occur only in crystal structures having tight frameworks with covalent bonding because the number of atoms changes before and after the substitution.

2.4.2 Type 2: coupled substitution within a rare earth site

The substitutions of this type, $2RE^{3+} = Ca^{2+} + Th^{4+}$ and $2Ca^{2+} = RE^{3+} + Na^+$, are the simple ones occurring at an independent site. They are the commonest substitutions observed in the crystal structures of rare earth minerals because they do not require any other concomitant substitution. Such examples are observed in aeschynite-(Y) $[(Y,Ca,Th)(Ti,Nb)_2(O,OH)_6]$ (Aleksandrov, 1962) and apatite $[(Ca,RE,Na)_5(PO_4)_3(F,OH)]$ (Hughes, Cameron and Mariano, 1991). The solid solution range of this type is generally wide.

2.4.3 Type 3: coupled substitution at two independent sites

In this case coupled substitutions occur at a rare earth site and another cation site, $RE^{3+} + M^{n+} = Ca^{2+} + M^{(n+1)+}$, or an anion site, RE^{3+}

$+ A^{n-} = Ca^{2+} + A^{(n-1)-}$, and these are different from the type 2, which occurs within one rare earth site. This substitution can be observed at the (RE,Sr) and (Na,Ca) sites in the crystal structure of carbocernaite [(Sr,Ce,Ba)(Ca,Na)(CO$_3$)$_2$] (Shi, Ma and Peng, 1982). Britholite-(Ce) [(Ce,Ca,Na)$_5$[(Si,P)O$_4$]$_3$(OH,F)] (Kalsbeek, Larsen and Rønsbo, 1990) and gadolinite-(Y) [(Y,Ca)$_2$Fe(Be,B)$_2$Si$_2$O$_{10}$] (Miyawaki, Nakai and Nagashima, 1984) shows this type of substitution at the RE and tetrahedral sites. Substitutions of this type can also be observed at octahedral and RE sites in the crystal structures of euxenite-(Y) [(Y,Ca,Th)(Ti,Nb)$_2$O$_6$] (Weitzel and Schröcke, 1980) and hibonite [(Ca,Ce)(Al,Ti,Mg)$_{12}$O$_{19}$] (Utsunomiya et al., 1988) combined with the other types of the coupled substitution.

Ancylite-(Ce) [(Ce,Ca)(CO$_3$)(OH,H$_2$O)] (Dal Negro, Rossi and Tazzoli, 1975) shows a coupled substitution, $RE^{3+} + (OH)^- = Ca^{2+} + H_2O$. This type, accompanied with charge compensation by replacing $(OH)^-$ with H$_2$O for the substitution of Ca^{2+} for Ce^{3+}, is able to occur in crystal structures containing isolated hydroxyl ions. As these anions are rarely present in such minerals, this is a rather unusual example. The other coupled substitutions at the RE and anion sites, $RE^{3+} + O^{2-} = Ca^{2+} + (OH,F)^-$, is frequently observed in many minerals such as Ca-rich agardite [(Ca,Y)Cu$_6$[(AsO$_3$(O,OH)]$_3$(OH)$_6$·3H$_2$O] (Aruga and Nakai, 1985) and fersmite [(Ca,RE,Na)(Nb,Ta,Ti)$_2$(O,OH,F)$_6$] (Cummings and Simonsen, 1970).

2.4.4 Type 4: substitution accompanying valence variation

This substitution, accompanied with valence variation of iron ions, $RE^{3+} + Fe^{2+} = Ca^{2+} + Fe^{3+}$, is one proposed by Ito and Hafner (1974) for solid solutions of the gadolinite series. Calciogadolinite [YCaFe^{3+}Be$_2$Si$_2$O$_{10}$] can be derived from gadolinite-(Y) [Y$_2$Fe^{2+}Be$_2$Si$_2$O$_{10}$] with this substitution mechanism.

2.5 Crystal chemistry of gadolinite–datolite group and related minerals

The structures of gadolinite–datolite group minerals, e.g. gadolinite-(Y) [Y$_2$FeBe$_2$Si$_2$O$_{10}$] (Miyawaki, Nakai and Nagashima, 1984), hingganite-(Y) [YBeSiO$_4$(OH)] (Yakubovich et al., 1983), datolite [CaBSiO$_4$(OH)] (Foit, Phillips and Gibbs, 1973), homilite [Ca$_2$FeB$_2$Si$_2$O$_{10}$] (Miyawaki, Nakai and Nagashima, 1985), hydroxylherderite [CaBePO$_4$(OH)] (Lager and Gibbs, 1974), are isomorphous to each other and consist of sheets formed by the corner-sharing linkages of SiO$_4$, PO$_4$, BeO$_4$ and/or BO$_4$ tetrahedra. These minerals are related to each other by the isomorphous substitutions shown in Figure 2.11. Naturally occurring specimens of these minerals are solid solutions resulting from these substitutions (Ito and Hafner, 1974; Miyawaki and Nakai, 1993a; Demartin et al., 1993). The crystal structure of semenovite-

CRYSTAL CHEMISTRY OF GADOLINITE–DATOLITE GROUP

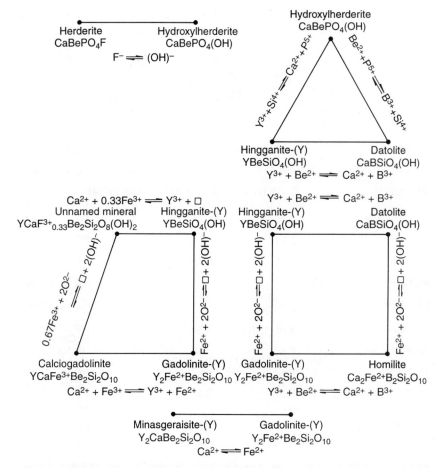

Figure 2.11 Phase relationships with isomorphous substitutions among the gadolinite–datolite group minerals.

(Ce) [(Ce,Na,□)$_2$(Na,□,Ce)$_2$(Fe,Mn,□)(□,Fe,Mn)(Ca,Na)$_8$Si$_8$ (Si,Be)$_6$O$_{40}$-(OH,O)$_2$(OH,F)$_6$] (Mazzi et al., 1979), nordite-(La) [(La,Ca)(Sr,Ca)Na$_2$-(Na,Mn)(Zn,Mg)Si$_6$O$_{17}$] (Bakakin et al., 1970), hellandite-(Y)[(Ca,Y)$_4$-Y$_2$(Al,Fe)B$_4$Si$_4$O$_{20}$(OH)$_4$] (Mellini and Merlino, 1977), tadzhikite-(Ce) [(Ca,Ce)$_4$Ce$_2$(Ti,Fe,Al)B$_4$Si$_4$(O,OH)$_2$O$_{22}$] (Chernitsova, Pudovkina and Pyatenko, 1982) and cappelenite-(Y) [BaY$_6$Si$_3$B$_6$O$_{24}$F$_2$] (Shen and Moore, 1984) consist of sheets made up of the tetrahedra of Si and B similar to the datolite–gadolinite group. In these structures large cations such as rare earths, calcium, sodium, barium and iron are situated between the sheets and connect the sheets to form three-dimensional structures. The intervals of the sheets are within the range 4.6–5.2 Å. In each sheet the tetrahedra

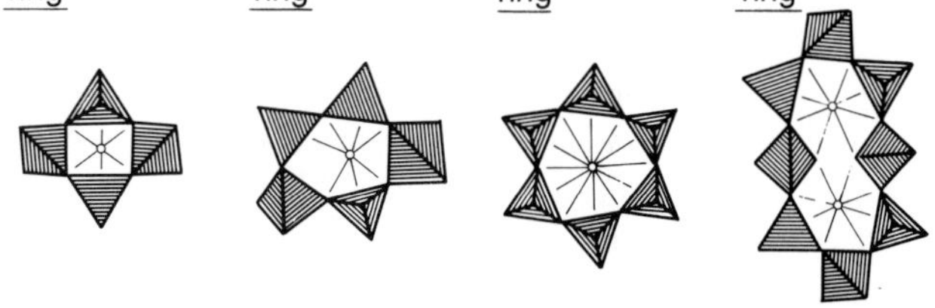

CRYSTAL CHEMICAL ASPECTS

	4-membered ring	5-membered ring	6-membered ring	8-membered ring
Interlayer cation	Fe,Mn,Al,Na	Na,Ca,RE	Ba	RE,Ca,Sr,Na
M–X distance[1] (Å)	1.97–2.38	2.47–2.52	2.96	2.38–2.60
Example	Interlayer cation and M–X distance[1] (Å)			
Gadolinite-(Y)[2]	Fe 2.18			RE 2.44
Hingganite-(Y)[3]	☐ 2.21			RE 2.42
Datolite[4]	☐ 2.16			Ca 2.49
Homilite[5]	Fe,Mn 2.15			Ca 2.49
Hydroxylherderite[6]	☐ 2.22			Ca 2.49
Nordite-(La)[7]	Na,Mn 2.38	Na 2.50		RE 2.50 Sr,Ca 2.60
Semenovite-(Ce)[8]	Fe,Mn 2.22 Fe,Mn 2.32	Na 2.52 Ca,Na 2.51		RE,Na 2.51 Na,RE 2.55
Hellandite-(Y)[9]	Al,Fe 1.97	Ca,RE 2.47		RE 2.44 Ca,RE 2.46
Tadzhikite-(Ce)[10]	Ti,Fe 1.98	Ca 2.47		RE 2.44 Ca 2.46
Cappelenite-(Y)[11]			Ba 2.96	RE 2.38 RE 2.43

Figure 2.12 Chemical compositions of the cation sites between the rings of tetrahedral sheets in the crystal structures of gadolinite and related minerals.

[1] Mean interatomic distance between the cation and coordination anions.
[2] $Y_2Fe^{2+}Be_2Si_2O_{10}$ (Miyawaki, Nakai and Nagashima, 1984).
[3] $YBeSiO_4(OH)$ (Yakubovich et al., 1983).
[4] $CaBSiO_4(OH)$ (Foit, Phillips and Gibbs, 1973).
[5] $Ca_2Fe^{2+}B_2Si_2O_{10}$ (Miyawaki, Nakai and Nagashima, 1985).
[6] $CaBePO_4(OH)$ (Lager and Gibbs, 1974).
[7] $LaSrNa_3ZnSi_6O_{17}$ (Bakakin et al., 1970).
[8] $Ce_2(Na,Ca)_9Fe(Si,Be)_{20}[O,(OH),F]_{48}$ (Mazzi et al., 1979).
[9] $(Ca,Y)_4Y_2(Al,Fe)B_4Si_4O_{20}(OH)_4$ (Mellini and Merlino, 1977).
[10] $(Ca,Ce)_4Ce_2(Ti,Fe,Al)B_4Si_4O_{22}[O,(OH)]_2$ (Chernitsova, Pudovkina and Pyatenko, 1982).
[11] $BaY_6Si_3B_6O_{24}F_2$ (Shen and Moore, 1984).

form 4-, 5-, 6- or 8-membered rings. Thus the kinds of cations occupying the sites between the rings depend on the volumes of the rings (Figure 2.12): i.e. large barium atoms occupy the sites between the 6-membered rings, while rare earth, calcium, strontium and sodium atoms occupy the sites between the 5- or 8-membered rings, and the relatively small space between the 4-membered rings tends to be occupied by relatively small atoms such as iron, manganese, aluminum and rarely sodium atoms.

2.6 Summary and conclusion

The rare earth minerals can be classified into six classes based on the arrangement of anionic groups. The rare earth atoms exhibit seven values of coordination number, from six to 12, in the crystal structure of rare earth minerals. The coordination numbers of the Ce-group rare earths are similar to those of calcium and are generally higher than those of Y-group rare earths. Consequently, isomorphous substitution between Ce-group rare earths and calcium is more commonly observed than that between Y-group rare earths and calcium. Even if the chemical formulae of the two minerals are similar to each other, the difference in the size of cations between the Y- and Ce-group rare earths results in different crystal structures, when these structures contain isolated anionic groups. The isomorphous substitutions between rare earths and the other heterovalent ions found in the crystal structures of rare earth minerals are coupled substitutions with charge compensation mechanisms, and can be summarized into the four types.

References

Aleksandrov, V. B. (1962) The crystal structure of aeschynite. *Dokl. Akad. Nauk SSSR*, **142**, 107–9.

Aruga, A. and Nakai, I. (1985) Structure of Ca-rich agardite, $(Ca_{0.40}Y_{0.31}Fe_{0.09}Ce_{0.06}La_{0.04} Nd_{0.01})Cu_{6.19}[(AsO_4)_{2.42}(HAsO_4)_{0.49}](OH)_{6.38} \cdot 3H_2O$. *Acta Crystallogr.*, **C41**, 161–3.

Bakakin, V. V., Belov, N. V., Borisov, S. V. and Solovyeva, L. P. (1970) The crystal structure of nordite and its relationship to melilite and datolite–gadolinite. *Am. Mineral.*, **55**, 1167–81.

Batalieva, N. G. and Pyatenko, Yu. A. (1972) Artificial yttrialite (y-phase) – a representative of a new structure type in the rare earth diorthosilicate series. *Sov. Phys. Crystallogr.*, **16**, 786–9.

Baudracco-Gritti, C., Quartieri, S., Vezzalini, G. *et al.* (1987) Une wakefieldite-(Ce) non plombifère: nouvelles données sur l'espèce minérale correspondant à l'orthovanadate de cérium. *Bull. Minéral.*, **110**, 657–63.

Bayliss, P. and Levinson, A. A. (1988) A system of nomenclature for rare earth mineral species; revision and extension. *Am. Mineral.*, **73**, 422–3.

Bevan, D. J. M., Strähle, J. and Greis, O. (1982) The crystal structure of tveitite, an ordered yttrofluorite mineral. *J. Solid State Chem.*, **44**, 75–81.

Chernitsova, N. M., Pudovkina, Z. V. and Pyatenko, Yu. A. (1982) Crystal structure of tadzhikite $\{(Ca,TR)_4(Y,TR)_2(Ti,Fe,Al)(O,OH)_2[Si_4B_4O_{22}]\}$. *Sov. Phys. Dokl.*, **27**, 367–8.

Cummings, J. P. and Simonsen, S. H. (1970) The crystal structure of calcium niobate ($CaNb_2O_6$) (fersmite). *Am. Mineral.*, **55**, 90–7.

Dal Negro, A., Rossi, G. and Tazzoli, V. (1975) The crystal structure of ancylite, $(RE)_x (Ca,Sr)_{2-x}(CO_3)_2(OH)_x(2 - x)H_2O$. *Am. Mineral.*, **60**, 280–4.
Dal Negro, A., Rossi, G. and Tazzoli, V. (1977) The crystal structure of lanthanite. *Am. Mineral.*, **62**, 142–6.
Davis, R. J. and Smith, G. W. (1971) Yttrotungstite. *Mineral. Mag.*, **38**, 261–85.
Demartin, F., Pilati, T., Diella, V. *et al.* (1993) A crystal-chemical investigation of alpine gadolinite. *Can. Mineral.*, **30**, 127–36.

Effenberger, H., Kluger, F., Paulus, H. and Wölfel, E. R. (1985) Crystal structure refinement of burbankite. *N. Jb. Mineral. Monatsh.*, **1985**, 161–70.

Foit, F. F., Phillips, M. and Gibbs, G. V. (1973) A refinement of the crystal structure of datolite, $CaBSiO_4(OH)$. *Am. Mineral.*, **58**, 909–14.

Gatehouse, B. M., Grey, I. E. and Kelly, P. R. (1979) The crystal structure of davidite. *Am. Mineral.*, **64**, 1010–7.
Ghose, S. and Wan, C. (1979) Agrellite, $Na(Ca,RE)_2Si_4O_{10}F$: a layer structure with silicate tubes. *Am. Mineral.*, **64**, 563–72.
Ghose, S., Sen Gupta, P. K. and Campana, C. F. (1987) Symmetry and crystal structure of montregianite, $Na_4K_2Y_2Si_{16}O_{38}\cdot 10H_2O$, a double-sheet silicate with zeolitic properties. *Am. Mineral.*, **72**, 365–74.

Hess, H. (1983) Die Kristallstruktur des Chlorotils, $SECu_6(AsO_4)_3(OH)_6\cdot H_2O$. *N. Jb. Mineral. Monatsh.*, **1983**, 385–92.
Hughes, J. M., Cameron, M. and Mariano, A. N. (1991) Rare-earth-element ordering and structural variations in natural rare-earth-bearing apatites. *Am. Mineral.*, **76**, 1165–73.

Ito, J. and Hafner, S. S. (1974) Synthesis and study of gadolinites. *Am. Mineral.*, **59**, 700–8.

Kalsbeek, N., Larsen, S. and Rønsbo, J. G. (1990) Crystal structures of rare earth elements rich apatite analogues. *Z. Kristallogr.*, **191**, 249–63.
Karpov, O. G., Pushcharovskii, D. Yu., Khomyakov, A. P. *et al.* (1980) Vitusite – a mineral with a disordered structure. *Sov. Phys. Crystallogr.*, **25**, 650–3.
Kinzhibalo, L. N., Trunov, V. K., Evdokimov, A. A. and Krongauz, V. G. (1982) Refinement of the crystal structure of fergusonite. *Sov. Phys. Crystallogr.*, **27**, 22–5.
Krstanović, I. (1965) Redetermination of oxygen parameters in xenotime, YPO_4. *Z. Kristallogr.*, **121**, 315–6.

Lager, G. A. and Gibbs, G. V. (1974) A refinement of the crystal structure of herderite $CaBePO_4OH$. *Am. Mineral.*, **59**, 919–25.
Levinson, A. A. (1966) A system of nomenclature for rare-earth minerals. *Am. Mineral.*, **51**, 152–8.

Mazzi, F., Ungaretti, L., Dal Negro, A. *et al.* (1979) The crystal structure of semenovite. *Am. Mineral.*, **64**, 202–10.
Mellini, M. and Merlino, S. (1977) Hellandite: a new type of silicoborate chain. *Am. Mineral.*, **62**, 89–99.
Mineev, D. A., Lavrischeva, T. I. and Bykova, A. V. (1970) Yttrium bastnaesite – a product of gagarivite alteration. *Zap. Vses. Mineral. Obshch.* **99**, 328–32.

REFERENCES

Miyawaki, R. and Nakai, I. (1987) Crystal structures of rare-earth minerals. *Rare Earths*, No. 11, 1–134.

Miyawaki, R. and Nakai, I. (1988) Crystal structures of rare-earth minerals, 1st supplement. *Rare Earths*, No. 13, 1–42.

Miyawaki, R. and Nakai, I. (1989) Crystal structures of rare-earth minerals, 2nd supplement. *Rare Earths*, No. 15, 1–12.

Miyawaki, R. and Nakai, I. (1990) Crystal structures of rare-earth minerals, 3rd supplement. *Rare Earths*, No. 17, 1–9.

Miyawaki, R. and Nakai, I. (1991) Crystal structures of rare-earth minerals, 4th supplement. *Rare Earths*, No. 19, 1–10.

Miyawaki, R. and Nakai, I. (1992) Crystal structures of rare-earth minerals, 5th supplement. *Rare Earths*, No. 21, 1–18.

Miyawaki, R. and Nakai, I. (1993a) Crystal structures of rare earth minerals. In K. A. Gschneidner, Jr. and L. Eyring (eds) *Handbook on the Physics and Chemistry of Rare Earths*, Vol. 16, Chapter 108, pp. 249–518.

Miyawaki, R. and Nakai, I. (1993b) Crystal structures of rare earth minerals, 6th supplement *Rare Earths*, No. 23, 1–21.

Miyawaki, R., Kuriyama, J. and Nakai, I. (1993) The redefinition of tengerite-(Y), $Y_2(CO_3)_3 \cdot 2-3H_2O$, and its crystal structure. *Am. Mineral.*, **78**, 425–32.

Miyawaki, R., Nakai, I. and Nagashima, K. (1984) A refinement of the crystal structure of gadolinite. *Am. Mineral.*, **69**, 948–53.

Miyawaki, R., Nakai, I. and Nagashima, K. (1985) Structure of homilite, $Ca_{2.00}(Fe_{0.90}Mn_{0.03})B_{2.00}Si_{2.00}O_{9.86}(OH)_{0.14}$. *Acta Crystallogr.*, **C41**, 13–5.

Moore, P. B. and Shen, J. (1983) Cerite, $RE_9(Fe^{3+},Mg)(SiO_4)_6(SiO_3OH)(OH)_3$: its crystal structure and relation to whitlockite. *Am. Mineral.*, **68**, 996–1003.

Moore, P. B., Sen Gupta, P. K., Schlemper, E. O. and Merlino, S. (1987) Ashcroftine, ca. $K_{10}Na_{10}(Y,Ca)_{24}(OH)_4(CO_3)_{16}(Si_{56}O_{140}) \cdot 16H_2O$, a structure with enormous polyanions. *Am. Mineral.*, **72**, 1176–89.

Nadezhina, T. N., Pushcharovskii, D. Yu. and Khomyakov, A. P. (1987) Refinement of crystal structure of belovite. *Mineral. Zh.*, **9**, 45–8.

Nagashima, K., Wakita, H. and Mochizuki, A. (1973) The synthesis of crystalline rare earth carbonates. *Bull. Chem. Soc. Jpn*, **46**, 152–6.

Ni, Y., Hughes, J. M. and Mariano, A. N. (1993) The atomic arrangement of bastnäsite-(Ce), $Ce(CO_3)F$, and structural elements of synchysite-(Ce), röntgenite-(Ce), and parisite-(Ce) *Am. Mineral.*, **78**, 415–8.

Peacor, D. R. and Dunn, P. J. (1988) Dollaseite-(Ce) (magnesium orthite redefined): structure refinement and implications for $F + M^{2+}$ substitutions in epidote-group minerals. *Am. Mineral.*, **73**, 838–42.

Pecora, W. T. and Kerr, J. H. (1953) Burbankite and calkinsite, two new carbonate minerals from Montana. *Am. Mineral.*, **38**, 1169–83.

Pertlik, F. and Preisinger, A. (1983) Crystal structure of sahamalite $(Mg,Fe)RE_2(CO_3)_4$. *Tschermaks Mineral. Petrol. Mitt.*, **31**, 39–46.

Qian, J., Fu, P., Kong, Y. and Gong, G. (1982) Determination of superstructure of huanghoite. *Acta Phys. Sinica*, **31**, 577–84.

Segalstad, T. V. and Larsen, A. O. (1978) Gadolinite-(Ce) from Skien southwestern Oslo region, Norway. *Amer. Mineral.*, **63**, 188–95.

Shannon, R. D. and Prewitt, C. T. (1969) Effective ionic radii in oxides and fluorides. *Acta Crystallogr.*, **B25**, 925–45.

Shannon, R. D. and Prewitt, C. T. (1970) Revised values of effective ionic radii. *Acta Crystallogr.*, **B26**, 1046–8.

Shen, J. and Moore, P. B. (1984) Crystal structure of cappelenite, $Ba(Y,RE)_6[Si_3B_6O_{24}]F_2$: a silicoborate sheet structure. *Am. Mineral.*, **69**, 190–5.

Shi, N., Ma, Z. and Peng, Z. (1982) The crystal structure of carbocernaite. *Kexue Tongbao*, **27**, 76–80.

Shinn, D. B. and Eick, H. A. (1968) The crystal structure of lanthanum carbonate octahydrate. *Inorg. Chem.*, **7**, 1340–5.

Shumyatskaya, N. G., Voronkov, A. A. and Pyatenko, Yu. A. (1980) Sazhinite, $Na_2Ce[Si_6O_{14}(OH)] \cdot nH_2O$: a new representative of the dalyite family in crystal chemistry. *Sov. Phys. Crystallogr.*, **25**, 419–23.

Szymanski, J. T. and Scott, J. D. (1982) A crystal-structure refinement of synthetic brannerite, UTi_2O_6, and its bearing on rate of alkaline-carbonate leaching of brannerite in ore. *Can. Mineral.*, **20**, 271–9.

Terada, Y., Nakai, I. and Kawashima, T. (1993) Crystal structure of bastnaesite (Ce,La,Nd,Sm,Gd) CO_3F. *Analytical Sciences*, **9**, 561–2.

Ueda, T. (1967) Reexamination of the crystal structure of monazite. *J. Jpn Assoc. Mineral. Petrol. Econ. Geol.*, **58**, 170–9.

Utsunomiya, A., Tanaka, K., Morikawa, H. *et al.* (1988) Structure refinement of $CaO \cdot 6Al_2O_3$. *J. Solid State Chem.*, **75**, 197–200.

Voronkov, A. A., Shumyatskaya, N. G. and Pyatenko, Yu. A. (1962) Crystal structure of gagarinite. *J. Struct. Chem. SSSR*, **3**, 665–9.

Weitzel, H. and Schröcke, H. (1980) Kristallstrukturverfeinerungen von Euxenit, $Y(Nb_{0.5}Ti_{0.5})_2O_6$, und M-fergusonit, $YNbO_4$. *Z. Kristallogr.*, **152**, 69–82.

Yakubovich, O. V., Matvienko, E. N., Voloshin, A. V. and Simonov, M. A. (1983) The crystal structure of hingganite-(Yb), $(Y_{0.51}TR_{0.36}Ca_{0.13}) \cdot Fe_{0.065}Be[SiO_4](OH)$. *Sov. Phys. Crystallogr.*, **28**, 269–71.

CHAPTER THREE
Perovskites: a revised classification scheme for an important rare earth element host in alkaline rocks

R. H. Mitchell

3.1 Introduction

Compounds that have the perovskite structure are of great scientific and economic importance. From a geophysical and mineralogical viewpoint they are of tremendous significance as it has been recognized recently that $MgSiO_3$-perovskite is the most abundant mineral in the Earth (Liu, 1975; Hazen, 1988; Navrotsky and Weidner, 1989). Many synthetic perovskite-structured compounds are ceramics which exhibit a wide array of electrical properties ranging from insulators through semiconductors to 'high-temperature' superconductors (Galasso, 1969; Hazen, 1988; Smyth, 1989; Wold and Dwight, 1993; Bourdillon and Bourdillon, 1994). Consequently, there is an extensive and rapidly increasing body of literature in the fields of materials science and solid state chemistry describing their synthesis, structure and properties.

Naturally occurring perovskite group minerals were first recognized by Gustav Rose (1839) in metamorphic rocks from the Ural Mountains, Russia. The name commemorates the Russian mineralogist Count Lev Aleksevich von Perovski. Subsequently, perovskite was found to be a common mineral in undersaturated alkaline rocks such as nepheline syenites, kimberlites, melilitites and carbonatites and to occur in meteorites and contact metamorphic or metasomatic rocks. Perovskites are one of the most important mineral hosts for the rare earth elements (REE) and all natural perovskites contain rare earth elements in concentrations ranging from trace through minor to major element status. Of these perovskites only the REE-rich variety, named loparite, strictly qualifies as a distinct rare earth mineral. However, many naturally occurring perovskites are complex solid solutions between loparite and other REE-free compositions. These solid solutions

Rare Earth Minerals: Chemistry, origin and ore deposits. Edited by Adrian P. Jones, Frances Wall and C. Terry Williams. Published in 1996 by Chapman & Hall. ISBN 0 412 61030 2

are discussed in detail in this work, as such perovskites are important hosts for the REE. Despite the ubiquity of perovskite in alkaline rocks there have been few modern investigations of its compositional variation or structure. Notable earlier works include those of Goldschmidt (1926), Nickel and McAdam (1963) and Vlasov, Kuzmenko and Eskova (1966). The structures of synthetic perovskites have, of course, been intensively investigated and are of use in understanding those of natural perovskites.

This work outlines a revised classification of perovskite group minerals based upon the structures of natural and synthetic perovskites and presents guidelines for describing the extensive compositional variation of naturally occurring rare earth-bearing perovskites. The phase relations and structures of silicate perovskites are not discussed in this work as they do not contain significant quantities of rare earth elements.

3.2 Perovskite structure

3.2.1 Ideal perovskite

Perovskites have the general structural formula ABX_3, where A and B are cations and X represents anions. The anions are oxygen in naturally occurring perovskites and the majority of ceramics but may be halogens in synthetic compounds such as $LiBaF_3$ and $CsCdBr_3$. The perovskite structure has an unusual ability to tolerate structural distortions within its framework. Consequently, a wide range of cations may occupy the A and B sites. Currently, over 20 different elements are known to be able to occupy the A site and over 50 elements may adopt the B site (Hazen, 1988; Galasso, 1969). The A-site cations, e.g. Na^+, Sr^{2+} or La^{3+}, are typically larger than the B-site cations, e.g. Cr^{3+}, Ti^{4+} or Nb^{5+}.

The ideal structure of perovskite (Náray-Szabó, 1943) may be viewed in several ways, depending upon whether the large A cations, B cations or oxygens are placed at the corners of a cubic unit cell. Figure 3.1a shows the basic structure of perovskite in which the A cation is placed at the centre of the unit cell. In this orientation, eight B cations occur at the corners of the cube and 12 oxygens are found at the midpoints of the cube edges. Figure 3.1a depicts the location of two corner-sharing TiO_6-polyhedra and Figure 3.1b shows that the structure may be visualized as consisting of eight corner-linked TiO_6-polyhedra grouped around a single A cation.

Figure 3.2a depicts two views of the perovskite structure with either A cations or oxygens at the corners of the cubic unit cell. In these representations, which are commonly used by material scientists, the structure is viewed as layers of atoms, rather than the polyhedron models favoured by mineralogists. This representation is useful in describing layered derivatives of the perovskite structure (Figure 3.2b), e.g. the Ruddlesden–Popper phases, $Sr_3Ti_2O_7$ (Ruddlesden and Popper, 1957, 1958) or $K_2La_2Ti_3O_{10}$

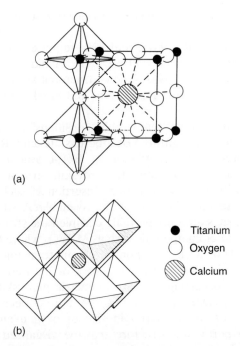

Figure 3.1 Structure of perovskite ABX$_3$, showing (a) the locations of atoms with Ti placed at the corners of the cubic unit cell. Two of the TiO$_6$ octahedral coordination polyhedra are shown (b) polyhedral model of the unit cell. (After Náray-Szabó, 1943 and Hazen, 1988.)

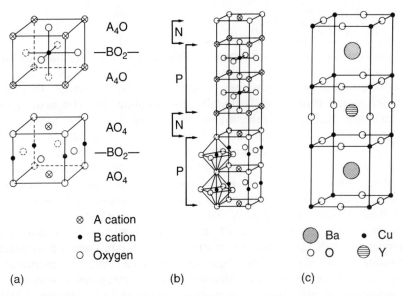

Figure 3.2 Structure of layered perovskites. (a) Ideal perovskite with the large cations or oxygens placed at the corners of the cubic unit cell (after Smyth, 1989), (b) the Ruddlesden–Popper compound Sr$_3$Ti$_2$O$_7$, and (c) the high-temperature superconductor YBa$_2$Cu$_3$O$_7$.

(Gopalakrishnan and Bhat, 1987), and high temperature superconducting compounds (Figure 3.2c), e.g. $YBa_2Cu_3O_7$ (Wold and Dwight, 1993; Bourdillon and Bourdillon, 1994). These structures are of direct relevance to complex naturally occurring REE-, Nb- and Sr-bearing perovskites (see below).

The ideal structure of perovskite may be considered as a derivative of the ReO_3 (or AlF_3) structure. Both ABO_3-perovskite and ReO_3 compounds consist of a corner-sharing framework of BO_6-octahedra located at the corners of a cubic unit cell. Relative to perovskite structure the unit cell of ReO_3 lacks a body-centred large cation. Insertion of such a cation results in a perovskite structure when all of the available A sites are occupied. However, if the A sites are only partially occupied the structure is known as a tungsten bronze, after compounds such as M_xWO_3 where $0 < x < 1$ and M is K, Na, Ba, Pb, Tl or REE. The non-stoichiometry is tolerated because of the fundamental similarites in the framework of the perovskite and ReO_3 structure. Naturally occurring non-stoichiometric perovskites analagous to tungsten bronzes are clearly possible but have not yet been recognised.

Finally, because of the similarity in size of oxygen anions and the A cations, the ideal perovskite structure may be visualized as a hexagonal close-packed array of A and O atoms with B cations occupying the octahedral holes. The close-packed array is parallel to (111) and consists of layers composed of oxygen atoms alone which alternate with layers composed of equal amounts of oxygen and the A cation.

3.2.2 Distorted perovskite structures

Many natural and synthetic perovskite structures deviate significantly from cubic symmetry at room temperature and pressure. Depending upon temperature, distorted perovskites commonly exhibit symmetries that range from tetragonal through orthorhombic to rhombohedral (Galasso, 1969). For example $BaTiO_3$ is cubic above 120°C but inverts to a tetragonal polymorph below this temperature. This phase remains stable until about 5°C, below which an orthorhombic phase is formed. This polymorph is stable until −90°C, when it inverts to a rhombohedral phase (Kay and Vousden, 1949). A similar pattern of phase transformations of decreasing symmetry with decreasing temperature has been recognized for many other perovskite compositions, e.g. $NaNbO_3$ and $SrTiO_3$ (Galasso, 1969). The phase transitions invariably result in the development of transformation twinning. Growth, glide and transformation twins are common in natural perovskites (Vlasov, 1966; Hu, Wenk and Sinitsyna, 1992). Transformation twins attest to the inversion of high-temperature tetragonal polymorphs to low-temperature orthorhombic derivatives (Wang and Liebermann, 1993).

Most naturally occurring $CaTiO_3$-rich perovskites are unlikely ever to have crystallized initially as cubic phases because of the high temperatures

(>1200°C) required for their stability (Liu and Liebermann, 1993; Guyot et al., 1993). In contrast NaNbO$_3$, which is cubic above 640°C (Glazer and Megaw, 1972), and SrTiO$_3$, which is cubic at ambient temperatures, undoubtedly initially crystallize from alkaline magmas as cubic phases.

Distorted perovskite structures may be produced in several ways:

1. Rotation of the octahedral polyhedra (Figure 3.3). These rotations result from the A cation being too small or too large for the 12-fold site. Tilt of the octahedra may be about any of the orthogonal crystallographic axes (Glazer, 1972, 1975). Figures 3.4a and 3.4b show how an original cubic cell is distorted to tetragonal symmetry by rotation of the octahedra about a tetrad [001], the ω angle of Megaw (1973) or the ϕ angle of Zhao et al. (1993). Further, tilting about another orthogonal diad axis [110], the φ angle of Megaw (1973) or the θ angle of Zhao et al. (1993), reduces the symmetry to orthorhombic (Figure 3.4c). Reduction in symmetry results in a larger unit cell ($a \approx c \approx \sqrt{2}/2b$; $Z = 4$), compared to the cubic cell ($Z = 1$). Rotations about [001] and [110] are equivalent to a single rotation (Φ) about [111] triad axis (Figure 3.3).

Glazer (1972, 1975), Aleksandrov (1976, 1978), Sasaki, Prewitt and Liebermann (1983) and Hyde and Andersson (1989) have discussed in detail the nature and classification of tilted octahedra. Megaw (1973), O'Keefe and Hyde (1977) and Zhao et al. (1993) give equations for assessing tilt angles from lattice parameters. Sasaki, Prewitt and Liebermann (1983) give relations between bond length distortions and Goldschmidt tolerance factors (see below), which are of use in predicting the structures of synthetic distorted perovskites. CaTiO$_3$-perovskite is an example of distorted perovskite structure formed by tilting of TiO$_6$-octahedra (Figure 3.5).

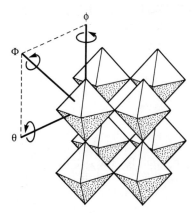

Figure 3.3 The octahedral framework of ideal perovskite showing the directions about which octahedra may be rotated to give orthorhombic and tetragonal derivatives.

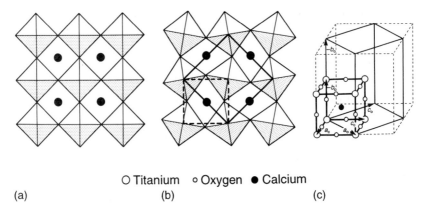

○ Titanium ○ Oxygen ● Calcium
(a) (b) (c)

Figure 3.4 (a) [001] projection of the octahedral framework of ideal perovskite. (b) Rotation of the octahedra about [001] showing how the cubic unit cell (dashed line) is transformed into a tetragonal unit cell (heavy solid line). (c) geometrical relationships between cubic and orthorhombic unit parameters (after Hu, Wenk and Sinitsyna, 1992).

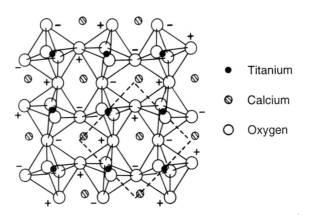

● Titanium
⊘ Calcium
○ Oxygen

Figure 3.5 Structure of orthorhombic perovskite. Positive and negative signs adjacent to apical oxygens indicate position of the octahedra above and below the plane of the illustration. (After Megaw, 1973.)

2. Displacement of the B cation from the centre of the BO_6-polyhedron due to the cation being too small for the octahedral site. $BaTiO_3$ below 120°C provides an example of this type of distortion, in which the symmetry is reduced to tetragonal and octahedra are not tilted. Brown (1992) has described in detail this type of distortion in terms of a bond-valence model.
3. Distortion of the BO_6-polyhedra due to pronounced covalency in the

B–O and/or A–O bonds, or Jahn–Teller effects. The latter type of distortion is commonly observed in Mn^{3+}-bearing perovskites (Goodenough et al., 1961).

Goldschmidt (1926) observed that the perovskite structure is stable if a tolerance factor, t, defined by $(R_A + R_O) = t\sqrt{2}(R_B + R_O)$ lies in the range $0.7 < t < 1.0$, where R refers to the ionic radii of the A, B and O atoms. If t lies outside these limits other structures exist. Only compounds for which t is close to unity adopt the cubic space group $Pm3m$, e.g. $SrTiO_3$ ($t = 1.002$ using Shannon and Prewitt (1969) ionic radii). The tolerance factor predicts that many naturally occurring Ca-rich perovskites will not be cubic at ambient temperatures, e.g. orthorhombic $CaTiO_3$ ($t = 0.970$; section 3.4). Discussion of tolerance factors may be found in Galasso (1969), Megaw (1973) and Sasaki, Prewitt and Liebermann (1983).

Sleight and Ward (1962) have modified the Goldschmidt tolerance factor to include complex perovskites such as $A(B^{3+}_{0.5},B^{5+}_{0.5})O_3$ (section 3.4.2). Thus

$$t = (R_A + R_O)/\sqrt{2}[(R^{3+}_B + R^{5+}_B)/2 + R_O] \quad (3.1)$$

Analogous equations may be derived for $(A^+_{0.5}A^{3+}_{0.5})B^{4+}O_3$ perovskites or complex solid solutions.

3.3 Perovskite space

Smyth (1989) has introduced a useful concept termed 'perovskite space' to illustrate variations in perovskite (*sensu lato*) compositions. Figure 3.6 shows the compositional relations of perovskites which are oxygen deficient or contain excess oxygen relative to ideal ABO_3 perovskite.

In this diagram the horizontal axis reflects variations in the oxygen/A cation ratio, which may range from 2.5 to 3.5. Compounds plotting on this line result from substitution of cations of either greater of lesser charge than that of the ion they replace. Thus, the substitution of Nb^{5+} for Ti^{4+} at B sites in $CaTiO_3$ results in the compound $Ca_2Nb_2O_7$ that has oxygen excess relative to ideal perovskite. Substitution of Fe^{3+} for Ti^{4+} at B sites in $CaTiO_3$ results in the oxygen deficient compound $Ca_2Fe_2O_5$. Compounds with excess or deficient oxygen typically form layered derivative structures of the ideal perovskite structure (Smyth, 1989; Hazen, 1988). The case of excess oxygen is particularly relevant to some naturally occurring perovskites and is discussed further below.

The vertical axis of Figure 3.6 includes perovskites with constant oxygen contents (oxygen/A-cation ratio = 3) and describes self-compensational or coupled cationic substitutions at the A and B sites, such that the sum of the nominal cation charge always totals six. Three groups of simple perovskites

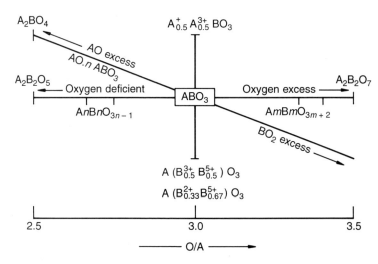

Figure 3.6 Perovskite compositional space. O/A = oxygen/A-site cation ratio. (After Smyth, 1989.)

are recognized: $A^+B^{5+}O_3$, $A^{2+}B^{4+}O_3$ and $A^{3+}B^{3+}O_3$. In many instances A^{2+} may be replaced by equal amounts of A^+ and A^{3+}, e.g. $(Na_{0.5},Ce_{0.5})TiO_3$, while B may be replaced by $0.5B^{3+}$ and $0.5B^{5+}$, e.g. $Ca(Fe^{3+}_{0.5},Nb^{5+}_{0.5})O_3$, $0.33B^{2+}$ and $0.66B^{5+}$, e.g. $Ca(Fe^{2+}_{0.33},Nb^{5+}_{0.66})O_3$ or $0.5B^{2+}$ and $0.5B^{6+}$, e.g. $Sr(Fe_{0.5}U_{0.5})O_3$. All of these compounds are termed complex perovskites by Galasso (1969). Many of them have the $(NH_4)_3FeF_6$ variant of the perovskite structure (Figure 3.7). Cation ordering occurs in many of these compounds (Filipev and Fesenko, 1965; Galasso, 1969). Complex substitutions of a similar type account for the majority of the compositional variation observed in natural perovskites (see below).

The inclined axis in Figure 3.6 describes perovskites with changes in the A/B ratios and variations in oxygen/A-cation ratio and includes Ruddlesden–Popper structures, e.g. $A_{n+1}B_nO_{3n+1}$ or AO-nABO$_3$ (Ruddlesden and Popper, 1957, 1958; Smyth, 1989). Perovskites with excess A cations of this type, such as $Sr_3Ti_2O_7$ ($n = 2$), consist of layers of SrO between n layers of SrTiO$_3$-perovskite (Figure 3.2b). Naturally occurring perovskites belonging to this series have not yet been described. Compounds with excess B cations do not adopt the perovskite structure.

3.4 Naturally occurring perovskites

3.4.1 Compositional variation and potential end-member compounds

The compositional variation of naturally occurring perovskites is principally along the vertical axis (oxygen/A-cation ratio = 3) of Figure 3.6, and along

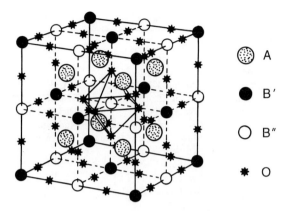

Figure 3.7 Structure of complex perovskites. (After Pauling, 1924.)

the horizontal axis towards perovskites containing excess oxygen (oxygen/A-cation ratio = 3.5). This restricted compositional range relative to synthetic perovskites is to be expected, given the relatively oxidizing character and limited range of compositional environments in which they form. Although oxygen-deficient Fe^{3+}-bearing perovskites such as brownmillerite ($Ca_2Fe_2O_5$) can form in air, it is considered unlikely that any oxygen-deficient perovskites will be found in common alkaline rocks, although such perovskites should be sought in highly reducing environments.

In natural perovskites A-site cations include Na^+, K^+, Ca^{2+}, Sr^{2+}, rare earth elements $(REE)^{3+}$, Pb^{2+} and Ba^{2+}, while B-site cations include Ti^{4+}, Nb^{5+}, Fe^{3+}, Fe^{2+}, Ta^{5+}, Th^{4+} and Zr^{4+}. Natural perovskites are invariably complex perovskites and represent complex solid solutions between several end-member compositions. Potential and actually recognized end-member perovskite compositions which best describe the compositional variation of natural perovskites are listed in Table 3.1. However, given the vast number of potential complex perovskites, other compounds may also be considered as potential end members, e.g. $Ca(La_{0.5}Nb_{0.5})O_3$.

Figure 3.8 depicts a modified version (this work) of the 'perovskite space' of Smyth (1989), which demonstrates that most naturally occurring perovskites have compositions which may be expressed in terms of only seven end-member compounds. The majority of these are cubic and orthorhombic perovskites (see below) whose compositions fall in the quaternary system $NaNbO_3$ (lueshite)–$(Na_{0.5}Ce_{0.5})TiO_3$ (loparite)–$SrTiO_3$ (tausonite)–$CaTiO_3$ (perovskite). For ease of graphical representation, compositions in terms of molecular percentages may be depicted in three ternary subsystems of the quaternary system: (1) lueshite–loparite–perovskite, (2) perovskite–loparite–tausonite and (3) lueshite–loparite–tausonite. Natural perovskites

Table 3.1 End-member perovskite compositions

Composition	Name	Type locality	Reference
$CaTiO_3$	Perovskite	Ural Mtns, Russia	Rose (1839)
$(Na_{0.5}Ce_{0.5})TiO_3$	Loparite	Khibina, Kola, Russia	Kuznetsov (1925)
$NaNbO_3$	Lueshite	Lueshe, Zaire	Safiannikoff (1959)
$Ca(Fe_{0.5}Nb_{0.5})O_3$	Latrappite	Oka, Quebec	Nickel (1964)
$PbTiO_3$	Macedonite	Crni Kaman, Macedonia	Radusinović and Markov (1971)
$SrTiO_3$	Tausonite	Murun complex, Russia	Vorobyev et al. (1984)
$KNbO_3$		Synthetic	
$BaTiO_3$		Synthetic	
$Ce_2Ti_2O_7$		Synthetic	
$Ca_2Nb_2O_7$		Synthetic	
$CaZrO_3$		Synthetic	
$CaThO_3$		Synthetic	

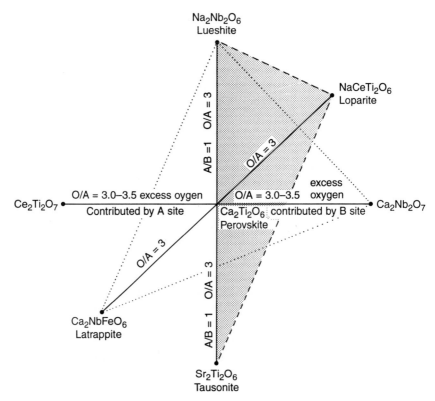

Figure 3.8 Compositional space of naturally occurring perovskites. O/A = oxygen/A-site cation ratio.

belonging to the ternary system perovskite–tausonite–lueshite have not yet been recognized.

Perovskites which are rich in REE^{3+}, Fe^{3+} and Nb^{5+}, but relatively poor in Na, are considered to represent primarily quaternary solid solutions between distorted perovskites such as lueshite, latrappite and $CaTiO_3$ and the compounds, $Ce_2Ti_2O_7$ and $Ca_2Nb_2O_7$, which contain excess oxygen. The latter are chosen as end members because they have structures which are layered derivatives of the perovskite structure, e.g. $Ce_2Ti_2O_7$ (Preuss and Gruehn, 1994), $La_2Ti_2O_7$ (Gasperin, 1975; Williams et al., 1991), $Ca_2Nb_2O_7$ (Scheunemann and Muller-Buschbaum, 1974). Other layered perovskites which have structures of this type which might be relevant to substitutions in natural perovskites include $Sr_2Nb_2O_7$ (Ishizawa et al., 1975), $Sr_2Ta_2O_7$ (Smyth, 1989) and $NaLaNb_2O_7$ (Sato, Jin and Uematsu, 1993). Such compounds have a greater probability of forming a solid-solution series with ideal perovskites than a compound such as $CaNb_2O_6$ (fersmite), which has a very different structure (Aleksandrov, 1960).

Thus, in Figure 3.8 the horizontal compositional axis is divided into compounds which gain their excess oxygen from either A- or B-site cations. Two ternary subsystems, $Ce_2Ti_2O_7$–latrappite–perovskite and $Ca_2Nb_2O_7$–latrappite–perovskite may be used to illustrate the compositional variation of such perovskites.

The light REE-dititanates, $REE_2Ti_2O_7$, where REE = La, Ce, Pr, Nd and Sm, are all layered perovskite compounds with monoclinic unit cells ($P2_1$; $a = 7.8–7.6$; $b = 5.55–5.43$; $c = 13.01–12.98$; $\beta \approx 98°$) at ambient temperatures (Preuss and Guehn, 1994). An orthorhombic modification of $La_2Ti_2O_7$ exists above 800°C (Williams et al., 1991) which is identical in structure to orthorhombic $Ca_2Nb_2O_7$ ($Pn2_1a$; $a = 25.457$; $b = 5.501$; $c = 7.692$; Scheunemann and Muller-Buschbaum, 1974). Other isomorphs undoubtedly exhibit similar phase relations.

The structures of the layered REE-dititanates and $Ca_2Nb_2O_7$ may be visualized by regarding the perovskite structure as cut apart at regular intervals parallel to (100), such that the corners of BO_6 octahedra which normally share oxygens are separated. Extra oxygens are in effect then added at these corners, thus creating the oxygen excess relative to the ABO_3 structure. The structure is now represented by slabs of perovskite of n layers of octahedra in thickness separated by planes in which the excess oxygen is accommodated (Figure 3.9a and b). Crystallographic shear takes place along the planes of excess oxygen by an amount equal to $a/2$, and the octahedra within the layers may be rotated to give monoclinic variants (Figure 9c and d). The structure represents a homologous series $A_{n+1}B_{n+1}O_{3n+5}$. As n tends to infinity the amount of excess oxygen which has to be accommodated decreases, i.e. the slabs become thicker as the number of octahedral units increases, and the simple perovskite structure is approached.

Clearly, REE-dititanates (or other members of the homologous series)

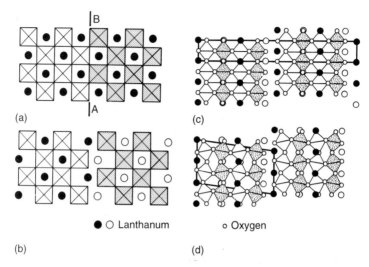

Figure 3.9 Structure of layered light-REE dititanates and $Ca_2Nb_2O_7$ illustrated by $La_2Ti_2O_7$. (a) Framework of ideal perovskite viewed down [001] with (100) plane (A–B) along which separation of slabs of TiO_6 octahedra occurs. (b) [001] projection showing separated slabs four TiO_6 octahedra in thickness and the displacement of La cations relative to the ideal perovskite structure. Filled and open circles depict La atoms differing in height along [100] by $a/2$. (c) [010] projection showing the structure of orthorhombic compounds. (d) [010] projection of monoclinic compounds. Stippled octahedra and filled circles depict polyhedra and La atoms at the same height in the projection. (After Williams *et al.*, 1991.)

and $Ca_2Nb_2O_7$ are able to form intergrowths with the orthorhombic perovskites that comprise the major solid-solution series in natural perovskites. These features will be enhanced at high temperatures when they probably exist as orthorhombic variants. Thus, it postulated that accommodation of Nb and REE in Na-deficient perovskites takes place by the intercalation of slabs of layered perovskites. Such slabs should be readily imaged by HRTEM in compounds such as latrappite. Compounds composed of such slabs have been reported in studies of the $Ca_2Nb_2O_7$–$NaNbO_3$ system (Portier *et al.*, 1974).

Note that heavy REE-dititanates do not adopt the layered perovskite structure and form pyrochlore-group compounds (Brixner, 1964). This may be one of the underlying reasons why natural perovskites are typically light REE-enriched compounds (section 3.9).

3.4.2 Structure

Surprisingly little is known about the structure of naturally occurring oxide perovskites as opposed to synthetic variants. Some data are available for end-member compositions but refinements of the structures of the complex solid solutions that are typically found have not been undertaken.

Perovskite ($CaTiO_3$) is orthorhombic ($t = 0.970$) at ambient temperatures and may be considered to belong to the space groups *Pnma* ($a = 5.4405$; $b = 7.6436$; $c = 5.3812$; JCPDS File 22-153) or *Pbnm* ($a = 5.3796$; $b = 5.4423$; $c = 7.6401$; Sasaki et al., 1987). The latter orthorhombic setting corresponds to the lanthanide orthoferrite or $GdFeO_3$ structure. All of the octahedra are tilted (Náray-Szabó, 1943; Kay and Bailey, 1957; Sasaki et al. 1987; Hyde and Andersson, 1989), with $\Phi = 10.16°$, and may be distorted (Liu and Liebermann, 1993).

Synthetic tausonite ($SrTiO_3$) is cubic (*Pm3m*; $a = 3.9050$) at ambient temperatures (Galasso, 1969). Tausonite is the only naturally occurring cubic perovskite ($t = 1.002$). Tausonites are typically extensively compositionally zoned and their crystal structure has not been accurately determined. Substitution of Na, REE at the A site undoubtedly results in a reduction in symmetry and thus tausonite–loparite solid solutions are expected to be orthorhombic (see Sr-loparite below).

Synthetic loparite ($NaLaTi_2O_6$) is orthorhombic (*Pnma*; $a = 5.476$; $b = 7.777$; $c = 5.481$; pseudocubic cell $a_p = 3.872$; $t = 0.959$; $\Phi = 5.80°$; this work). Natural loparites deviate significantly from the ideal end-member composition and contain significant amounts of Ca, Sr and Nb. Such loparites are also orthorhombic, e.g. Sr-loparite from Sarambi ($a = 5.497$; $b = 7.788$; $c = 5.494$; $a_p = 3.886$; $\Phi = 2.87°$; Haggerty and Mariano, 1983). Although there are as yet insufficient data to prove conclusively, it is possible that in the system $NaLaTi_2O_6-Sr_2TiO_6$ the parameter a_p increases and Φ angle decreases with increasing tausonite content. Sr-poor, Nb-bearing loparites from the Kola Peninsula region, Russia (Vlasov, 1966) are also orthorhombic ($a = 5.491$; $b = 7.728$; $c = 5.457$; $a_p = 3.8605$). Ordering of A cations in loparites has not been investigated. However if the observations of Galasso (1969) for B-cation ordering also apply to the A site, it can be expected that ordering will be present given the very different atomic numbers of Na and La.

Lueshite ($NaNbO_3$) at ambient temperatures has the space group (*$P222_1$*; $a = 5.512$; $b = 5.557$; $c = 15.540$; $a_p = 3.91$; $t = 0.942$; JCPDS File 14-603) and thus solid solutions with loparite are also expected to be orthorhombic.

Latrappite (Ca_2NbFeO_6) is a complex perovskite ($t = 0.952$) which Filipev and Fesenko (1965) consider to be monoclinic ($a = 3.89$; $b = 3.88$; $c = 3.89$; $\beta = 91.2°$). However, natural perovskites rich in the latrappite component are orthorhombic (*Pbnm*; $a = 5.4479$; $b = 5.5259$; $c = 7.1575$; Hawthorne and Mitchell, unpublished data).

3.5 Nomenclature of naturally occurring perovskites

Figures 3.10–3.13 illustrate perovskite compositions plotted as molecular percentages of end-member components in standard ternary compositional

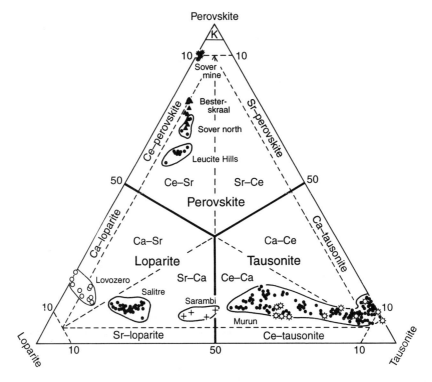

Figure 3.10 Compositions of perovskites in the system perovskite–loparite–tausonite. Data sources given in the text.

diagrams. Note that the existing terminology of perovskite is preserved, although some new, as yet unnamed, end members (Table 3.1) are introduced for non-stoichiometric Ce and Nb-rich, and zirconium-bearing, perovskites.

For these ternary systems major compositional divisions are as recommended by Nickel (1992). Thus, individual perovskite solid solutions are initially named on the basis of the '50% rule' for dominant end-member molecules. However, experience has shown that a simple name is inadequate for describing the complex compositional variation exhibited by natural perovskites (Mitchell and Vladykin, 1993; Mitchell and Platt, 1993; Platt, 1994). Therefore it is recommended here that prefixes be added to the basic name to reflect the dominant cations present in the ternary solid solution. Such compound names permit description of the compositional variation found within particular perovskite solid solutions, e.g. the tausonite, calcian tausonite, cerian calcian tausonite series (Mitchell and Vladykin, 1993). These perovskites of very different composition would all be uninformatively termed tausonite using the unmodified ternary nomenclature diagram of

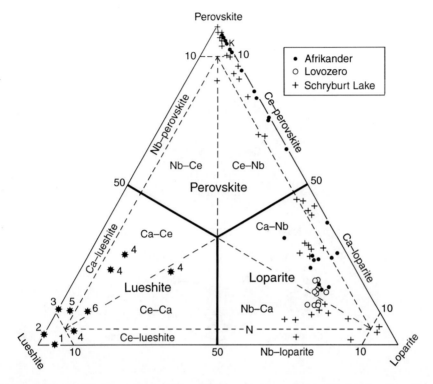

Figure 3.11 Compositions of perovskites in the system perovskite–loparite–lueshite. Data sources given in the text. K = knopite, N = niobolparite. Numbered compositions of lueshites (*) are from: 1, Lueshe (this work); 2, Lueshe (Safiannikoff, 1959); 3, Kovdor (Kukharenko *et al.*, 1965); 4, Kovdor (Lapin and Kazakova, 1966); 5, Siberian carbonatite (Bagdasarov *et al.*, 1962); 6, natroniobite, Lesnaya Varaka, Kola (Bulakh *et al.*, 1962).

Nickel (1992). Prefixes should be assigned in order of decreasing abundance of elements. 'Pure' end members are considered to be those perovskites which contain more than 90 mol.% of a particular end member.

The last revision of perovskite terminology was undertaken by Nickel and McAdam (1963) who recommended that most perovskite compositions could be represented in terms of the ternary system lueshite–loparite–perovskite. Note that Nickel and McAdam (1963) also recommended the use of appropriate modifiers to the three end members, e.g. niobian perovskite, to reduce the proliferation of names. However, as knowledge of the compositional variation of perovskite has increased, this classification has been shown to be inadequate. This is especially important with regard to the ternary diagrams utilized by Nickel and McAdam (1963) to illustrate perovskite compositional variation. These are based upon structural formulae calculated on the basis of three oxygens. As a consequence of

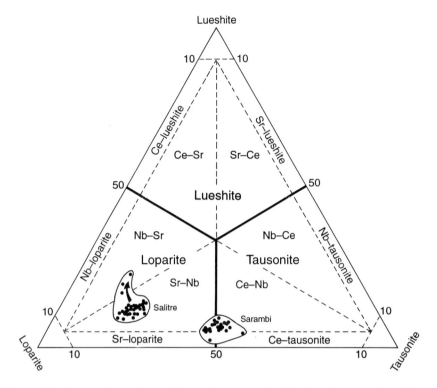

Figure 3.12 Compositions of perovskites in the system lueshite–loparite–tausonite. All data this work.

non-stoichiometry many Fe^{3+}, Nb and REE-rich compositions plot in the 'forbidden region' in the ternary system Na–Ca–REE (Haggerty and Mariano, 1983; Mitchell and Steele, 1992). This is a consequence of not considering perovskites as defect structures.

Names which are considered, either in this work and or by Nickel and McAdam (1963), to be obsolete include igdloite, nioboloparite, knopite, dysanalyte and irinite.

Dano and Sörenson (1959) have described a sodium niobate replacing eudialyte from the Igdlunguaq nepheline syenite (Greenland) and named this material 'igdloite'. However, the mineral is inadequately described and is probably merely lueshite containing inclusions of other minerals.

Nioboloparite (Tikonenkov and Kazakova, 1957) and knopite (Holmqvist, 1894) have been suggested as varietal names for Nb and REE-bearing perovskites respectively. Knopite is merely a perovskite (Figure 3.11) containing small amounts of REE (<10 mol.% loparite) The name has been previously considered as unnecessary by Nickel and McAdam (1963). Nioboloparite is better described as niobian loparite (Figure 3.11), for which

STRUCTURAL FORMULAE

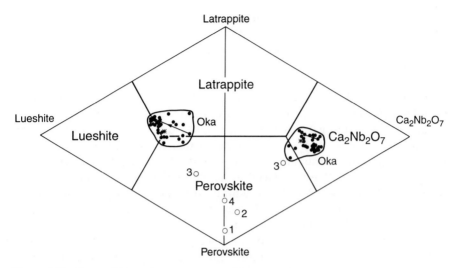

Figure 3.13 Compositions of perovskites in the systems lueshite–latrappite–perovskite and latrappite–perovskite–$Ca_2Nb_2O_7$. All data for perovskites from La Trappe, Oka, Quebec are from this work with the exception of the bulk composition of 'latrappite' (*) from Nickel and McAdam (1963). Numbered compositions of Nb-bearing perovskites or 'dysanalytes' (o) from: 1, Oka (this work); 2, Magnet Cove (Lupini, Williams and Woolley, 1992); 3, Kaiserstuhl (Meigen and Hugel, 1913); 4, Kaiserstuhl (Hauser, 1908).

a separate name is not required. Reasons for the elimination of irinite and dysanalyte as useful names are given in sections 3.8.4 and 3.8.5 respectively.

A mineral termed 'metaloparite' has been reported from the Lovozero complex by Gerasimovskii (1941). The mineral occurs as an alteration product of loparite and is considered by Borodin (1954) to have the composition $(Ce_{0.34},Ce_{0.15})(Ti_{0.87},Nb_{0.13})O_{2.46}(OH)_{0.53}$. The mineral is reported to contain 3.49 wt% H_2O and to be poor in Na_2O (0.23 wt%). The conversion of loparite to metaloparite is believed to involve the loss of Na and introduction of water, resulting in elevated Ti, Nb and REE contents (Vlasov, 1966). The status of metaloparite as a mineral is unclear as there have been no modern investigations of its composition or paragenesis.

The mineral termed 'natroniobate' is considered to be the monoclinic dimorph of lueshite by Bulakh et al. (1962). In terms of its composition it is actually calcian cerian lueshite (Figure 3.11).

3.6 Structural formulae and end-member compositions

Calculation of structural formulae and end-member compositions on the basis of two cations and three oxygens/unit cell is satisfactory for the majority of perovskite solid solutions falling in the quaternary system lueshite–loparite–perovskite–tausonite where A/B = 1 and A/O = 3.

Perovskites containing substantial amounts of the $Ce_2Ti_2O_7$ and $Ca_2Nb_2O_7$ components are by definition non-stoichiometric with respect to ABO_3 and individual end members may be calculated only from oxide molecular equivalents.

Molecular recalculation or end-member component schemes have experienced considerable popularity in mineralogy and have proven to be extremely useful in describing the compositional variation found in such diverse mineral groups as pyroxenes, garnets and spinels. Unfortunately, in any recalculation scheme one is presented with the problem that the calculated composition depends upon the sequence of assignment of cations to particular end members. This problem is reduced if the end members are formed from the least to the most abundant. Structural formulae and end-member compositions presented in this paper were calculated using a recalculation scheme written as an APL computer program.

Vector representation of perovskite compositions as suggested by Burt (1989) is another manner of illustrating the compositional variation of complex solid solutions. With respect to perovskites, this method assumes stoichiometry and does not provide an easy means of recognizing the proportions of end members present or describing compositional variation. Thus, the vector $NaCeCa_{-2}$, describing solid solutions between loparite and perovskite, provides no significant improvement over conventional terminology ($Na^+, Ce^{3+} = 2Ca^{2+}$) for coupled solid solutions. Consequently, there are no compelling mineralogical grounds to adopt this method of representation in preference to well-established plots of cation ratios or end-member recalculation schemes.

In the recalculation procedures employed in this work all REE are treated as forming compounds which are isomorphous with loparite. Note that all loparites so far analysed are properly termed, using the system of nomenclature of REE minerals advocated by Levinson (1966), loparite-(Ce). This suffix, for reasons of brevity and clarity, and because most perovskites are REE-bearing minerals rather than REE minerals, is omitted in all references to loparite in this work.

The oxidation state of iron in perovskites has been insufficiently studied. Commonly, total Fe is calculated as FeO when abundances are low (Mitchell, 1972, 1986) or as Fe_2O_3 in cases where abundances are high (Lupini, Williams and Woolley, 1992; Nickel and McAdam, 1963). It should be realized that these procedures are quite arbitrary. Mössbauer studies of perovskites are required to determine the oxidation state of iron. Calculation of the oxidation state of iron on the basis of stoichiometry is clearly not possible for non-stoichiometric Nb- and iron-rich perovskites.

3.7 Paragenesis

Perovskites are a common accessory mineral in a wide variety of undersaturated alkaline rocks and carbonatites (Nickel and McAdam, 1963; Vlasov, 1966; Kapustin, 1980; Haggery and Mariano, 1983; Mitchell, 1986; Ulrych *et al.*, 1988; McCallum, 1989; Michell and Bergman, 1991; Mitchell and Vladykin, 1993). Perovskites are commonly complexly twinned (Vlasov, 1966) as a consequence of the phase transitions which occur during cooling (Hu, Wenk and Sinitsyna, 1992).

Perovskites typically occur as discrete crystals, although they are also commonly found intergrown with other oxide minerals. Thus, mantles of perovskite may occur on earlier-formed ilmenite as a consequence of reaction of the latter with groundmass-forming fluids. In turn perovskites themselves are commonly mantled by other minerals which form as a consequence of changes in oxygen fugacity, silica and/or fluorine activity in the magma, e.g. perovskite mantled by titanite, titanite mantled by perovskite, perovskite decomposing to rutile, perovskite overgrown by pyrochlore, perovskite mantled by Ba-hollandite.

In alkaline rocks perovskites occur as:

1. Single-phase, early-crystallizing discrete euhedral to subhedral crystals. In kimberlites, olivine lamproites, melilitites and alnoites these may be only very weakly zoned and are commonly relatively poor in REE (1–5 wt% total REE_2O_3). In perovskite pyroxenites, ultrapotassic syenites and carbonatites these may be REE-rich and strongly and complexly zoned (Mitchell and Vladykin, 1993; Platt, 1994).
2. Discrete early-forming euhedral to subhedral REE- and Nb-poor crystals mantled by subhedral overgrowths of a second generation of REE and Na-rich perovskite. These are common in alkaline plutonic rocks but have not been observed in kimberlites or lamproites. The formation of the second-generation perovskite is attributed to reaction of earlier-formed perovskite with residual (deuteric or hydrothermal) fluids (Platt, 1994).
3. Poikilitic late-stage groundmass plates. REE- and Sr-rich perovskites exhibiting this habit have been recognized only in some lamproites (Mitchell and Steele, 1992) and orangeites (Mitchell, 1995).

3.8 Major element compositional variation of perovskite in alkaline rocks

3.8.1 Analysis of perovskite

Initial studies of the major element composition of perovskite were undertaken by gravimetric or X-ray fluorescence methods of bulk crystals. Recent studies utilizing electron backscattered imagery and electron microprobe methods have demonstrated that these techniques are inappropriate for most natural perovskites which exhibit pronounced compositional zoning. Thus, many of the compositions of loparite, lueshite and latrappite presented by Vlasov (1966), Lapin and Kazakova (1966), Tikhonenkov and Kazakova (1957) and Nickel and McAdam (1963) are undoubtedly average compositions. The structural formulae of these minerals suggest that many of these data are erroneous, especially with regard to the abundances of the alkali elements, titanium and niobium. Reinvestigation of many of these perovskites by modern techniques is being undertaken by Mitchell and Chakmouradian (University of St Petersburg).

Since about 1970 wavelength- and energy-dispersive electron microprobe techniques have provided the only suitable methods of analysis of complexly zoned perovskite crystals. In many of the initial studies only partial compositions are presented due to lack of suitable REE standards. In examples for which REE were determined (Ilupin, Khomyakov and Balashov, 1971; Blagulkina and Tarnovskaya, 1975; Boctor and Boyd, 1979, 1982), there are unfortunately serious analytical errors arising from a combination of poor choice of analytical lines and backgrounds, inadequate assessment of interferences and less than optimum sensitivity for the beam current and operating voltages employed (Jones and Wyllie, 1984; Mitchell and Reed, 1988). Suitable analytical protocols for wavelength-dispersive electron microprobe analysis of the REE and other elements in perovskite are given by Jones and Wyllie (1984), Mitchell and Vladykin (1993), Dawson, Smith and Steele (1994) and Williams (Chapter 13). Energy-dispersive methods of analysis have been utilized by Mitchell and Steele (1992) and Platt (1994).

Perovskite compositional variation is described below with reference to three ternary systems and one quaternary system as studies have shown that perovskites in alkaline rocks fall into four compositional groups:

1. Nb-poor, Ca-, Sr- and REE-rich perovskites which represent solid solutions between perovskite, loparite and tausonite.
2. Sr-poor, Nb-, Ca- and REE-rich perovskites which represent solid solutions between perovskite, loparite and lueshite.
3. Ca-poor, Na-, REE-, Nb- and Sr-rich perovskites which represent solid solutions between lueshite, loparite and tausonite.
4. Sr-poor, Na-, Ca- and Nb-rich perovskites which represent solid solutions between lueshite, latrappite, perovskite and the non-stoichiometric layered perovskite $Ca_2Nb_2O_7$.

3.8.2 Ternary system perovskite–loparite–tausonite

Figure 3.10 illustrates the composition of Nb-poor, Sr-bearing perovskites from Leucite Hills lamproites, the Sover and Besterskraal orangeites (also known as group 2 kimberlites; Mitchell, 1994, 1995) and ultrapotassic syenites from the Little Murun alkaline complex.

Perovskite in the leucite phlogopite transitional madupitic lamproite of Middle Table Mountain, Leucite Hills (Wyoming) occurs as poikilitic groundmass plates. Individual crystals are homogeneous and little intergrain compositional variation is present. Figure 3.10 and Table 3.2 show that these perovskites are cerian strontian perovskites.

Perovskites occur as discrete subhedral crystals and as poikilitic groundmass minerals in orangeites (Boctor and Boyd, 1982; Mitchell, 1994, 1995). Figure 3.10 and Table 3.3 indicate that the former are perovskite (and cerian perovskite) and the latter are cerian strontian perovskites. The evolutionary trend of composition is from $CaTiO_3$ towards loparite with slightly increasing tausonite contents.

Perovskites occurring in the ultrapotassic syenites from the Little Murun complex, Siberia, contain the highest Sr contents of all perovskites so far examined (Table 3.4). The complex is the type locality for tausonite (Vorobyev et al., 1984). Mitchell and Vladykin (1993) and Vorobyev, Konyev and Maklyshok (1987) have shown that these perovskites are complexly zoned and range in composition from tausonite through calcian tausonite to cerian calcian tausonite (Table 3.4; Figure 3.10). Evolution is from tausonite towards Sr-Ca loparite with slightly increasing perovskite contents.

Table 3.2 Representative compositions of Sr-bearing perovskites from lamproite

	Wt% oxides		Cations/3 oxygens			Mol.% components		
	1	2		1	2		1	2
La_2O_3	3.78	3.94	La	0.036	0.038	Lueshite	–	0.07
Ce_2O_3	8.12	8.41	Ce	0.076	0.080	Loparite	28.80	30.45
Pr_2O_3	0.81	0.82	Pr	0.008	0.008	$BaTiO_3$	0.28	0.28
Nd_2O_3	2.51	2.62	Nd	0.023	0.024	Tausonite	9.74	9.85
Sm_2O_3	0.18	0.17	Sm	0.002	0.002	Perovskite	61.18	59.35
SrO	6.45	6.48	Sr	0.096	0.097			
CaO	22.1	21.4	Ca	0.606	0.592			
BaO	0.28	0.28	Ba	0.003	0.003	% Cations not assigned		
Na_2O	2.86	3.01	Na	0.142	0.151			
			A =	0.992	0.995	A site	2.53	0.74
						B site	2.09	2.11
Nb_2O_5	0.90	0.93	Nb	0.010	0.011			
TiO_2	51.2	50.7	Ti	0.986	0.985			
FeO	0.53	0.53	Fe	0.011	0.011			
	99.72	99.29	B =	1.007	1.007			

Total Fe calculated as FeO. Compositions 1 and 2 are perovskites from leucite phlogopite lamproite, Middle Table Mountain, Leucite Hills, Wyoming (Mitchell and Steele, 1992). Both samples contain 0.03 wt.% ZrO_2. Mg, Mn, Ta and Th were not determined.

Table 3.3 Representative compositions of perovskites from orangeites

	Wt% oxides		Cations/3 oxygens		Mol.% components		
	1	2	1	2		1	2
ThO_2	n.d.	0.18	Th –	0.001	Lueshite	0.36	0.0
La_2O_3	1.51	3.38	La 0.013	0.031	Loparite	3.20	19.61
Ce_2O_3	3.54	8.00	Ce 0.031	0.083	Tausonite	1.03	4.97
Pr_2O_3	n.d.	0.74	Pr –	0.007	Perovskite	95.41	75.42
Nd_2O_3	0.81	2.18	Nd 0.007	0.020			
Sm_2O_3	n.d.	0.10	Sm –	0.001			
SrO	0.70	3.27	Sr 0.009	0.048			
CaO	35.39	26.86	Ca 0.894	0.724			
BaO	n.d.	n.d.	Ba –	–	% Cations not assigned		
Na_2O	1.06	1.93	Na 0.048	0.094			
			A = 1.002	1.009	A site	2.99	3.79
					B site	3.34	4.73
Nb_2O_5	0.32	0.65	Nb 0.003	0.007			
TiO_2	54.60	50.75	Ti 0.968	0.961			
FeO	2.64	1.25	Fe 0.052	0.026			
	100.57	99.29	B = 1.023	0.994			

Total Fe calculated as FeO. Composition 1 is subhedral early-forming perovskite, Sover Mine, South Africa. Composition 2 is groundmass poikilitic perovskite, Besterskraal, South Africa, n.d. = not detected. All data this work.

Figure 3.10 also illustrates the compositions of Sr-rich perovskites from the Sarambi and Salitre fenites (see below). These perovskites are rich in Nb and do not strictly fall within this ternary system. Compositions are projected into this system from lueshite to show that a continuous solid solution between tausonite and loparite appears to be possible.

Figure 3.10 shows that natural perovskites with compositions along the binary join perovskite–tausonite have not yet been reported. Phase equilibrium studies have shown that there is complete solid solution between $CaTiO_3$ and $SrTiO_3$ at temperatures above 1300°C (Čeh, Kolar and Golic, 1987). The absence of natural $(Sr,Ca)TiO_3$ may reflect either low-temperature immiscibility in this system or bulk compositional control of perovskite compositions.

Many of the perovskites which plot in Figure 3.10 coexist with Ba- and K-titanates and titanosilicates such as K-Ba-hollandite, barium pentatitanate, shcherbakovite and barytolamprophyllite. Their BaO contents are typically low (<1.0 wt% BaO), suggesting that the complete solid solution which has been observed in the system $BaTiO_3$–$SrTiO_3$ at high temperature (Wechsler and Kirby, 1992), is not present at low temperatures. Alternatively, partition coefficients for Ba between perovskite and other titanates in natural silicate systems may not favour incorporation of Ba in the perovskite structure.

Table 3.4 Representative compositions of tausonite and strontian loparite

	Wt% oxides		Cations/3 oxygens		Mol.% components		
	1	2	1	2		1	2
ThO_2	n.d.	2.00	Th –	0.013	Lueshite	0.21	0.15
La_2O_3	0.68	6.34	La 0.007	0.069	$BaTiO_3$	0.68	0.58
Ce_2O_3	1.23	9.59	Ce 0.014	0.103	Loparite	5.05	37.41
Pr_2O_3	0.06	0.54	Pr 0.001	0.006	Perovskite	7.27	11.04
Nd_2O_3	0.31	1.30	Nd 0.003	0.014	Tausonite	86.79	49.83
Sm_2O_3	n.d.	0.04	Sm –	–			
SrO	49.40	27.78	Sr 0.862	0.475			
CaO	2.24	3.33	Ca 0.072	0.105	% Cations not assigned		
BaO	0.57	0.48	Ba 0.007	0.006			
Na_2O	0.59	3.31	Na 0.034	0.189	A site	0.73	2.80
			A = 1.000	0.980	B site	1.10	5.02
Nb_2O_5	0.15	0.79	Nb 0.002	0.011			
Ta_2O_5	n.d.	0.05	Ta –	–			
TiO_2	44.03	44.48	Ti 0.997	0.986			
FeO	0.19	0.26	Fe 0.005	0.006			
	99.48	100.29	B = 1.003	1.003			

Total Fe calculated as FeO. Compositions 1 and 2 from the Little Murun ultrapotassic complex, Aldan Shield, Siberia (Mitchell and Vladykin 1993). n.d. = not detected.

Naturally occurring perovskites containing $BaTiO_3$ as the dominant end-member molecule have not yet been reported.

3.8.3 Ternary system lueshite–loparite–perovskite

Perovskites whose compositions may be represented by this system (Figure 3.11) comprise the majority of perovskites occurring in alkaline undersaturated rocks, perovskite pyroxenites and carbonatites.

Sr- and REE-poor perovskites in kimberlites and melilitites (McCallum, 1989; Ulrych et al., 1988) plot at the perovskite apex of this system as they typically contain more than 90 mol.% $CaTiO_3$. The evolutionary trend of compositions in this system has been established from studies of the Lovozero and Khibina nepheline syenite complexes, the Kovdor and Afrikanda (also spelt Afrikander) (Vlasov, Kuzmenko and Eskova, 1966; Vlasov, 1966; Chakmouradian, 1994; this work), Sayan (Chernysheva et al., 1990), and Schryburt Lake (Platt, 1994) carbonatite complexes, and Brazilian perovskite pyroxenite complexes (Mariano and Mitchell, 1991). It should be particularly noted that the data reported by Vlasov, Kuzmenko and Eskova (1966) and Vlasov (1966) are average bulk compositions and greater compositional variation than illustrated on Figure 3.11 is undoubtedly present. Perovskites evolve from perovskite (*sensu stricto*) through cerian perovskite towards calcian niobian loparite and niobian calcian loparite and then towards niobian loparite. Representative compositions are given in Table 3.5. Loparites from the Afrikanda (this work) and Schryburt Lake (Platt, 1994) carbonatite–ultramafic rock complexes are poor in SrO (<0.32 wt%) relative to loparites from Lovozero nepheline syenites (0.6–4.4 wt%) SrO, 1–10 mol.% $SrTiO_3$; Figure 3.6; Vlasov, Kuzmenko and Eskova, 1966).

Loparitic perovskites are considered to evolve ultimately during differentiation towards lueshite. However, perovskites having the composition of lueshite (Table 3.5) are extremely rare (Figure 3.11). Apart from the type locality (Lueshe, Congo; Safiannikoff, 1959), they have only been reported from the Kovdor carbonatite (Kukharenko *et al.*, 1965; Lapin and Kazakova, 1966), the Lesnaya Varaka and Sallanlahti carbonatites, Kola Peninsula (Bulakh *et al.*, 1962), unspecified Siberian carbonatites (Bagdasarov *et al.*, 1962; Kapustin, 1980) and the Gem Park (Colorado) carbonatite (Parker and Sharp, 1970). The majority of the reported compositions are in error due to inaccurate Na determinations. Some compositions approach that of

Table 3.5 Representative compositions of cerian perovskite, loparite and lueshite

	Wt% oxides				Cations/3 oxygens				Mol.% components		
	1	2	3		1	2	3		1	2	3
ThO_2	0.62	1.94	0.55	Th	0.004	0.013	0.003	Tausonite	0.43	0.33	–
La_2O_3	4.46	8.53	0.44	La	0.042	0.089	0.005	Lueshite	2.55	12.76	94.55
Ce_2O_3	11.12	21.64	1.12	Ce	0.103	0.225	0.011	Perovskite	64.20	17.42	0.85
Pr_2O_3	0.89	1.66	0.22	Pr	0.008	0.017	0.002	Loparite	32.81	69.49	4.60
Nd_2O_3	3.16	4.68	0.27	Nd	0.029	0.048	0.003				
Sm_2O_3	0.30	0.17	n.d.	Sm	0.003	0.002	–				
SrO	0.29	0.18	n.d.	Sr	0.004	0.003	–				
CaO	22.86	5.21	1.05	Ca	0.619	0.159	0.031	% Cations not assigned			
BaO	n.d.	n.d.	0.02	Ba	–	–	–				
Na_2O	3.73	7.85	15.81	Na	0.183	0.433	0.855	A site	2.88	6.50	2.60
				A =	0.993	0.987	0.910	B site	3.30	0.02	13.11
Nb_2O_5	2.15	9.03	76.10	Nb	0.025	0.116	0.959				
Ta_2O_5	0.07	0.02	0.35	Ta	0.001	–	0.003				
TiO_2	50.98	40.47	2.35	Ti	0.969	0.865	0.049				
FeO	0.14	0.25	n.d.	Fe	0.003	0.006	–				
	100.77	101.63	98.28	B =	0.997	0.987	1.011				

Total Fe calculated as FeO. Compositions 1 and 2 are cerian perovskite and calcian niobian loparite, respectively, Afrikanda complex, Kola, Russia. Composition 3 is lueshite, Lueshe, Zaire. All data this work. n.d. = not detected.

ideal NaNbO$_3$ and, apart from the Kovdor examples, very little solid solution towards loparite or perovskite is present. Lueshites may contain 1–2 wt% K$_2$O, presumably reflecting very limited solid solution towards KNbO$_3$ (Kukharenko et al., 1965; Lapin and Kazakova, 1966). Lueshites from Kovdor are considered by Lapin and Kazakova (1966) to contain 0.9–4.20 wt% Ta$_2$O$_5$. These exceptionally high contents of Ta may reflect analytical errors. Accurate analyses for Na in lueshite are required to determine whether or not some of these perovskites are actually non-stoichiometric niobium bronzes.

Figure 3.11 also shows that perovskites with compositions falling along the binary join perovskite–lueshite have not yet been recognized.

3.8.4 Ternary system lueshite–loparite–tausonite

Perovskites occurring in rocks which are considered to be rheomorphic fenites (Haggerty and Mariano, 1983) are Sr- and REE-rich and contain significant amounts of Nb. Figures 3.10 and 3.12 show that each 'fenite' contains perovskite of distinct composition (Table 3.6). Perovskites from Sarambi (Paraguay) range in composition from strontian niobian loparite to

Table 3.6 Representative compositions of strontian loparite and niobian tausonite

	Wt% oxides			Cations/3 oxygens				Mol.% components		
	1	2	3		1	2	3		1	2	3
ThO$_2$	0.09	0.70	6.19	Th	–	0.005	0.041	Lueshite	7.09	11.39	20.60
La$_2$O$_3$	8.60	10.12	8.23	La	0.095	0.109	0.088	Perovskite	10.79	9.78	10.44
Ce$_2$O$_3$	12.32	16.24	12.74	Ce	0.135	0.174	0.135	Loparite	38.02	59.00	56.07
Pr$_2$O$_3$	0.72	1.01	0.78	Pr	0.008	0.011	0.009	Tausonite	44.11	19.80	12.95
Nd$_2$O$_3$	1.00	2.31	1.89	Nd	0.011	0.024	0.019				
Sm$_2$O$_3$	n.d.	0.09	0.12	Sm	–	0.001	0.002				
SrO	23.34	11.07	6.95	Sr	0.406	0.187	0.116				
CaO	3.09	2.96	3.03	Ca	0.099	0.093	0.094				
BaO	n.d.	n.d.	n.d.	Ba	–	–	–	% Cations not assigned			
Na$_2$O	4.13	6.84	8.09	Na	0.240	0.387	0.453				
				A =	0.996	0.996	0.955	A site	7.50	4.40	5.97
								B site	7.47	4.95	10.89
Nb$_2$O$_5$	4.81	8.04	14.17	Nb	0.065	0.106	0.185				
Ta$_2$O$_5$	n.d.	0.21	n.d.	Ta	–	0.002	–				
TiO$_2$	40.13	40.15	37.75	Ti	0.906	0.882	0.819				
FeO	0.97	0.27	0.17	Fe	0.024	0.006	0.004				
	99.20	100.01	100.11	B =	0.995	0.990	1.008				

Total Fe calculated as FeO. Composition 1 is cerian niobian tausonite, Sarambi, Paraguay. Compositions 2 and 3 are strontian niobian loparite and Th-bearing, niobian strontian loparite, respectively, Salitre, Brazil. All data this work. n.d. = not detected.

cerian niobian tausonite. At Salitre-I (Brazil) perovskites are primarily strontian niobian loparites.

Perovskites from Salitre are in some examples zoned towards niobian strontian loparites which contain significant amounts of ThO_2 (4–6 wt%). Similar high Th contents are found in 'niobioloparites' from Khibina (Mitchell and Chakmouradian, unpublished). Perovskites containing 13.0 wt% ThO_2, termed 'irinite' have been reported from pegmatites in the Khibina complex (Kola Peninsula, Russia) by Borodin and Kazakova (1954). These and the material from Salitre do not contain Th as the dominant B-site cation and are best termed thorian loparites. Nickel and McAdam (1963) have previously recommended that the name irinite be abandoned. The Th-rich material from Khibina should be re-examined.

3.8.5 Quaternary system lueshite–latrappite–perovskite–$Ca_2Nb_2O_7$

In 1963 Nickel and McAdam described a new variety of perovskite from the Oka carbonatite complex (Quebec). This was initially termed niobian perovskite but was subsequently named 'latrappite' by Nickel (1964) on the basis that the Nb content was greater than the Ti content. Note that the 'latrappite' composition as determined by Nickel and McAdam (1963) did not represent an end-member composition. The end-member complex perovskite, ($Ca_2Nb^{5+}Fe^{3+}O_6$), had previously been recognized by Tikhonenkov and Kazakova (1957) as a major component of 'dysanalyte' (Knop, 1877; Hauser, 1908) from the Kaiserstuhl complex, but was not given a specific name.

Nickel and McAdam's (1963) composition was obtained by gravimetric methods and represented an average bulk composition. Table 3.7 and Figure 3.13 present new compositional data for 'latrappite' from the type locality. These data show clearly that latrappite is a complex solid solution in the quaternary system lueshite–latrappite–perovskite–$Ca_2Nb_2O_7$, and that the layered perovskite $Ca_2Nb_2O_7$ is the dominant end-member component.

The term 'dysanalyte' is imprecisely defined. Compositions of perovskites, termed 'dysanalyte', summarized by Vlasov (1966) vary widely in terms of their Nb, REE and Fe contents and analytical totals. Recalculation of the data suggests significant errors in the determination of Ti. The majority of these perovskites appear to be niobian cerian loparites. Only dysanalyte from the type locality (Meigen and Hugel, 1913) is closely similar in composition to Oka 'latrappite' (Figure 3.13).

Although latrappites from the type locality do not have end-member compositions, it is suggested that the name be retained for the Ca_2NbFeO_6 component. This recommendation is based on the grounds that $Ca_2Nb_2O_7$ is a synthetic end member and 'dysanalyte' is not adequately characterized. Nickel and McAdam (1963) have previously recommended that the term dysanalyte be abandoned. However, it is recognized and could be argued

Table 3.7 Representative compositions of lueshite–perovskite–latrappite–$Ca_2Nb_2O_7$ solid solutions

	Wt% oxides		Cations/3 oxygens		Mol.% components		
	1	2		1	2		1	2
ThO_2	n.d.	0.07	Th	–	–	Tausonite	0.27	0.32
La_2O_3	0.70	0.45	La	0.007	0.004	Loparite	4.41	2.73
Ce_2O_3	1.66	0.98	Ce	0.016	0.009	Lueshite	11.31	20.90
Pr_2O_3	0.10	0.05	Pr	0.001	–	Latrappite	23.48	18.21
Nd_2O_3	0.36	0.26	Nd	0.003	0.002	Perovskite	24.44	19.54
Sm_2O_3	0.07	0.01	Sm	0.001	–	$Ca_2Nb_2O_7$	35.87	38.31
SrO	0.19	0.22	Sr	0.003	0.003			
CaO	29.82	26.97	Ca	0.848	0.766	% Mol. equiv. not assigned		
BaO	n.d.	n.d.	Ba	–	–			
Na_2O	2.33	3.96	Na	0.120	0.204	A site	8.52	0.04
			A =	0.999	0.991	B site	1.80	14.9
Nb_2O_5	40.53	47.91	Nb	0.486	0.574			
Ta_2O_5	n.d.	n.d.	Ta	–	–			
TiO_2	11.71	8.89	Ti	0.234	0.177			
Fe_2O_3	8.92	6.89	Fe	0.198	0.153			
MnO	0.27	0.34	Mn	0.006	0.008			
MgO	1.54	1.89	Mg	0.061	0.074			
	99.19	99.66	B =	0.985	0.987			

Total Fe calculated as Fe_2O_3. Compositions 1 and 2 are core and rim of zoned euhedral crystal, Oka, Quebec. All data this work. n.d. = not detected.

that, because of the dominance of the $Ca_2Nb_2O_7$ molecule in latrappite, this name be given to this end member and dysanalyte be retained for Ca_2NbFeO_6 on the grounds of historical precedence.

3.8.6 Other perovskites

Some perovskites from kimberlites, carbonatites and alnöites are characterized by low or negligible Na coupled with significant Nb and REE contents (Table 3.8). Solid solutions in such perovskites cannot be towards either loparite or lueshite. Recalculation of the compositions suggests that these perovskites are characterized by the presence of limited solid solution towards the $Ce_2Ti_2O_7$, $Ca_2Nb_2O_7$ and latrappite end members. These perovskites have been insufficiently investigated.

Radusinović and Markov (1971) have reported the existence of the end member $PbTiO_3$, as euhedral crystals within an amazonite granite pegmatite and named the mineral 'macedonite'. The material from the type locality is characterized by the presence of significant amounts of Bi replacing Pb at

Table 3.8 Representative compositions of Na-poor, REE-bearing perovskites from kimberlite and alnöite

	Wt% oxides		Cations/3 oxygens			Mol.% components		
	1	2		1	2		1	2
ThO_2	1.02	0.11	Th	0.006	0.001	Tausonite	0.28	0.71
La_2O_3	1.11	2.39	La	0.010	0.022	Lueshite	0.28	0.0
Ce_2O_3	2.41	3.69	Ce	0.021	0.033	Latrappite	1.27	5.42
Pr_2O_3	0.18	0.15	Pr	0.002	0.001	$Ca_2Nb_2O_7$	0.0	0.48
Nd_2O_3	1.94	1.43	Nd	0.017	0.013	$Ce_2Ti_2O_7$	2.35	3.56
Sm_2O_3	n.d.	n.d.	Sm	–	–	Perovskite	95.81	89.83
SrO	0.23	0.58	Sr	0.003	0.008			
CaO	35.89	36.35	Ca	0.928	0.951	% Mol. equiv. not assigned		
BaO	n.d.	n.d.	Ba	–	–			
Na_2O	0.06	n.d.	Na	0.003	–	A site	0.42	18.47
			A =	0.989	1.028	B site	0.04	15.46
Nb_2O_5	0.60	1.98	Nb	0.007	0.022			
Ta_2O_5	n.d.	n.d.	Ta	–	–			
TiO_2	53.66	49.90	Ti	0.974	0.916			
FeO	0.87	2.42	Fe	0.018	0.049			
	97.97	99.00	B =	0.998	0.988			

Total calculated as FeO. Compositions: 1, euhedral perovskite in kimberlite, Frank Smith Mine, South Africa; 2, euhedral perovskite in alnöite, Oka, Quebec. All data this work; n.d. = not detected.

the A site (70.55 wt% PbO, 2.20 wt% Bi_2O_3, 2.69 wt% FeO, 25.07 wt% TiO_2).

Pb-bearing perovskites are apparently very rare. The only other known occcurrence is in ultrapotassic syenites from the Little Murun complex, Siberia (Vorobyev *et al.*, 1987). Note that not all perovskites in this complex are Pb-bearing. These perovskites are essentially tausonites (76–85 mol.%) that exhibit significant amounts of solid solution towards macedonite (0.6–8.6 mol.% $PbTiO_3$, corresponding to 0.71–10.2 wt% PbO), loparite (4.2–7.4 mol.%) and perovskite (9.2–11.0 mol.%).

Perovskites of unusual composition have been described recently from the Polino carbonatite by Lupini, Williams and Woolley (1992). These are characterized by the presence of 2.8–3.3 wt% ZrO_2 and are the only known occurrence of Zr-rich perovskite. The minerals have high CaO, TiO_2 and Fe_2O_3 (5.7 wt%) contents but low Nb_2O_5 (0.22 wt%) and Na_2O (0.12 wt%) contents, and hence are essentially perovskites (*sensu stricto*) exhibiting limited solid solution towards the $CaZrO_3$ (2.8 mol.%), loparite (1.7 mol.%), $Ce_2Ti_2O_7$ (1.4 mol.%) and latrappite (0.8 mol.%) end members.

Lueshite containing 1.03 wt% ZrO_2 has been reported by Lapin and Kazakova (1966) from the Kovdor carbonatite.

3.9 Rare earth element distribution patterns

Very few reliable data are available on the relative distribution of REE in perovskites. This is because REE abundances quoted in earlier works (Vlasov, 1966; Blagulkina and Tarnovskaya, 1975; Boctor and Boyd, 1979, 1982; Boctor and Yoder, 1980) are either inaccurate and/or the analytical techniques were not sufficiently sensitive to allow the determination of the heavy REE. However, most recent studies (Jones and Wyllie, 1984; Mitchell and Reed, 1988; Mitchell and Steele, 1992; Dawson, Smith and Steele, 1994) are in agreement that all perovskites are enriched in the light REE (La–Gd) relative to the heavy REE (Tb–Lu).

Figure 3.14 illustrates chondrite-normalized REE distribution patterns for perovskites from diverse alkaline rocks. Light REE enrichments are so great that La/Yb ratios of perovskites typically exceed 2000 (Mitchell and Reed,

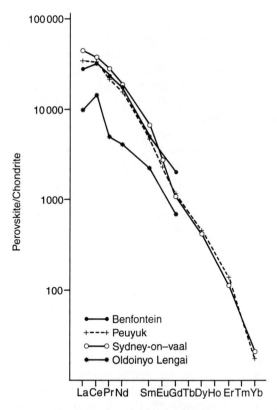

Figure 3.14 Representative chondrite-normalized REE distribution patterns for perovskites. 1. Benfontein kimberlite (Jones and Wyllie, 1984); 2, Peuyuk kimberlite (Mitchell and Reed, 1988); 3, Sydney-on-Vaal orangeite (Mitchell and Reed, 1988); 4, Oldoinyo Lengai Jacupirangite (Dawson, Smith and Steele, 1994).

1988). La_{CN} may be greater or less than Ce_{CN}. Negative Eu anomalies have only been recognized by Eby (1975) in perovskites from the Oka carbonatite. This material requires re-examination as Eu anomalies were not found in perovskites from alnöites of the same complex (Mitchell and Reed, 1988).

3.10 Conclusions

This review has shown that the major element compositions of naturally occurring perovskites may be represented by only seven end-member compositions. The majority of these perovskites have compositions within the quaternary system lueshite–loparite–perovskite–tausonite. Perovskites with ternary or quaternary compositions in this system do not occur and most compositions cluster near the binary joins loparite–tausonite, perovskite–loparite and loparite–lueshite. This observation suggests that there are significant, but as yet unknown, geological or thermodynamic (i.e. low-temperature immiscibility) controls on the compositions of perovskites that crystallize from alkaline undersaturated magmas.

Perovskites are one of the principal hosts for REE in alkaline rocks. They are typically strongly enriched in the light REE. Few reliable data are available on the distribution of REE in perovskites from a wide range of parageneses. Reinvestigation of the trace-element geochemistry of the REE and other elements (Sc, Y, Zr, Pb, Th, U, etc.) known to be concentrated in perovskites is desirable.

Perovskites from la Trappe (Oka, Quebec) and the Kaiserstuhl (Germany) carbonatites are unusual in that they are members of a quaternary solid-solution series involving lueshite, latrappite, perovskite and the postulated non-stoichiometric layered perovskite compound, $Ca_2Nb_2O_7$. Limited solid solution towards this end member and $Ce_2Ti_2O_7$ occurs in some Na-deficient perovskites in kimberlites and alnöites.

Perovskites containing significant contents of K, Ta, Zr, Th and Ba do not occur in undersaturated alkaline rocks. In these rocks perovskites form in environments that are commonly rich in these elements, and the reason for their paucity in perovskite is unknown. Note that Sitnin and Leonova (1961) claim that loparites in albitized and greisenized granites contain 1.95 wt% Ta_2O_5 and 0.86 wt% SnO_2. These minerals should be re-examined to determine whether they really are perovskites and whether or not the high Ta contents are artifacts of the analytical methods.

Detailed modern mineralogical investigations of the composition of perovskites from all parageneses are required in conjunction with phase equilibrium studies. The latter should be designed so as to lead to an understanding of the physico-chemical controls on the composition of perovskites crystallizing in undersaturated silicate and carbonatite magmas. Initial studies of this type have been undertaken by Veksler and Teptelev (1990), Jago (1990) and Jongejan and Wilkins (1972).

Although this review is concerned primarily with the composition of naturally occurring REE-bearing perovskites, it is evident that many other aspects of perovskite mineralogy have been insufficiently investigated. Useful crystallographic studies that might be undertaken include the nature of cation ordering and lattice defects in complex solid solutions. Mossbäuer, infrared, visible light and cathodoluminescence spectrometric studies of perovskites are also required.

Acknowledgements

This research is supported by the Natural Sciences and Engineering Research Council of Canada and Lakehead University. Garth Platt and Anton Chakhmouradian are thanked for discussions on the occurrence and composition of perovskites. Alexandra Navrotsky, Linda Campbell and Frances Wall are thanked for reviews of the initial draft of this paper.

References

Aleksandrov, K. S. (1976) The sequences of structural phase transitions in perovskites. *Ferroelectrics*, **14**, 801–5.

Aleksandrov, K. S. (1978) Mechanisms of the ferroelectric and structural phase transitions. Structural distortions in perovskites. *Ferroelectrics*, **20**, 61–7.

Aleksandrov, V. B. (1960) Kristallicheskaya struktura fersmita. (Crystal structure of fersmite.) *Dokl. Akad. Nauk SSSR*, **132**, 669–72 (in Russian).

Bagdasarov, Y. A., Gaidukova, V. S., Kuznetsova, N. N. and Sidorenko, G. A. (1962) Nakhodka lueshita v karbonatitakh Sibiri. (Discovery of lueshite in Siberian carbonatites.) *Dokl. Akad. Nauk SSSR*, **147**, 1168–71 (in Russian).

Blagulkina, V. H. and Tarnovskaya, A. N. (1975) O perovskite iz kimberlitov Yakutii. (Perovskite from Yakutian kimberlites.) *Zap. Vses. Mineral. Obshch.*, **104**, 703–10 (in Russian).

Boctor, N. Z. and Boyd, F. R. (1979) Distribution of rare earth elements in perovskite from kimberlites. *Carnegie Inst. Washington Yearbook*, **78**, 572–4.

Boctor, N. Z. and Boyd, F. R. (1982) Petrology of kimberlite from the DeBruyn and Martin Mine, Bellsbank, South Africa. *Am. Mineral.*, **67**, 917–25.

Boctor, N. Z. and Yoder, H. S. (1980) Distribution of rare earth elements in perovskite from the Oka carbonatite, Quebec. *Carnegie Inst. Washington Yearbook*, **79**, 304–7.

Borodin, L. S. (1954) O vkhozhdenii vody v kristallicheskuyu reshetku mineralov gruppy perovskita. (Water in the crystal lattice of minerals of the perovskite group.) *Dokl. Akad. Nauk SSSR*, **95**, 873–5 (in Russian).

Borodin, L. S. and Kazakova, M. E. (1954) Irinit – novii mineral iz gruppy perovskita. (Irinite, a new member of the perovskite group.) *Dokl. Akad. Nauk SSSR*, **97**, 725–8 (in Russian).

Brixner, L. H. (1964) Preparation and properties of the $Ln_2Ti_2O_7$-type rare earth titanates. *Inorg. Chem.*, **3**, 1065–7.

Bourdillon, A. and Bourdillon, T. (1994) *High Temperature Superconductors: Processing and Science*, Academic Press, San Diego.

Brown, I. D. (1992) Chemical and steric constraints in inorganic solids. *Acta Cryst.*, **B48**, 553–72.

Bulakh, A. G., Kukharenko, A. A., Knipovich, Y. N. *et al.* (1962) Nekotorye novyi minerali v karbonatitakh Kolskogo Polusostrova. (Some new minerals in carbonatites from the Kola Peninsula.) Ann. Rpt VSEGEI Ministry of Geology, 1959, Leningrad, 114–16. Abstracted in *Zap. Vses. Mineral. Obshch.*, **91**, 190 (in Russian).

Burt, D. M. (1989) Compositional and phase relations among rare earth element minerals, in *Geochemistry and Mineralogy of Rare Earth Elements* (eds B. R. Lipin and G. A. McKay,) 259–307. Mineralogical Society of America, Reviews in Mineralogy, **21**.

Čeh, M., Kolar, D. and Golič, L. (1987) The phase diagram $CaTiO_3$–$SrTiO_3$. *J. Solid State Chem.*, **68**, 68–72.

Chakhmouradian, A. (1994) Perovskites: their behaviour during the formation of carbonatite-bearing massifs. Geological Association of Canada – Mineralogical Association of Canada, Annual Meeting, Waterloo, Program with Abstracts.

Chernysheva, E. A., Nechelyustov, G. N., Kvitko, T. D. and Weiss, B. T. (1990) Evolutsiya sostava perovskita v shchelochnikh porodakh Nizhnesayanskogo karbonatitovogo kompleksa. (Evolution of perovskite composition in alkaline rocks of the Lower Sayan carbonatite complex.) *Geokhimiya*, **1990**, 1330–7.

Dano, M. and Sörensen, H. (1959) An examination of some rare minerals from the nepheline syenites of southwest Greenland. *Medd. Gronland*, **162**, 1–35.

Dawson, J. B., Smith, J. V. and Steele, I. M. (1994) Trace element distribution between coexisting perovskite, apatite and titanite from Oldoinyo Lengai, Tanzania. *Chem. Geol.*, **117**, 285–90.

Eby, G. N. (1975) Abundance and distribution of the rare earths and yttrium in the rocks and minerals of the Oka carbonatite complex. *Geochim. Cosmochim. Acta*, **39**, 597–620.

Filipev, V. S. and Fesenko, E. G. (1965) Preparation and structures of complex perovskites of type $Ca_2B'B''O_6$. *Sov. Phys. Crystallogr.*, **10**, 243–7.

Galasso, F. S. (1969) *Structure, Properties and Preparation of Perovskite Type Compounds*, Pergamon Press, London.

Gasperin, M. (1975). Dititanate de lanthane. *Acta Crystallogr.*, **B31**, 2129–30.

Gerasimovskii, V. I. (1941) Novii mineral Lovozerskikh massiv – metaloparit. (Metaloparite, a new mineral of the Lovozero Massif.) *Dokl. Akad. Nauk SSSR*, **33**, 61–3 (in Russian).

Glazer, A. M. (1972) The classification of tilted octahedra in perovskites. *Acta Crystallogr.* **B28**, 3384–92.

Glazer, A. M. (1975) Simple ways of determining perovskite structures. *Acta Crystallogr.*, **A31**, 756–62.

Glazer, A. M. and Megaw, H. D. (1972) The structure of sodium niobate (T_2) at 600°C, and the cubic–tetragonal transition in relation to soft-phonon modes. *Philosoph. Mag.*, **26**, 1119–35.

Goldschmidt, V. M. (1926) *Geochemische Verteilungsgesetze der Elemente VII*. Skrifter Norske Videnskaps Akademi Klasse 1 Matematisk Naturvidensakaplig Klasse, Oslo.

Goodenough, J. B., Wold, A., Arnott, R. J. and Menyuk, N. (1961) Relation between crystal symmetry and magnetic properties of ionic compounds containing Mn^{+++}. *Phys. Rev.*, **124**, 373–84.

Gopalakrishnan, J. and Bhat, V. (1987) $A_2Ln_2Ti_3O_{10}$ (A = K or Rb; Ln = La or rare earth): a new series of layered perovskites exhibiting ion exchange. *Inorg. Chem.*, **26**, 4299–301.

Guyot, F., Richet, P., Courtial, P. and Gillet, P. (1993) High-temperature heat capacity and phase transitions of $CaTiO_3$ perovskite. *Phys. Chem. Minerals.*, **20**, 141–6.

Haggerty, S. E. and Mariano, A. N. (1983) Strontian-loparite and strontio-chevkinite: two new

REFERENCES

minerals in rheomorphic fenites from the Parana Basin carbonatites, South America. *Contr. Mineral. Petrol.*, **84**, 365–81.

Hauser, O. (1908) Über den sogenannten Dysanalyt von Vogtsburg im Kaiserstuhl. *Z. Anorg. Chem.*, **60**, 237–41.

Hazen, R. M. (1988) Perovskites. *Sci. Am.*, **258**, 74–81.

Holmquist, P. J. (1894) Knopit, ett perowskit narstaende, nytt mineral fran Alno. *Geol. Foren. Stocklholm Forhandl.*, **16**, 73.

Hu, M., Wenk, H. R. and Sinitsyna, D. (1992) Microstructures in natural perovskites. *Am. Mineral.*, **77**, 359–73.

Hyde, B. G. and Andersson, S. (1989) *Inorganic Crystal Structures*, J. Wiley & Sons, New York.

Ilupin, I. P., Khomyakov, A. P. and Balashov, Y. A. (1971) Redkiye zemli v aktsessornykh mineralov kimberlitov Yakutii. (Rare earths in accessory minerals of Yakutian kimberlites.) *Dokl. Akad. Nauk SSSR*, **201**, 1214–17 (in Russian).

Ishizawa, N., Marumo, F., Kawamura, T., and Kimura, M. (1975) The crystal structure of $Sr_2Nb_2O_7$, a compound with perovskite-type slabs. *Acta Crystallogr.* **B31**, 1912–15.

Jago, B. C. (1990) The role of fluorine in the evolution of alkali-bearing carbonatite magmas and the formation of carbonate-hosted apatite and pyrochlore deposits. Unpublished Ph.D Thesis, University of Toronto, Ontario, Canada.

Jones, A. P. and Wyllie, P. J. (1984) Minor elements in perovskite from kimberlites and the distribution of the rare earth elements. *Earth Planet. Sci. Lett.*, **69**, 128–40.

Jongejan, A. and Wilkins, A. L. (1972) Liquidus determinations in the 25% SiO_2 plane of the quaternary system $CaO-Nb_2O_5-TiO_2-SiO_2$. *J. Less Common Metals*, **29**, 349–60.

Kapustin, Yu. L. (1980) *Mineralogy of Carbonatites*, Amerind Pub. Co. Pvt. Ltd, New Delhi.

Kay, H. F. and Bailey, P. C. (1957) Structure and properties of $CaTiO_3$. *Acta Crystallogr.* **10**, 219–26.

Kay, H. F. and Vousden, P. (1949) Symmetry changes in barium titanate at low temperatures and their relation to its ferroelectric properties. *Philosoph. Mag.*, **40**, 1019–40.

Knop, A. (1877) Dysanalyt, ein pyrochloratiges Mineral. (Fruher Perowskit von Vogtsburg im Kaiserstuhl). *Z. Kristallogr.*, **1**, 284–96.

Kukharenko, A. A., Orlova, M. P., Bulakh, A. G. *et al.* (1965) *Kaledonskii kompleks ultraosnovnykh shchelochnykh porod i karbonatitov Kolskogo Poluostrova i Severnoi Karelii.* (Caledonian complexes of alkaline ultrabasic rocks and carbonatites of the Kola Peninsula and Northern Karelia.) Nedra, Leningrad (in Russian).

Kuznetsov, I. G. (1925) Loparit – novyi redkozemelnyi mineral Kibinskikh tundr. (Loparite, a new rare earth element from the Khibina tundra.) *Izvestiya Geologicheskogo Komiteta*, **44**, 663–82 (in Russian).

Lapin, A. V. and Kazakova, M. Y. (1966) O titanovom lueshit iz Kovdorskogo massiva i ozomorfizme v gruppy perovskita. (Titanium, lueshite from the Kovdor massif and isomorphism in the perovskite group.) *Dokl. Acad. Sci. SSSR*, **171**, 956–9 (in Russian).

Levinson, A. A. (1966) A system of nomenclature for rare earth minerals. *Am. Mineral.*, **51**, 152–8.

Liu, L. G. (1975) Post oxide phase of forsterite and enstatite. *Geophys. Res. Lett.*, **2**, 284–96.

Liu, X. and Liebermann, R. C. (1993) X-ray powder diffraction study of $CaTiO_3$ perovskite at high temperatures. *Phys. Chem. Minerals*, **20**, 171–5.

Lupini, L., Williams, C. T. and Woolley, A. R. (1992) Zr-rich garnet and Zr- and Th-rich perovskite from the Polino carbonatite, Italy. *Mineral. Mag.*, **56**, 581–6.

Mariano, A. N. and Mitchell, R. H. (1991) Mineralogy and geochemistry of perovskite-rich

pyroxenites. Proc. 5th Internat. Kimberlite Conf. Araxa, Brazil. Extended Abstracts, pp. 251–3.
McCallum, M. E. (1989) Oxide minerals in the Chicken Park kimberlite, northern Colorado, in *Kimberlites and Related Rocks* (ed. J. Ross). *Geol. Soc. Austr. Spec. Publ.*, **14**, 241–63.
Megaw, H. D. (1973) *Crystal Structures: A Working Approach*. W.B. Saunders Co., Philadelphia.
Meigen, W. and Hugel, E. (1913) Chemical composition of dysanalyte from Vogtsburg, Kaiserstuhl. *Z. Anorg. Chem.*, **82**, 2424–48.
Mitchell, R. H. (1972) Composition of perovskite in kimberlites. *Am. Mineral.*, **57**, 1748–53.
Mitchell, R. H. (1986) *Kimberlites: Mineralogy, Geochemistry and Petrology*, Plenum Press, New York.
Mitchell, R. H. (1994) Suggestions for revisions to the terminology of kimberlites and lamprophyres from a genetic viewpoint, in Proceedings of the 5th International Kimberlite Conference (eds H. O. A. Meyer and O. A. Leonardos), 15–26. Companhia de Pesquisa de Recursos Minerais, Brasilia, Spec. Publ. 1A.
Mitchell, R. H. (1995) *Kimberlites, Orangeites and Related Rocks*, Plenum Press, New York.
Mitchell, R. H. and Bergman, S. C. (1991) *Petrology of Lamproites*, Plenum Press, New York.
Mitchell, R. H. and Platt, R. G. (1993) Compositional variation of rare earth, strontium and niobium-bearing perovskites from alkaline rocks and carbonatites. *Rare Earth Minerals: chemistry, origin and ore deposits*. Mineralogical Society Spring Meeting, Natural History Museum, London, UK, Extended Abstracts, 84–5.
Mitchell, R. H. and Reed, S. J. B. (1988) Ion microprobe determination of rare earth elements in perovskites from kimberlites and alnoites. *Mineral. Mag.*, **52**, 331–9.
Mitchell, R. H. and Steele, I. (1992) Potassian zirconium and titanium silicates and strontian cerian perovskite in lamproites from the Leucite Hills, Wyoming. *Can. Mineral.* **30**, 1153–9.
Mitchell, R. H. and Vladykin, N. V. (1993) Rare earth element-bearing tausonite and potassium barium titanates from the Little Murun potassic alkaline complex, Yakutia, Russia. *Mineral. Mag.*, **57**, 651–4.

Náray-Szabó, S. V. (1943) Der strukturtyp des Perowskits ($CaTiO_3$). *Naturwiss*, **31**, 202–3.
Navrotsky, A. and Weidner, D. J. (1989) *Perovskite: A Structure of Great Interest to Geophysics and Materials Science*. Am. Geophys. Union Monograph, **45**, Washington, DC
Nickel, E. H. (1964) Latrappite – a proposed new name for the perovskite-type calcium niobate mineral from the Oka area of Quebec. *Can. Mineral.*, **8**, 121–2.
Nickel, E. H. (1992) Solid solutions in mineral nomenclature. *Mineral. Mag.*, **56**, 127–30.
Nickel, E. H. and McAdam, R. C. (1963) Niobian perovskite from Oka, Quebec, a new classification of the perovskite group. *Can. Mineral.*, **7**, 683–97.

O'Keefe, M. and Hyde, B. G. (1977) Some structures topologically related to cubic perovskite ($E2_1$), ReO_3 (DO_9) and Cu_3Au ($L1_2$). *Acta Cryst.*, **B33**, 3802–13.

Parker, R. L. and Sharp, W. N. (1970) Mafic–ultramafic igneous rocks and associated carbonatites of the Gem Park complex, Custer and Fremont Counties, Colorado. *U.S. Geol. Surv. Prof. Paper*, **649**, 1–24.
Pauling, L. (1924) Crystal structures of ammonium fluoferrate, fluoaluminate and oxyfluomolybdate. *J. Am. Chem Soc.*, **46**, 2738–51.
Portier, R., Fayard, M., Carpy, A. and Galy, J. (1974) Electron microscopic study of some members of the $(Na,Ca)_nNb_nO_{3n+2}$ series. *Materials Res. Bull.*, **9**, 371–8.
Platt, R. G. (1994) Perovskite, loparite and Ba–Fe hollandite from the Schryburt Lake carbonatite complex, northwestern Ontario, Canada. *Mineral. Mag.*, **58**, 49–57.
Preuss, A. and Gruehn, R. (1994). Preparation and structure of cerium titanates Ce_2TiO_5, $Ce_2Ti_2O_7$, and $Ce_4Ti_9O_{24}$. *J. Solid State Chem.*, **110**, 363–9.

REFERENCES

Radusinović, D. and Markov, C. (1971) Macedonite – lead titanate: A new mineral. *Am. Mineral.*, **56**, 387–94.

Rose, G. (1839) Beschreibung einiger neuer Mineralien vom Ural. *Pogendorff Annalen der Physik und Chemie*, **48**, 551–72.

Ruddlesden, S. N. and Popper, P. (1957) New compounds of the K_2NiF_4 type. *Acta Crystallogr.*, **10**, 508–39.

Ruddlesden, S. N. and Popper, P. (1958) The compound $Sr_3Ti_2O_7$ and its structure. *Acta Crystallogr.*, **11**, 54–5.

Safiannikoff, A. (1959) Un nouveau mineral de niobium. *Bull. Seances Acad. Roy. Sci. Outre-Mer*, **5**, 1251–5.

Sasaki, S., Prewitt, C. T. and Liebermann, R. C. (1983) The crystal structure of $CaGeO_3$ perovskite and the crystal chemistry of the $GdFeO_3$-type perovskites. *Am. Mineral.*, **68**, 1189–98.

Sasaki, S., Prewitt, C. T., Bass, J. D. and Schulze, W. A. (1987) Orthorhombic perovskite $CaTiO_3$ and $CdTiO_3$: structure and space group. *Acta Crystallogr.*, **C43**, 1668–74.

Sato, M., Jin, T. and Uematsu, K. (1993) Proton conduction of $MLaNb_2O_7$ (M = K,Na,H) with a layered perovskite structure. *J. Solid State Chem.*, **102**, 557–61.

Scheunemann, K. and Muller-Buschbaum, H. K. (1974). Zur kristallstruktur von $Ca_2Nb_2O_7$. *J. Inorg. Nuclear Chem.*, **36**, 1965–70.

Shannon, R. D. and Prewitt, C. T. (1969) Effective ionic radii in oxides and fluorides. *Acta Crystallogr.*, **B25**, 925–46.

Sitnin, A. A. and Leonova, T. N. (1961) Loparit – novii aktsessornii mineral albitizirovannikh i greizenizirovannikh granitov. (Loparite – a new accessory mineral in albitized and greisenized granite.) *Dokl. Akad. Nauk SSSR*, **140**, 1407–10 (in Russian).

Sleight, A. W. and Ward, R. (1962) Compounds of hexavalent and pentavalent uranium with the ordered perovskite structure. *Inorg. Chem.*, **1**, 790–3.

Smyth, D. M. (1989) Defects and structural changes in perovskites systems; from insulators to superconductors. *Cryst. Latt. Defects and Amorphous Mat.*, **18**, 355–75.

Tikhonenkov, I. P. and Kazakova, M. E. (1957) Nioboloparite, novyi mineral iz gruppy perovskita. (Nioboloparite, a new mineral of the perovskite group.) *Zap. Vses Mineral. Obshch.*, **86**, 641–4 (in Russian).

Ulrych, J., Piveč, E., Povondra, P. and Rutsek, J. (1988) Perovskite from melilite rocks, Osečná complex, northern Bohemia, Czechoslovakia. *N. Jb. Mineral. Monatsh.*, **1988**, 81–95.

Veksler, I. M. and Teptelev, M. P. (1990) Conditions for the crystallization and concentration of perovskite-type minerals in alkaline magmas. *Lithos*, **26**, 177–89.

Vlasov, K. A. (1966) *Geochemistry and Mineralogy of Rare Elements and Genetic Types of Their Deposits II. Mineralogy of Rare Elements*. Israel Program for Scientific Translations, Jerusalem.

Vlasov, K. A., Kuzmenko, M. V. and Eskova, E. M. (1966) *The Lovozero Alkai Massif*. Oliver & Boyd, Edinburgh.

Vorobyev, E. I., Konyev, A. A., Malyshok, Y. V. *et al.* (1984) Tausonit $SrTiO_3$, novyye mineraly v perovskit gruppa. (Tausonite $SrTiO_3$, a new mineral of the perovskite group.) *Zap. Vses. Mineral. Obshch.*, **113**, 83–9 (in Russian).

Vorobyev, E. I. Konyev, K. A., Malyshok, Y. V. *et al.* (1987) *Tausonit, geologicheskiye usloviya obrazovaniya i mineralnii paragenezisi. (Tausonite, geological conditions of formation and mineral paragenesis.)* Nauka Press, Novosibirsk (in Russian).

Wang, Y. and Leibermann, R. C. (1993) Electron microscopy study of domain structure due to phase transitions in natural perovskite. *Phys. Chem. Minerals.*, **20**, 147–58.

Wechsler, B. A. and Kirby, K. W. (1992) Phase equilibria in the system barium titanate – strontium titanate. *J. Am. Ceram. Soc.*, **75**, 981–84.

Wold, A. and Dwight, K. (1993) *Solid State Chemistry: Synthesis, Structure and Properties of Selected Oxides and Sulphides*, Chapman & Hall, New York.

Williams, T., Schmalle, H., Reller, A. *et al.* (1991) On the crystal structures of $La_2Ti_2O_7$ and $La_5Ti_5O_{17}$: high resolution electron microscopy. *J. Solid State Chem.*, **93**, 534–48.

Zhao, Y., Weidner, D. J., Parise, J. C. and Cox, D. (1993) Thermal expansion and structural distortions of perovskites: data for $NaMgF_3$ perovskite (I). *Phys. Earth Planet. Interiors*, **76**, 1–16.

CHAPTER FOUR
Rare earth elements in carbonate-rich melts from mantle to crust

P. J. Wyllie, A. P. Jones and J. Deng

4.1 Introduction

The rare earth elements (REE) are widely distributed as trace elements in mantle rocks and mantle-derived magmas, and in crustal rocks. There is a huge literature on their geochemistry, but much less has been written about how they become concentrated from trace concentrations in the mantle to high percentages in ore deposits. Carbonatite is one host rock for REE deposits, and this review is concerned with the behaviour of REE in carbonate-rich melts in environments ranging from the Earth's mantle to shallow level magmatic carbonatite intrusions and hydro-carbo-thermal veins.

A high percentage of the trace REE elements in the upper mantle may be stored in a small percentage of discrete titanate minerals (Haggerty, 1983; Jones, 1989), and most would dissolve in carbonate-rich melts generated in the mantle at depths greater than about 70 km; abundances would remain at the trace element levels, however. The immiscible separation of carbonate-rich magma from silicate magma is expected to cause enrichment in REE (Wendlandt and Harrison, 1979), but the experiments of Hamilton, Bedson and Esson (1989) did not reveal the kind of enrichment required to explain the REE in many carbonatites. In the Mountain Pass carbonatite, bastnäsite constitutes up to 15 vol.% of the ore body (Olson *et al.*, 1954; Heinrich, 1966, p. 357). Furthermore, experiments on silicate–carbonate liquid immiscibility indicate that although immiscibility is likely to occur within the crust, it appears to be improbable in the lower lithosphere. The high concentrations of REE are probably caused by fractional crystallization of a carbonatite magma already somewhat enriched in REE, and we have conducted experiments to test this proposal.

We are approaching the problem in two ways. The first is to build from simple to more complex phase diagrams, in order to establish precisely the behaviour of the rare earth elements in carbonate-rich melts, as a guide for

Rare Earth Minerals: Chemistry, origin and ore deposits. Edited by Adrian P. Jones, Frances Wall and C. Terry Williams. Published in 1996 by Chapman & Hall. ISBN 0 412 61030 2

interpretation of the more complex systems and of the rocks themselves. We present here the results for the simplest model of a rare earth-carbonatite magma, the composition join $CaCO_3-Ca(OH)_2-La(OH)_3$ through the system $CaO-La_2O_3-CO_2-H_2O$ (Jones and Wyllie, 1986), followed by joins to synthetic hydroxylbastnäsite (Deng and Wyllie, in preparation). These experiments outline conditions under which bastnäsite and calcite can be coprecipitated from melts at moderate temperatures; this provides a simple analogue for the primary formation of RE-carbonatite from a magma.

The second approach is to melt complex mixtures approximating the composition of the ore body at Mountain Pass, to follow paths of crystallization and to locate the parageneses and conditions for the precipitation of bastnäsite. The system studied includes baryte and fluorite. Preliminary results have been published (Jones and Wyllie, 1983; Wyllie and Jones, 1985). These experiments show that with fractional crystallization, residual melts will contain high concentrations of REE (18 wt % $La(OH)_3$).

This experimental review outlines likely conditions for the formation of primary RE minerals from carbonatite magmas. Carbonatites are amongst the lowest temperature terrestrial magmas known. If crystallized at shallow depths in the upper crust, primary bastnäsite group minerals could form at temperatures as low as ~550°C. Subsolidus changes, which may occur frequently in natural systems, are not considered here. However, some examples of possible carbothermal or hydrothermal replacement reactions are given by Gieré in Chapter 5.

4.2 Formation of carbonate-rich melts in the mantle

The possible existence of immiscible carbonate-rich melts in mantle peridotite was suggested by Irving and Wyllie (1973, 1975) and Koster van Groos (1975), and the coexistence of carbonate-rich melt with synthetic harzburgite was demonstrated by Huang and Wyllie (1974). According to Huang and Wyllie (1975, 1976) dolomite forms in peridotite with CO_2 at pressures above about 20 kbar (decreases with temperature), and the near-solidus melt in dolomite–lherzolite–CO_2 at higher pressures is carbonatitic with about 40 wt% CO_2. This was challenged by Eggler, Holloway and Mysen (1976) who concluded that the liquid 'contained at most 20 per cent CO_2'. On the basis of these experimental results and those of Eggler (1974, 1976) in model systems, Wyllie (1977) concluded that near-solidus melt in peridotite–H_2O–CO_2 at pressures higher than the carbonation reaction corresponds to that of a dolomitic carbonatite with Ca/Mg > 1 (0.7, Wyllie *et al.*, 1983, Figure 5), with about 10–15% dissolved silicates (5–10% SiO_2), and enriched in alkali carbonates to the extent that alkalis were available. It is now agreed that the near-solidus liquid from model dolomite-lherzolite (magnesite-lherzolite at somewhat higher pressures) is carbonatitic (Eggler, 1978, 1987, 1989; Wyllie, 1987, 1989).

The results from the model systems have since been confirmed by investigations using natural rock compositions. The significant point on the solidus where the system changes from CO_2- to dolomite-peridotite was located at 2.2 GPa and 1090°C by Rutter and Wyllie (1986; see Wyllie, 1987), and at 2.05 GPa and 1050°C for dolomite-pyrolite by Falloon and Green (1989). Wallace and Green (1988), using a fertile natural peridotite composition, reported a liquid corresponding to a carbonatite of 'sodic dolomitic character': Na_2O, 5%; CaO, 21%; MgO, 14%; FeO, 4.6%. Baker and Wyllie (1989; manuscript in preparation) used recombined natural mineral assemblages, obtaining similar results for dolomite-lherzolite. Thibault, Edgar and Lloyd (1992), using carbonated phlogopite peridotite, measured an alkaline dolomitic melt: Na_2O, 5%; K_2O, 7%; CaO, 22%; MgO, 15%; FeO, 4.6%. Dalton and Wood (1993), using depleted natural lherzolite with dolomite, measured similar melt compositions but with only 0.08% Na_2O and 0.01% K_2O. Sweeney (1994) worked with two rocks corresponding to (1) sodic oceanic peridotites and (2) more potassic lithospheric (continental) peridotites, and confirmed that the percentages of Na_2O and K_2O in the calcic dolomitic melts were a direct reflection of their contents in the source rocks: (1) 5% Na_2O; 2.2% K_2O, and (2) 2.1% Na_2O; 5.4% K_2O.

4.3 Rare earth elements in metasomatized mantle rocks

Many mantle xenoliths have been metasomatized by aqueous vapours, silicate melts (Menzies and Hawkesworth, 1987) or carbonate-rich melts (Menzies and Wass, 1983; Meen, Ayers and Fregeau, 1989; Haggerty, 1989). Menzies and Chazot (1994) presented a detailed review and classification of these different types of mantle metasomatism in different tectonic environments. Relative enrichment in LREE and Sr over HREE and HFSE appears to be characteristic of mantle carbonates and carbonate-bearing peridotites. Many of the silicate-incompatible trace elements enriched in metasomatized mantle rocks and the distinctive MARID suite of kimberlite xenoliths are incorporated in a small group of titanate minerals, dominated by members of the crichtonite series (Jones, Smith and Dawson, 1982; Haggerty, 1983). Haggerty (1989) discussed the distribution of these minerals (and CO_2 and H_2O) in mantle metasomes occurring at different depths, identified from kimberlite xenolith suites, which correspond closely to three depth levels proposed independently by Wyllie (1980, 1989) on the basis of phase relationships. Jones (1989) noted that approximately 1 part in 1000 titanate, occurring as discrete grains, would double the content of Ba, LREE (La, Ce), Rb, Sr and U in average mantle. He suggested that the agent responsible for metasomatism and crystallization of the titanates in the upper mantle could have been small-volume hydrous carbonatitic melts, which would be efficient scavengers, and could derive most of their incompatible elements, including the REE, from the titanates. There is growing appreciation for the metasomatizing properties of mantle-derived carbonatitic liquids (Green and Wallace, 1988; Jones, 1989; Hunter and McKenzie, 1990; Brenan and Watson, 1991; Yaxley, Crawford and Green, 1991; Hauri *et al.*, 1993). Signatures of carbonate-related metasomatism may be high Sr/Sm, Sm/Hf, La/Nb, Zr/Hf and Nb/Ta (Ionov *et al.*, 1993).

Experimental studies have demonstrated that carbonatitic melts are very mobile at very low melt fractions (Hunter and McKenzie, 1989; White and Wyllie, 1992), largely due to their very low viscosities (Wolff, 1994; Jones, Dobson and Genge, 1995). Experimental studies of trace element distributions between mantle minerals and carbonatitic liquids are beginning to permit quantification of some metasomatic processes, but data are at present sparse (Brenan and Watson, 1991; Sweeney *et al.*, 1991; Green, Adam and Sie, 1992; Baker and Wyllie, 1992). The trace element contents of the melts are controlled by the solubility of trace element-rich accessory minerals (e.g. apatite, rutile), crystal–liquid partition coefficients and elemental diffusion rates in the various solids. Brenan and Watson (1991) showed that carbonate melt in contact with depleted lherzolite would raise the levels of LIL elements in clinopyroxene with only subtle mineralogical change which is one form of mantle metasomatism commonly observed (e.g. Yaxley, Crawford and Green, 1991).

Haggerty (1989) concluded that the incompatible elements from the exotic LIL-enriched titanates in the metasomes become concentrated more strongly into carbonatites than into kimberlites. Similarly, Jones (1989) suggested that the trace element enrichment in carbonatites is largely controlled by LREE-enriched titanate minerals, and that the REE and other incompatible elements were derived originally from titanates in mantle peridotite.

4.4 REE in carbonatites

Anomalously high REE content is a distinctive aspect of carbonatites, as featured elsewhere in this book (see especially Chapter 8). According to Heinrich (1966, p. 173), bastnäsite is both the most abundant and the most widely distributed of the dozen or so discrete rare earth carbonate minerals reported from carbonatites; volumetrically, it is typically of accessory or minor status. Mineral hosts containing more than trace amounts of the REE in carbonatites can be divided into three main groups: (1) oxides, (2) phosphates and (3) carbonates. The common representatives are: for group (1), pyrochlore and perovskite; for group (2), apatite and monazite; and for group (3), bastnäsite-series REE carbonates such as synchysite, bastnäsite, parisite and roentgenite.

Pecora (1956) distinguished two varieties of carbonatite, the apatite–magnetite and the rare earth mineral varieties. Heinrich (1966, p. 157), noted that although this is an oversimplification of the geochemistry of carbonatites, there is a real distinction in the paragenesis of these assemblages.

BASTNÄSITE SYNTHESIS AND STABILITY RANGE

Many carbonatites of the apatite–magnetite type do contain REE, commonly in monazite. Heinrich added that if the term 'rare-earth carbonate type' is substituted, the schism is more pronounced and paragenetically more significant. The most abundant RE carbonate in carbonatites, bastnäsite, contains up to about 70 wt% REE (see Taylor and Pollard, Chapter 7). Light REE are concentrated in carbonatites (Heinrich, 1966; Loubet et al., 1972; Eby, 1975) through more than one process. Eby (1975) presented a study of the REE distribution in minerals from the Oka carbonatite, whereas secondary REE mobility in the Fen carbonatite was considered by Andersen (1984).

Mariano (1989) gave a detailed review of carbonatites as major sources of Nb, phosphate and REE, classifying RE mineralization in carbonatites into three categories:

1. primary magmatic crystallization; Mountain Pass deposit is the only known example, and bastnäsite and parisite are the only ore minerals;
2. hydrothermal mineralization with ancylite, monazite, or bastnäsite-type minerals forming replacement veins in carbonatite and related rocks, usually accompanied by baryte, fluorite, quartz and strontianite;
3. supergene mineralization produced by weathering of carbonatite bodies; the source is from chemical breakdown of calcite, dolomite and apatite, not from primary REE minerals.

The rare earth minerals in most carbonatites are precipitated from hydrothermal solutions. The distinction between magmatic carbonatite dykes and veins with similar mineralogy is not always obvious. Naldrett and Watkinson (1981) pointed out that the later stage, lower-temperature carbonatites, often found with hematite, baryte, fluorite and other relatively oxidized phases, are more favourable for prospecting because they are more likely to be repositories of incompatible elements such as U, Th and rare earth elements.

The RE carbonate type of carbonatite usually represents the last in a series of carbonatitic differentiates, and is normally subordinate to earlier carbonatite (Heinrich, 1966, p. 234). Although this stage is represented in nearly all carbonatites, it is characterized by the significant development of RE carbonate species only in some. The RE-carbonatite at Mountain Pass, California, is unusual in that there are no observed earlier carbonatite differentiates.

4.5 Bastnäsite synthesis and stability range

Bastnäsite was originally discovered as a new cerium-bearing mineral in skarn deposits at Bastnäs, Sweden. Sporadic occurrences and chemical analyses were compiled by Glass and Smalley (1945) who gave the formula, $CeFCO_3$, where Ce represents the light REE dominated by Ce. Natural bastnäsite group minerals can be represented by the general formula $nXYCO_3 \cdot mCaCO_3$, where X = LREE, Y = (F, OH), and where $m = 0$ (bastnäsite) or 1 (synchysite, parisite, röntgenite), and $n = 1$ (bastnäsite, synchysite), 2 (parisite) or 3 (röntgenite). Additional natural mixed-layer compounds in this series with compositions between synchysite and bastnäsite have been identified by Van Landuyt and Amelinckx (1975). Bastnäsite has been reported as a common minor or accessory mineral associated with several carbonatites from Africa (Deans, 1966; Le Bas, 1977) and the USSR (Kapustin, 1980). Heinrich (1966, pp. 258–61) and Mariano (1989) reviewed REE deposits in carbonatites. Bastnäsite is the main host for REE at Mountain Pass carbonatite (Olson et al. 1954).

Figure 4.1 shows phases known in the system $La_2O_3-CO_2-H_2O$, with hydroxylbastnäsite at the point B. Addition of CaO as a component introduces additional mineral compositions in Figure 4.2, and shows the triangular composition join CC–CH–LH described below. Hydroxyl end members of the RE carbonate minerals röntgenite (R), parisite (P) and synchysite (S) plot within the tetrahedron between hydroxylbastnäsite (B, on the front face) and calcite (CC, on a rear edge of the tetrahedron). This line does not coincide with the triangle CC–CH–LH, but it is not far removed from it.

There have been several hydrothermal experiments on the synthesis and

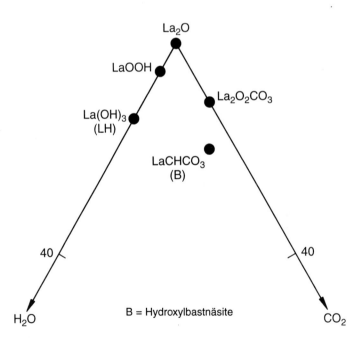

Figure 4.1 Known phases of interest in the system $La_2O_3-CO_2-H_2O$ (Jones and Wyllie, 1986).

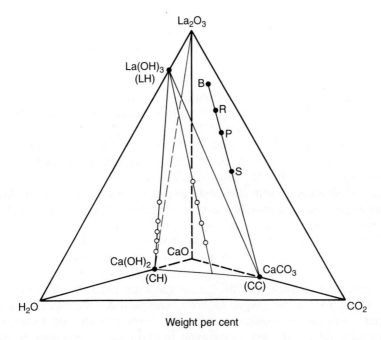

Figure 4.2 Location of the triangle $CaCO_3$–$Ca(OH)_2$–La_2O_3 in the quaternary system CaO–La_2O_3–CO_2–H_2O. The mixtures studied are given by open circles along the two joins CH–LH and $CC_{55}CH_{45}$–LH. Hydroxylbastnäsite as $La(OH)CO_3$ plots on the front face of the tetrahedron at B. Additional bastnäsite group minerals plot along the join between B and CC, with R = roentgenite, P = parisite, S = synchysite (Jones and Wyllie, 1986).

stability of REE compounds containing H_2O, CO_2 and Cl. Caro and co-workers described low temperature hydrated RE carbonates of the type $RE_2(CO_3)x(OH)_{2(3-x)} \cdot nH_2O$ (Caro, 1966; Caro, Lemaitre-Blaise and Trombe, 1968; Caro and Lemaitre-Blaise, 1969). Aumont et al. (1971) gave X-ray data for hydrothermally synthesized Cl-bearing bastnäsite $La(OH)_{0.84}Cl_{0.16}CO_3$, for La, Nd, Sm and Gd end members. Further characterization of synthesized RE-hydroxylcarbonates and their analogy with hydroxylbastnäsite were made by Christensen (1973) and Haschke and Eyring (1971). Haschke (1975) subsequently reported that the continuous series between $La(OH)CO_3$ and $LaFCO_3$ (chemical analogues of bastnäsite) differed in detail, with lattice parameters decreasing in proportion of F/OH. Chai and Mroczkowski (1978) performed a more systematic study for each of several RE oxides and examined their behaviour with varying CO_2 and H_2O, at temperatures of 250–550°C and pressures of 1–2 kbar. Chai and Mroczkowski (1978) produced a series of isothermal phase diagrams for the generalized system RE_2O_3–CO_2–H_2O at 11 temperatures from 900 to 1000°C. Kutty, Viswanathiah and Tareen (1978) adopted a more rigorous experimental approach and produced the $T-X_{CO_2}$ phase diagram for the system $Nd_2O_3-H_2O-CO_2$ at 1.5 kbar, and documented the X-ray characteristics of each of four phases encountered at temperatures up to 900°C: $Nd(OH)CO_3$, $Nd(OH)_3$, $Nd_2O_3CO_3$ and NdOOH. Subsequently, Tareen and Kutty (1980) repeated the same $T-X_{CO_2}$ diagram at 0.5 kbar and established the decrease in T-stability of $Nd(OH)CO_3$ with decreasing pressure: 575° at 1.5 kbar, 540°C at 0.5 kbar. This is at least 25°C higher than that for the analogous $RE(OH)CO_3$ (<550°C) given in the generalized scheme of Chai and Mroczkowski (1978). Some of the above studies have also been summarized recently by Williams-Jones and Wood (1992).

Hsu (1992) noted that the previous experimental syntheses of bastnäsite had provided little information regarding the physicochemical conditions, and he presented a comprehensive investigation of the stability fields of fluorocarbonates and hydroxylcarbonates of Ce and La, and explained why the former are much more common than the latter in rocks. He used vapour-phase compositions with H_2O, CO_2 and HF. Hydroxylbastnäsite-(La) is stable up to 810°C, and fluobastnäsite-(La) is stable up to 860°C. The corresponding minerals with Ce are stable up to 650 ± 10°C and 780 ± 20°C, depending on the oxygen fugacity. The petrogenetic conclusions from the study were that bastnäsites can be stable from hydrothermal to magmatic conditions, that F-enriched species can form in an environment relatively low in F content, and that OH species are rare and occur only in low-temperature environments essentially devoid of fluorine.

4.6 Phase equilibrium studies in the system $CaO-La_2O_3-CO_2-H_2O$

Figures 4.2, 4.3 and 4.4 show the minerals, composition joins and mixtures used in our experiments. This quaternary system contains the low-temperature synthetic carbonatite magmas reported by Wyllie and Tuttle (1960) in $CaCO_3-Ca(OH)_2$, with lanthanum added to represent the LREE. We have determined phase relationships in selected parts of the system for determination of the conditions for the precipitation of hydroxylbastnäsite and calcite from liquid.

4.6.1 $CaCO_3-Ca(OH)_2-La(OH)_3-H_2O$

Jones and Wyllie (1986) determined the liquidus relationships at 1 kbar for part of the join $CaCO_3-Ca(OH)_2-La(OH)_3$, using the mixtures shown in Figure 4.2 (open circles). The results are shown in Figure 4.4. The liquidus surface was followed to the side $CaCO_3-La(OH)_3$ by Deng and Wyllie (in preparation), with the field boundary from E-a extending to a eutectic liquid at 713°C containing 70% $La(OH)_3$. The phase relationships remain ternary to the highest temperature reached, 750°C.

The effect of a vapour phase was determined by Deng and Wyllie (in

THE CaO–La$_2$O$_3$–CO$_2$–H$_2$O SYSTEM

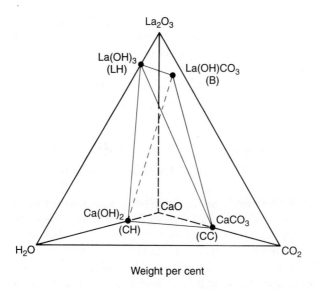

Figure 4.3 Location of the tetrahedron CaCO$_3$–Ca(OH)$_2$–La(OH)$_3$–La(OH)CO$_3$ in the quaternary system CaO–La$_2$O$_3$–CO$_2$–H$_2$O (Deng and Wyllie, in preparation).

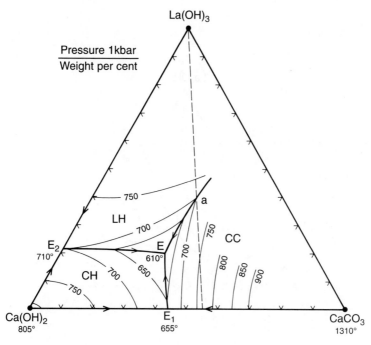

Figure 4.4 Liquidus surface in the join CaCO$_3$–Ca(OH)$_2$–La(OH)$_3$ at 1 kbar pressure, based on results from the mixtures plotted in Figure 4.2 (Jones and Wyllie, 1986).

preparation) by addition of 15% H_2O. Some CO_2 was released from the condensed phases, and the vapour phase changed composition from H_2O at $Ca(OH)_2$ with increase in CO_2/H_2O as the bulk composition changed toward $CaCO_3$. The effect of vapour is to lower the whole liquidus surface by 5–15°C. There are also minor changes in the positions of the field boundaries, as shown by Figure 4.5.

The two solidus reactions are:

$$CaCO_3 + Ca(OH)_2 + La(OH)_3 = \text{liquid} \qquad (4.1)$$

$$CaCO_3 + Ca(OH)_2 + La(OH)_3 + \text{vapour} = \text{liquid} \qquad (4.2)$$

Consider the field boundaries enclosing the liquidus surface for $La(OH)_3$: these show the solubility of $La(OH)_3$ in the synthetic carbonatite liquids, on either side of the eutectics E1 in Figures 4.4 and 4.5. When more than the amount represented by points on the field boundary is present, a lanthanum mineral becomes the primary phase. The solubility of $La(OH)_3$ in this synthetic carbonatite liquid increases with CO_2/H_2O (in bulk composition or vapour phase) from about 20% at the eutectic (585–600°C) to about 40% at point 'a' (665–700°C).

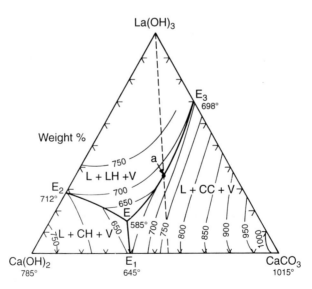

Figure 4.5 Phase diagram of $CaCO_3–Ca(OH)_2–La(OH)_3$ with 15 wt% H_2O at 1 kbar pressure (Deng and Wyllie, in preparation).

4.6.2 $CaCO_3-Ca(OH)_2-La(OH)CO_3$ (CC-CH-B)

There is no indication that a liquidus volume for hydroxylbastnäsite intersects the triangle $CaCO_3-Ca(OH)_2-La(OH)_3$, with or without a vapour phase present (Figures 4.4 and 4.5). In an effort to locate the liquidus volume for hydroxylbastnäsite, Deng and Wyllie (in preparation) synthesized the mineral from La_2O_3 and oxalic acid dihydrate, and made starting mixtures with this for a series of experiments in the joins CC-CH-B, CC-LH-B and CC-CH-LH-B (Figure 4.3). The join CC-CH-B is not ternary, as shown by the formation of phases $La(OH)_3$, $La_2O_2CO_3$ and a vapour, as well as B and calcite. The results suggest a quaternary solidus at 1 kbar and 555°C for the reaction:

$$CaCO_3 + Ca(OH)_2 + La(OH)_3 + La(OH)CO_3 = \text{liquid} \quad (4.3)$$

Although the liquid composition has not been located, it lies within the volume $CaCO_3-Ca(OH)_2-La(OH)_3-B$ in Figure 4.3. The results confirm that hydroxylbastnäsite and calcite can be coreciptiated from a La-enriched synthetic carbonatite magma between about 630°C and the solidus temperature at 550°C. This is consistent with the results of Hsu (1992) on hydroxylbastnäsite stability.

4.7 Mountain Pass synthetic carbonatite magma

The results outlined above in the quaternary system provide information on the behaviour of lanthanum (i.e. light REE) in carbonate-rich melts, and on the conditions for coprecipitation of calcite and hydroxylbastnäsite from melts. In parallel investigations Jones and Wyllie (1983, 1985) have selected compositions containing the main components of the Sulphide Queen ore body at Mountain Pass, California (Olson, 1954). According to Heinrich (1966, p. 357), the most abundant rock of the body consists of calcite (40–75%), baryte (15–50%), and bastnäsite 5–15%. The most abundant carbonatite dykes associated with the body contain calcite (up to 90%), baryte (5–30%) and bastnäsite (locally up to 30%). The rare earth elements are chiefly of the cerium subgroup, very enriched in the lightest rare earths.

Mariano (1989) classified the Mountain Pass deposit as the only known example of primary magmatic crystallization, whereas Bailey (1993) referred to the 'hydrothermal/carbothermal deposition' of the Mountain Pass carbonatite, excluding a magmatic origin. Le Bas (1977) referred to five generations of baryte in carbonatites, with the three main stages easily recognized. The first is early-stage carbonatite baryte associated with apatite, pyrochlore or perovskite; the second is late-stage carbonatite baryte associated with fluorite and bastnäsite; and the third is hydro-carbo-thermal vein baryte associated with fluorite, bastnäsite and quartz. Where these minerals become concentrated, they may form economic ore deposits (Kapustin, 1980; Naldrett and Watkinson, 1981; Semenov, 1974). Baryte–fluorite–quartz–calcite veins in many ore districts may be related to carbonatite magmas in a manner analogous to the relationship between silicate magmas and quartz, pegmatite and aplite veins (Kuellmer et al., 1966).

We have attempted to determine under what conditions calcite–baryte–fluorite–bastnäsite assemblages could be precipitated from a magma, rather than a hydrothermal solution. Following the approach illustrated by Figures 4.2–4.4, we determined first the phase relationships in the REE-free system, and then added rare earths as $La(OH)_3$. We build on the results of Kuellmer et al. (1966) in the system $CaCO_3$–$BaSO_4$–CaF_2 ± H_2O at 0.5 kbar.

4.7.1 The system $CaCO_3$–$BaSO_4$–CaF_2–$Ca(OH)_2$

The minerals calcite, baryte, fluorite and bastnäsite are represented in the system $CaCO_3$–$BaSO_4$–CaF_2; CaF_2 was added to provide fluorine for the fluocarbonate. $Ca(OH)_2$ was added to reduce the liquidus temperature. The liquidus relationships for several of the bounding systems are known: $CaCO_3$–$Ca(OH)_2$ (Wyllie and Tuttle, 1960), $CaCO_3$–CaF_2–$Ca(OH)_2$ (Gittins and Tuttle, 1964) and $CaCO_3$–$BaSO_4$–$Ca(OH)_2$ (Kuellmer et al., 1966). Figure 4.6 shows the arrangement of liquidus volumes, surfaces, field boundaries and eutectics in the quaternary system at 0.5 kbar, as reported by Kuellmer et al. (1966). The composition of the quaternary eutectic liquid, EQ, is 35% $CaCO_3$, 15% CaF_2, 8% $BaSO_4$ and 42% $Ca(OH)_2$:

$$\text{calcite} + \text{baryte} + \text{fluorite} + \text{portlandite} = \text{liquid} \qquad (4.4)$$

Deng and Wyllie (manuscript in preparation) determined the ternary eutectic reactions at up to 10 kbar for three bounding ternary systems, and

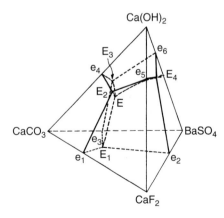

Figure 4.6 Phase diagram of $CaCO_3$–$BaSO_4$–CaF_2 at 0.5 kbar (Kuellmer et al., 1966).

for the quaternary system. They also located the equivalent solidus reactions with excess H_2O added. Results are shown in Figure 4.7. Note that addition of $Ca(OH)_2$ to the ternary system $CaCO_3-BaSO_4-CaF_2$ lowers the solidus temperature by about 250°C.

4.7.2 Phase relationships for Mountain Pass model composition

Figure 4.8 illustrates selection of the model mixture composition. The shaded area on the side $CaCO_3-BaSO_4-CaF_2$ represents the range of ore body mineralogy at Sulphide Queen. The mixture E was selected to be consistent with the ore body mineralogy, and $Ca(OH)_2$ was added to obtain a mixture with a low liquidus temperature, and to approximate the quaternary eutectic liquid composition EQ in Figure 4.6. The selection was made on the basis of the results of Gittins and Tuttle (1964) and Kuellmer et al. (1966). The mixture E has composition 39% $CaCO_3$, 16.9% CaF_2, 13% $BaSO_4$ and 31.2% $Ca(OH)_2$, which is listed in terms of components in Table 4.1, and compared with the electron microprobe analysis of the fused material.

Figure 4.9 shows the phase fields intersected by the join between the

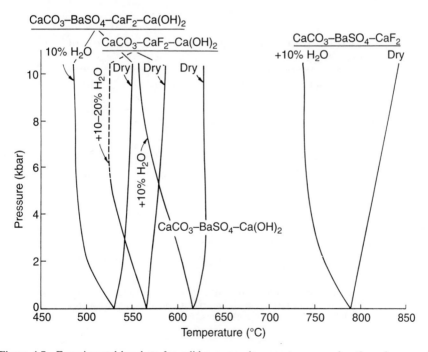

Figure 4.7 Experimental brackets for solidus curves (temperatures as a function of pressure) in the system $CaCO_3-BaSO_4-CaF_2 \pm H_2O$. Point K is the solidus temperature for the dry system at 0.5 kbar from Kuellmer et al. (1966). (Deng and Wyllie, in preparation).

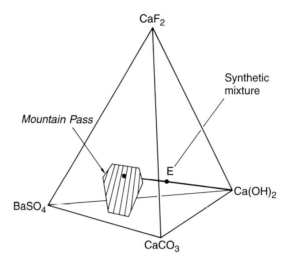

Figure 4.8 Mixture E located in the quaternary system $CaCO_3$–$Ca(OH)_2$–CaF_2–$BaSO_4$, and the estimated range of anhydrous bulk compositions (hatched) for Mountain Pass are on the $BaSO_4$–$CaCO_3$–CaF_2 face (i.e. recasting natural mineral assemblage into these equivalent components).

Table 4.1 Electron probe analysis[a] of carbonate glass compared with starting mixture

Component	Mixture wt.%	Glass[b]
CaO[c]	46.03	45.0
BaO	6.81	6.59
La_2O_3[d]	17.16	18.2
SO_3	3.55	3.08
F[e]	6.58	6.2[e]
CO_2	13.71	
H_2O	8.90	
Total – O = F	99.97	
Total – (CO_2, H_2O, O = F)	77.35	76.46

[a] Average of 9 spots, using defocused beam (~10 μm), 15 kV and 0.4 μA with routine wavelength dispersive procedures and $LaNi_5$ as La standard; [b] refractive index of glass = 1.588; [c] includes CaO equiv. to CaF_2, resolved in total; [d] La as La(OH)$_3$ in starting mix; [e] precision for F is much poorer (~±0.5%) than for other elements.

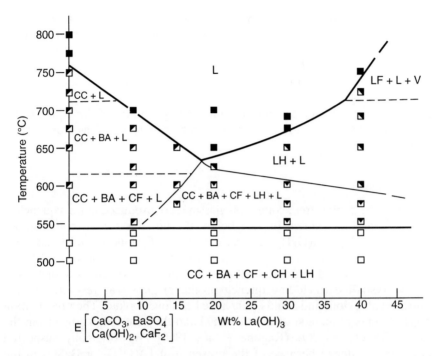

Figure 4.9 Isobaric phase diagram for the join E–LH at 1 kbar pressure. Heavy lines indicate liquidus and solidus. Abbreviations used are $CaCO_3$ = CC, $Ca(OH)_2$ = CH, CaF_2 = CF, $BaSO_4$ = BA, $La(OH)_3$ = LH, $LaOF$ = LF, L = liquid, V = vapour.

synthetic mixture 'E' and the light REE component $La(OH)_3$. The sub-solidus assemblage is calcite, portlandite, baryte, fluorite and lanthanum hydroxide. Addition of $La(OH)_3$ (LH) to the system $BaSO_4$–$CaCO_3$–CaF_2–$Ca(OH)_2$ reduces the solidus temperature by approximately 30°C from 576°C (at 0.5 kbar pressure; Kuellmer et al., 1966) to 543 ± 7°C (at 1 kbar pressure; Jones and Wyllie, 1983 and manuscript in preparation). The liquidus temperature for mixture E, without added $La(OH)_3$, was bracketed between runs at 750 and 775°C. There is a minimum on the liquidus located at a composition of $E_{82}LH_{18}$ and a temperature of 635°C. At temperatures below the liquidus, quenched liquid in the form of glass or quenched crystalline products coexists with various crystalline assemblages.

The appearance of minerals with decreasing temperature below the liquidus for pure E agrees with those predicted from the quaternary system. Calcite, present alone at the highest temperatures (>725°C), is joined by baryte at 700°C and then by fluorite at 600°C. The large field of LH + L to the right of the minimum on the liquidus in Figure 4.9 is terminated at approximately 710°C when LaOF (LF) becomes stable. The lanthanum oxyfluoride is readily distinguished by its isotropic character. The existence of prominent vesicles in the run product containing LF supports a possible reaction of the type

$$La(OH)_3 + F \rightarrow LaOF + nH_2O \qquad (4.5)$$

involving a net release of water (vapour). The geometry of the liquidus also fits with an inflexion where it is intersected by this reaction, as illustrated in Figure 4.9.

4.7.3 Glasses

Volatile-rich, bubble-free, colourless glasses were produced in runs quenched from temperatures above or just below the liquidus for all compositions studied, including $La(OH)_3$-free, pure E at 775°C. One of these has been described previously (Jones and Wyllie, 1983). Most runs between 700 and 625°C in Figure 4.9 produced glasses. In one or two charges that were mostly crystalline (including quench products) glass was restricted to the central axial region, and could easily have been overlooked. The preservation of glasses in carbonate systems is unusual and they are only known from the system K_2CO_3–$MgCO_3$ (Ragone et al., 1966). We originally suspected they were produced because of the presence of $La(OH)_3$ or $BaSO_4$ in the mixtures (Jones and Wyllie, 1983).

Seven glasses were analysed by electron microprobe for Ca, La, Ba, S and F. The analyses are compared with weighed starting mixtures in Figure 4.10. All plotted points represent the averages of several analyses for each different glass, and horizontal lines are drawn to show significant variations in results. Glasses from runs above the liquidus have uniform compositions and plot closest to the 1:1 slopes, indicating close agreement with starting mixtures. Glasses from runs below the liquidus are more variable and tend to stray from the 1:1 correlation lines. Overall, most of the probe analyses lie above the 1:1 lines, perhaps suggesting that the original mixtures were not perfectly dry; this is especially true for Ca.

Table 4.1 compares the analysis of the starting mixture E with the probe analysis of the glass quenched from just above the liquidus. The averaged probe analysis total is 76.5 wt% with the balance allotted to unanalysed H_2O and CO_2. This is consistent with the composition of the starting mixture, where $H_2O + CO_2 = 22.6\%$. Based on the close agreement between the probe analyses and the starting mixtures, the liquid E contained about 33% of dissolved volatile components, namely 13.7% CO_2, 8.9% H_2O, 6.6% F and 3.6% SO_3. This volatile-charged liquid has a composition suitable for complete crystallization into the subsolidus minerals of the original mixture, without evolution of a vapour phase.

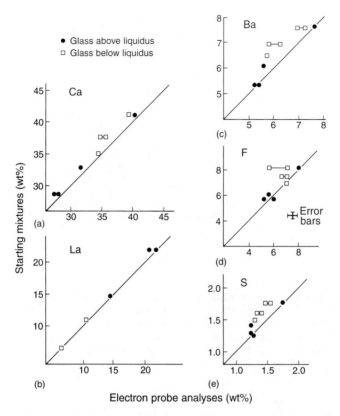

Figure 4.10 Glass compositions: electron probe analyses versus weighed starting mixtures (element wt%) for (a) Ca, (b) La, (c) Ba, (d) F and (e) S. Symbols differentiate between glasses from above and below liquidus (see text for details).

4.8 Silicate–carbonate liquid immiscibility

There is excellent petrological evidence that some carbonatites separate from silicate magmas by liquid immiscibility (Le Bas, 1977; Kjarsgaard and Peterson, 1991; Macdonald *et al.*, 1993), and experimental evidence confirms that liquid immiscibility does occur between silicates and carbonates across a wide range of compositions (Koster van Groos and Wyllie, 1966, 1973; Koster van Groos, 1975; Freestone and Hamilton, 1980). Figure 4.11 summarizes recent results in a generalized projection of experimentally determined miscibility gaps into the triangle $(SiO_2 + Al_2O_3 + TiO_2)$–$(CaO + MgO + FeO)$–$(Na_2O + K_2O)$, with sufficient CO_2 present to generate carbonates and saturate liquids. The silicate minerals and silicate liquids (Ls) with small amounts of dissolved CO_2 project into the area extending

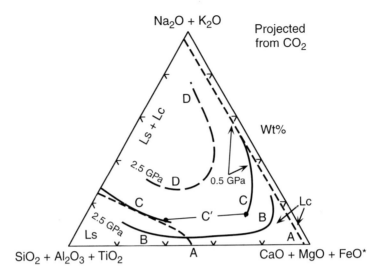

Figure 4.11 A variety of liquid miscibility gap boundaries determined for various silicate (Ls) and carbonate liquid (Lc) compositions at the pressures marked, projected from CO_2 into the triangle given. DD is for primitive magnesian melts under mantle conditions; the others are for Mg/Fe-free feldspar–carbonate mixtures; CC is a revised version of AA, consistent with the result BB. A, Kjarsgaard and Hamilton (1989); B, Lee and Wyllie (1992); C, Kjarsgaard and Hamilton (in preparation); D, Baker and Wyllie (1990). (Modified after Lee and Wyllie, 1994a; Kjarsgaard and Hamilton's revised boundaries CC were published in MacDonald et al., 1994.)

from the SiO_2 corner; carbonates and carbonate liquids (Lc) project in a band parallel to the opposite side, for alkali and cafemic oxides.

Baker and Wyllie (1990), using a primitive mantle-derived nephelinite and carbonate mixtures dominated by dolomite, located the limited miscibility gap DD at upper mantle pressures. Liquid compositions derived by partially melting carbonate-peridotites remain well separated from the miscibility gap, and it appears to be unlikely that immiscible carbonate melts can separate from mantle magmas in the lower lithosphere. Kjarsgaard and Hamilton (1988, 1989) and Brooker and Hamilton (1990) reported wide miscibility gaps in several feldspar–$CaCO_3$ systems at crustal pressures and high temperatures, mostly in the presence of CO_2 vapour, represented in Figure 4.11 by the dashed lines AA. Brooker and Hamilton (1990) added a second immiscible carbonate ($CaCO_3$-rich) liquid to this arrangement. Lee and Wyllie (1992a, 1992b, in preparation) studied the phase fields intersected by the join $NaAlSi_3O_8$–$CaCO_3$, concluding that the rounded $CaCO_3$ phase previously interpreted as immiscible liquid represented crystalline calcite (Lee and Wyllie, 1994a), and they defined the miscibility gap BB in Figure 4.11. Kjarsgaard and Hamilton (in Macdonald et al., 1994) revised their earlier result AA to CC, in conformity with the results of Lee and Wyllie

(1992a, 1992b, 1994b). The conjugate Ca-carbonate liquid boundary is quite far removed from pure $CaCO_3$.

Immiscibility does occur between silicate and carbonate liquids. The size of the miscibility gap varies as a function of pressure and of MgO/CaO (Lee and Wyllie, 1992a, 1992b, 1994b). It appears unlikely that melt paths in the upper mantle will intersect the miscibility gap, but with decreasing pressure and fractionation of alkaline silicate magmas, immiscible carbonatite magmas may be produced in the uppermost mantle, and certainly in the crust.

Wendlandt and Harrison (1979) demonstrated that in a system with immiscible silicate and carbonate liquids with CO_2 vapour, REE were fractionated into the vapour. Hamilton, Bedson and Esson (1989) determined the distribution of REE between immiscible silicate and carbonate liquids at crustal pressures, using phonolite and nephelinite systems. They found that enrichment of REE into the carbonate liquids was favoured by decreasing temperature, increasing pressure and increased polymerization of the co-existing silicate melt. In the nephelinite system the distribution coefficients were close to 1 in all conditions, indicating no significant enrichment of REE in the carbonate liquid. In the phonolite system enrichment in carbonate liquid was favoured below 1250°C, and in silicate liquid at higher temperatures; light REE were relatively more abundant in carbonate liquid than the heavy REE at fixed pressure and temperature, but extrapolation indicates that pressures much higher than 6 kbar would be required for enrichment in carbonate liquids. They concluded that if liquid immiscibility is controlling carbonatite genesis, the high level of REE enrichment commonly found in carbonatites must have occurred at pressures much greater than 6 kbar.

4.9 Conclusions and applications

The results reviewed indicate that the processes leading to high concentrations of REE in ore deposits are limited to the crust. Rare earth minerals may be precipitated through fractional crystallization of carbonatite magmas, but redistribution by fluids at subsolidus temperatures is clearly effective, although experimental data are few. Hamilton, Bedson and Esson (1989) preferred a multistage process (p. 425): 'an initial moderate concentration due to immiscibility followed by an increasing concentration in the carbonate melt due to the removal of large amounts of calcite (or dolomite) . . . and culminating in the precipitation and accumulation of phases which have high contents of the incompatible elements, e.g. bastnäsite, pyrochlore, and baddeleyite.' (Kjarsgaard and Hamilton, 1989). Our results in Figures 4.4, 4.5 and 4.9 support the process of fractionation as a process for concentrating REE to the stage of precipitation of bastnäsite, whether or not liquid immiscibility was involved in the petrogenesis of the carbonatite magma (cf. Bailey, 1993).

The system $CaCO_3-Ca(OH)_2$ is the simplest of synthetic carbonatite magmas. The precipitated product, calcite, is without question a good model for sövite. The magmatic model is too simple, lacking the alkalis which play such an important role in fenitization around carbonatites, and also problematic in that the residual liquid from the synthetic magma precipitates portlandite, whereas portlandite is not a normal component of carbonatites or fenites. The former aspect of carbonatites has been explored in many synthetic systems, and the latter problem has been eased by the discovery of an additional eutectic for calcite, dolomite and periclase, without portlandite, in the system $CaO-MgO-CO_2-H_2O$ (Fanelli, Cava and Wyllie, 1986; Wyllie and Jones, 1985).

The presence of a minimum on the liquidus at $E_{82}LH_{18}$ (Figure 4.9) shows the high solubility of $La(OH)_3$ in mixture E. Lanthanum, and presumably the other REE, can obviously play a role as major components rather than as trace components in carbonate-rich melts (Figures 4.4 and 4.5), if conditions are suitable for their concentration. The experimental results are entirely consistent with the observation that RE carbonates occur in more differentiated carbonatites (Heinrich, 1966, p. 234). Initial carbonatitic liquid compositions in Figures 4.4 and 4.5 could follow two different types of crystallization paths. An initial liquid could have more or less dissolved REE than values given for the line $CaCO_3-E$. A liquid with lower REE would first precipitate calcite. The liquid composition would change across the calcite liquidus surface until it reached the field boundary E1–E, where calcite would be joined by $Ca(OH)_2$, and the liquid would then change towards E where the rare earth mineral would be precipitated. A liquid with higher REE would also precipitate first calcite, but it would reach the field boundary E–E3 where calcite would be joined by $La(OH)_3$, and the two minerals would coprecipitate with further cooling down to E. In a more complex magma, with different bulk composition and possibly with higher CO_2/H_2O, the RE mineral could conceivably be one of the bastnäsite group minerals (compare Figures 4.5 and 4.2).

Given these results, one might expect that RE carbonatites would be more abundant. However, if the initial concentration of REE was small enough, this could be camouflaged in the major carbonates and other minerals, and the REE might not have the opportunity to grow as RE minerals. Given more than this minimum level of REE concentration, the biggest factor in precluding their concentration in the liquid to yield RE carbonatites is probably the efficient extraction of REE from the magma by higher temperature REE-bearing minerals such as perovskite (Mitchell, Chapter 3), apatite and monazite during crystallization of most carbonatite magmas. A carbonatite magma low in phosphorus may not precipitate the apatite and monazite characteristic of the magnetite–apatite type of carbonatite, leading to the concentration of REE in residual liquids and the formation of a rare earth carbonatite.

The highly soluble lanthanum contrasts with poorly soluble components in synthetic carbonatites. The solubility of silicate components (as defined by calcite field boundaries such as E1a in Figures 4.5 and 4.6) SiO_2, Mg_2SiO_4 (olivine), feldspars and feldspathoids is less than 5 wt% (for reviews see Wyllie, 1966, 1978; Wyllie and Jones, 1985). Other components such as CaS (Helz and Wyllie, 1979) and P_2O_5 (Wyllie and Biggar, 1966; Wyllie, 1966; Biggar, 1969) are similarly low. Addition of CaS to the system $CaCO_3$–$Ca(OH)_2$ reduces the solidus temperature by only about 3°C. A few per cent CaS or P_2O_5 in a carbonatite melt would be expected to crystallize early sulphides or apatite, respectively, whereas Figures 4.4, 4.5 and 4.9 indicate that $La(OH)_3$ would be concentrated in the residual liquid. Stabilities of RE minerals such as bastnäsite are strongly dependent on CO_2/H_2O ratios, as established in the previously published synthesis experiments described above. The low solidus temperature of this system E–LH (Figure 4.9) implies that addition of water to the Mountain Pass carbonatite is all that is required for it to be molten at approximately 543°C and 1 kbar pressure. This temperature is at least 30°C below the stability limit of bastnäsite given by Tareen and Kutty (1980; ~575°C) and more than 100°C below the stabilities given by Hsu (1992; >650–780°C depending on fO_2). Therefore it seems quite likely that, given suitable CO_2/H_2O ratios, bastnäsite could have crystallized at Mountain Pass as a primary phase from carbonatite magma similar to E in bulk composition (Jones and Wyllie, 1983).

This model predicts that natural RE-carbonatites should be an expected product of fractional crystallization in carbonatites, yet they are unusually rare. One apparent solution to this dilemma is that in natural carbonatites the REE partition strongly into an early-crystallizing phase. The phosphate minerals apatite and monazite are locally abundant in carbonatites and can carry considerable REE. Together with the expectedly low solubility of P_2O_5 in carbonatite magmas, extraction of the REE via early crystallization of apatite, for example, provides at least one reasonable mechanism for preventing sizeable volumes of late-stage RE-carbonatite in typical carbonatite complexes. Thus, the development of large RE carbonatites such as at Mountain Pass might have been related to crystal fractionation of P_2O_5-poor antecedants. This agrees with the empirical observation that the carbonatite at Mountain Pass is relatively poor in phosphate minerals (Jones and Wyllie, 1983).

Acknowledgements

This research was supported by the Earth Science section of the US National Science Foundation, grant EAR 9218806. Numerous colleagues in the 'carbonatite community' are thanked for their contributions to an enjoyable series of discussions over the past few years. APJ would like to thank, in particular, John Gittins, Mike LeBas and Alan Woolley for their help with earlier versions of this work.

References

Andersen, T. (1984) Secondary processes in carbonatites: petrology of 'rødberg' (hematite-calcite-dolomite carbonatite) in the Fen central complex, Telemark (South Norway). *Lithos.*, **17**, 227–45.

Aumont, R., Genet, F., Passaret, M. and Toudic, Y. (1971) Preparation par voie hydrothermale d'hydroxycarbonates d'elements des terres rares, et des produits correspondants obtenus par substitution de OH par Cl; determination des principaux caracteres cristallographiques. *C. R. Acad. Sci.*, **272(C)**, 314–17.

Bailey, D. K. (1993) Carbonate magmas. *J. Geol. Soc., London*, **150**, 637–51.

Baker, M. B. and Wyllie, P. J. (1989) Liquid compositions and Fe–Mg partitioning at 30 kbar in the systems olivine–dolomite and olivine–orthopyroxene–dolomite. Abstract. *EOS Trans. Geophys. Union*, **70**, 1394.

Baker, M. B. and Wyllie, P. J. (1990) Liquid immiscibility in a nephelinite–carbonate system at 25 kbar and implications for carbonatite origin. *Nature*, **346**, 168–70.

Baker, M. B. and Wyllie, P. J. (1992) High pressure apatite stability in carbonate-rich liquids: implications for mantle metasomatism. *Geochim. Cosmochim. Acta*, **56**, 3409–22.

Biggar, G. M. (1969) Phase relationships in the join $Ca(OH)_2-CaCO_3-Ca_3(PO_4)-H_2O$ at 1000 bars. *Mineral. Mag.*, **37**, 75–82.

Brenan, J. M. and Watson, E. B. (1991a) Partitioning of trace elements between carbonate melt and clinopyroxene and olivine at mantle P–T conditions. *Geochim. Cosmochim. Acta*, **55**, 2203–14.

Brenan, J. M. and Watson, E. B. (1991b) Partitioning of trace elements between olivine and aqueous fluids at high P–T conditions: implications for the effect of fluid composition on trace-element transport. *Earth Planet. Sci. Lett.*, **107**, 672–88.

Brooker, R. A. and Hamilton, D. L. (1990) Three-liquid immiscibility and the origin of carbonatites. *Nature*, **346**, 459–62.

Caro, M. P. (1966) Sur l'existence apparente d'un 'ion complexe' $Ln_2O_2^{2+}$ ou LnO^+ dans le nombreux des terres rares. *C. R. Acad. Sci. Paris*, Ser. C, 992–5.

Caro, P. and Lemaitre-Blaise, M. (1969) Hydroxycarbonates de terres rares $Ln_2(CO_3)_x (OH)_{2(3-x)} nH_2O$ (Ln = terres rares). *C. R. Acad. Sci.*, **269(C)**, 687–90.

Caro, P., Lemaitre-Blaise, M. and Trombe, F. (1968) Identification et solubilites des phases solides a l'equilibre sous une atmosphere de gaz carbonique dans les systemes ternaires oxydes de terres rares–gaz carbonique–eau. *C. R. Acad. Sci.*, **267(C)**, 1594–7.

Chai, H. T. and Mroczkowski, S. (1978) Synthesis of rare-earth carbonates under hydrothermal conditions. *J. Cryst. Growth*, **44**, 84–97.

Christensen, A. N. (1973) Hydrothermal preparation of rare earth hydroxy-carbonates. The crystal structure of $NdOHCO_3$. *Acta Chem. Scand.*, **27**, 2973–82.

Dalton, J. A. and Wood, B. J. (1993) The compositions of primary carbonate melts and their evolution through wallrock reaction in the mantle. *Earth Planet, Sci. Lett.*, **119**, 511–25.

Deans, T. (1966) Economic mineralogy of African carbonatites. In *Carbonatites* (ed. O. F. Tuttle and J. Gittins), Wiley, New York, pp. 385–413.

Eby, G. N. (1975) Abundance and distribution of the rare-earth elements and yttrium in the rocks and minerals of the Oka carbonatite complex, Quebec. *Geochim. Cosmochim. Acta,* **39**, 597–620.

Eggler, D. H. (1974) Effect of CO_2 on the melting of peridotite. *Carn. Inst. Wash. Yb.*, **73**, 215–24.

Eggler, D. H. (1976) Does CO_2 cause partial melting in the low-velocity layer of the mantle? *Geology,* **4**, 69–72.

Eggler, D. H. (1978) The effect of CO_2 upon partial melting of peridotite in the system $Na_2O-CaO-Al_2O_3-MgO-SiO_2-CO_2$ to 35 kb, with an analysis of melting in a peridotite–H_2O-CO_2 system. *Am. J. Sci.*, **278**, 305–43.

Eggler, D. H. (1987) Discussion of recent papers on carbonated peridotite, bearing on mantle metasomatism and magmatism: an alternative. *Earth Planet. Sci. Lett.*, **82**, 398–400.

Eggler, D. H. (1989) Carbonatites, primary melts, and mantle dynamics, in *Carbonatites; Genesis and Evolution* (ed. K. Bell), Unwin Hyman, London, pp. 561–79.

Eggler, D. H., Holloway, J. R. and Mysen, B. O. (1976) High CO_2 solubilities in mantle magmas: comment. *Geology,* **4**, 198–9.

Falloon, T. J. and Green, D. H. (1989) The solidus of carbonated fertile peridotite. *Earth Planet. Sci. Lett.*, **94**, 364–70.

Fanelli, M. F., Cava, N. and Wyllie, P. J. (1986) Calcite and dolomite without portlandite at a new eutectic in $CaO-MgO-CO_2-H_2O$, with applications to carbonatites, in *Morphology and Phase Equilibria of Minerals,* Proceedings 13th General Meeting International Mineralogical Association, Bulgarian Academy of Sciences, Sofia, pp. 313–22.

Freestone, I. C. and Hamilton, D. L. (1980) The role of liquid immiscibility in the genesis of carbonatites – an experimental study. *Contrib. Mineral. Petrol.* **73**, 105–17.

Gittins, J. and Tuttle, O. F. (1964) The system $CaF_2-Ca(OH)_2-CaCO_3$. *Am. J. Sci.*, **262**, 66–75.

Glass and Smalley (1945) Bastnasite. *Can. Mineral.*, **30**, 601–15.

Green, D. H. and Wallace, M. E. (1988) Mantle metasomatism by ephemeral carbonatite melts. *Nature*, **336**, 459–62.

Green, T. H., Adam, J. and Sie, S. H. (1992) Trace element partitioning between silicate minerals and carbonatite at 25 kbar and application to mantle metasomatism. *Mineral. Petrol.*, **46**, 179–84.

Haggerty, S. E. (1983) The mineral chemistry of new titanates from the Jagersfontein kimberlite, South Africa: implications for metasomatism in the upper mantle. *Geochim. Cosmochim. Acta,* **47**, 1833–54.

Haggerty, S. E. (1989a) Upper mantle opaque mineral stratigraphy and the genesis of metasomites and alkali-rich melts, in *Kimberlites and Related Rocks*, vol. 2, *Geol. Soc. Austr Spec. Publ.* **14**, Blackwell Scientific, pp. 687–99.

Haggerty, S. E. (1989b) Mantle metasomes and the kinship between carbonatites and kimberlites, in *Carbonatites: Genesis and Evolution* (Ed. K. Bell), Unwin Hyman, London, pp. 546–60.

Hamilton, D. L., Bedson, P. and Esson, J. (1989) The behaviour of trace elements in the evolution of carbonatites. In *Carbonatites: Genesis and Evolution* (ed. K. Bell), Unwin Hyman, London, pp. 405–27.

Haschke, J. M. (1975) The lanthanum hydroxide fluoride carbonate system: the preparation of synthetic bastnaesite. *J. Solid State Chem.*, **12**, 115–21.

Haschke, J. M. and Eyring, L. (1971) Hydrothermal equilibria and crystal growth of rare earth oxides, hydroxides, hydroxynitrates, and hydroxycarbonates. *Inorg. Chem.*, **10**, 2267–74.

Hauri, E. H., Shimizu, N., Dieu, J. J. and Hart, S. R. (1993) Evidence for hotspot-related carbonatite metasomatism in the oceanic upper mantle. *Nature,* **365**, 221–7.

Heinrich, E. W. (1966) *The Geology of Carbonatites*, Rand McNally.
Helz, G. R. and Wyllie, P. J. (1979) Liquidus relationships in the system $CaCO_3-Ca(OH)_2-$ CaS and the solubility of sulfur in carbonatite magmas. *Geochim. Cosmochim. Acta*, **43**, 259–65.
Hsu, L. C. (1992) Synthesis and stability of bastnaesites in a part of the system (Ce, La)–F–H–C–O. *Mineral. Petrol.*, **47**, 87–101.
Huang, W. L. and Wyllie, P. J. (1974) Eutectic between wollastonite II and calcite contrasted with thermal barrier in $MgO-SiO_2-CO_2$ at 30 kilobars, with applications to kimberlite-carbonatite petrogenesis. *Earth Planet. Sci. Lett.*, **24**, 305–10.
Hunter, R. H. and McKenzie, D. (1989) The equilibrium geometry of carbonate melts in rocks of mantle composition. *Earth Planet. Sci. Lett.*, **92**, 347–56.

Ionov, D. A., Dupuy, C., O'Reilly, S. Y. *et al.* (1993) Carbonated peridotite xenoliths from Spitsbergen: implications for trace element signature of mantle carbonate metasomatism. *Earth Planet. Sci. Lett.*, **119**, 283–97.
Irving, A. J. and Wyllie, P. J. (1973) Melting relationships in $CaO-CO_2$ and $MgO-CO_2$ to 36 kilobars, with comments on CO_2 in the mantle. *Earth Planet. Sci. Lett.*, **20**, 220–5.
Irving, A. J. and Wyllie, P. J. (1975) Subsolidus and melting relationships for calcite, magnesite, and the join $CaCO_3-MgCO_3$ to 36 kilobars. *Geochim. Cosmochim. Acta*, **39**, 35–53.

Jones, A. P. (1989) Upper-mantle enrichment by kimberlitic or carbonatitic magmatism. *Carbonatites: genesis and evolution* (ed. K. Bell), Unwin Hyman, London, pp. 448–63.
Jones A. P. and Wyllie, P. J. (1983) Low-temperature glass quenched from a synthetic, rare earth carbonatite: implications for the origin of the Mountain Pass deposit, California. *Econ. Geol.*, **78**, 1721–3.
Jones, A. P. and Wyllie, P. J. (1985) Paragenetic trends of oxide minerals in carbonate-rich kimberlites, with new analyses from the Benfontein sill, South Africa. *J. Petrol.*, **26**, 210–22.
Jones, A. P. and Wyllie, P. J. (1986) Solubility of REE in carbonatite magmas, indicated by the liquidus surface in $CaCO_3-Ca(OH)_2-La(OH)_3$ at 1 kbar pressure. *Appl. Geochem.*, **1**, 95–104.
Jones A. P., Dobson, D. and Genge, M. (1995) Comment on the physical properties of carbonatite magmas inferred from molten salt data, and application to extraction patterns from carbonatite–silicate magma chambers. *Geol. Mag.*, **132**, 121.
Jones, A. P., Smith, J. V. and Dawson, J. B. (1982) Mantle metasomatism in 14 veined peridotite xenoliths from Bulfontein Mine, South Africa. *J. Geol.*, **90**, 435–53.

Kapustin, Yu. L. (1980) *Mineralogy of Carbonatites*, Amerin. Pub. Co. Pvt. Ltd.
Kjarsgaard, B. A. and Hamilton, D. L. (1988) Liquid immiscibility and the origin of alkali-poor carbonatites. *Mineral. Mag.*, **52**, 43–55.
Kjarsgaard, B. A. and Hamilton, D. L. (1989) The genesis of carbonatites by immiscibility, in *Carbonatites: Genesis and Evolution* (ed. K. Bell), Unwin Hyman, London, pp. 388–404.
Kjarsgaard, B. A. and Petersen, D. (1991) Nephelinite–carbonatite liquid immiscibility at Shombole Volcano, East Africa: petrographic and experimental evidence. *Mineral. Petrol.*, **143**, 293–314.
Koster van Groos, A. F. (1975) The effect of high CO_2 pressures on alkalic rock and its bearing on the formation of alkalic ultrabasic rocks and the associated carbonatites. *Am. J. Sci.*, **275**, 163–85.
Koster van Groos, A. F. and Wyllie, P. J. (1966) Liquid immiscibility in the system $Na_2O-Al_2O_3-SiO_2-CO_2$ at pressures up to 1 kilobar. *Am. J. Sci.*, **265**, 234–55.
Koster van Groos, A. F. and Wyllie, P. J. (1973) Liquid immiscibility in the join $NaAlSi_3O_8-CaAl Si_2O_8-Na_2CO_3-H_2O$. *Am. J. Sci.*, **273**, 465–87.
Kuellmer, F. J. *et al.* (1966) Preliminary survey of the system barite–calcite–fluorite at 500 bars, in *Carbonatites* (eds O. F. Tuttle and J. Gittins), Interscience, pp. 353–64.

REFERENCES

Kutty, T. R. N., Viswanathiah, M. N. and Tareen, J. A. K. (1978) Hydrothermal equilibria in $Nd_2O_3-CO_2$ system. *Proc. Indian Acad. Sci*, **87A**, 69–74.

Le Bas, M. J. (1977) *Carbonatite–Nephelinite Volcanism*, Wiley, London.

Lee, W. J. and Wyllie, P. J. (1992a) New data on CO_2-rich immiscible liquids in $Na_2O-CaO-Al_2O_3-SiO_2-CO_2$ from 25 to 1 kb: carbonatite genesis (abstract). *EOS Trans. Am. Geophys. Union*, **73(14)**, 349–50.

Lee, W.-J. and Wyllie, P. J. (1992b) Liquid immiscibility between silicates and carbonates must intersect suitable liquidus field boundaries to have petrogenetic significance. 29th International Geological Congress, Kyoto, Abstracts, p. 571.

Lee, W.-J. and Wyllie, P. J. (1994a) Conditions for formation of immiscible carbonate-rich magmas from primitive (magnesian) nephelinite. Geol. Soc. Am. Abstr. Prog., **26**, A224, (abstract).

Lee, W.-J. and Wyllie, P. J. (1994b) The generation of Na-rich carbonate-rich magmas at crustal conditions. EOS *Trans. Am. Geophys. Union*, **75(44)**, 720, (abstract).

Loubet, M., Bernat, M., Javoy, M. and Allegre, C. J. (1972) Rare earth contents in carbonatites. *Earth Planet. Sci. Lett.*, **14**, 226–32.

Macdonald, R., Kjarsgaard, B. A., Skilling, P. *et al.* (1993) Liquid immiscibility between trachyte and carbonate in ash flow tuffs from Kenya. *Contrib. Mineral. Petrol.*, **114**, 276–87.

Mariano, A. N. (1989) Nature of economic mineralization in carbonatites and related rocks, in *Carbonatites: Genesis and Evolution* (ed. K. Bell), Unwin Hyman, London, pp. 149–76.

Meen, J. K., Ayers, J. C. and Fregeau, E. J. (1989) A model of mantle metasomatism by carbonated alkaline melts: trace-element and isotopic compositions of mantle source regions of carbonatite and other continental rocks, in *Carbonatites: genesis and evolution* (ed. K. Bell), Unwin Hyman, London, pp. 464–99.

Menzies, M. A. and Chazot, G. (1994) Mantle metasomatism – the transfer of silicate and nonsilicate melts in the Earth's mantle. In Invited Lectures, International Symposium on Physics and Chemistry of the Upper Mantle, Sao Paulo, Brazil, August 1994, pp. 117–36.

Menzies, M. A. and Hawkesworth, C. J. (1987) *Mantle Metasomatism*, Academic Press, London, 472 pp.

Menzies, M. A. and Wass, S. Y. (1983) CO_2- and LREE-rich mantle below eastern Australia: a REE and isotopic study of alkaline magmas and apatite-rich mantle xenoliths from the Southern Highlands Province, Australia. *Earth Planet. Sci. Lett.*, **65**, 287–302.

Naldrett, A. J. and Watkinson, D. H. (1981) Ore formation within magmas, in *Mineral Resources: Genetic Understanding for Practical Applications* (ed. P. B. Barton), Natl Acad., Washington.

Olson, J. C., Shawe, D. R., Pray, L. C. and Sharp, W. N. (1954) Rare earth mineral deposits of the Mountain Pass district, San Bernardino County, California. *US Geol Surv. Prof. Paper*, **261**.

Pecora, W. T. (1956) Carbonatites: a review. *Bull. Geol. Soc. Am.*, **67**, 1537–56.

Ragone S. E. (1966) The system potassium carbonate–magnesium carbonate. *J. Phys. Chem.*, **70(10)**, 3361.

Semenov, E. I. (1974) Economic mineralogy of alkaline rocks, in *The Alkaline Rocks* (ed. H. Sorensen), Wiley, pp. 543–52.

Sweeney, R. J. (1994) Carbonatite melt compositions in the Earth's mantle. *Earth Planet. Sci. Lett.*, **128**, 259–70.

Sweeney, R. J., Green, D. H. and Sie, S. H. (1992) Trace and minor element partitioning between garnet and amphibole and carbonatitic melt, *Earth Planet. Sci. Lett.*, **113**, 1–14.

Tareen, J. A. K. and Kutty, T. R. N. (1980) Hydrothermal phase equilibria in $Ln_2O_3–H_2O–CO_2$ system. *J. Crystal Growth*, **50**, 527–32.

Thibault, Y., Edgar, A. D. and Lloyd, F. E. (1992) Experimental investigation of melts from a carbonated phlogopite lherzolite: implications for metasomatism in the continental lithospheric mantle. *Am. Mineral.*, **77**, 784–94.

Van Landuyt and Amelinckx (1975) Multiple beam imaging of new mixed-layer compounds of the bastnaesite–synchisite series. *Am. Mineral.*, **60**, 351–8.

Wallace, M. E. and Green, D. H. (1988) An experimental determination of primary carbonatite magma composition. *Nature*, **335**, 343–6.

Wendlandt, R. F. and W. J. Harrison (1979) Rare earth partitioning between immiscible carbonate and silicate liquids and CO_2 vapor: results and implications for the formation of light rare earth-enriched rocks. *Contrib. Mineral Petrol.*, **69**, 409–19.

White, B. S. and Wyllie, P. J. (1992) Phase relations in synthetic lherzolite–H_2O–CO_2 from 20–30 kb, with applications to melting and metasomatism. *J. Volcan. Geotherm. Res.*, **50**, 117–30.

Williams-Jones, A. E. and Wood, S. A. (1992) A preliminary petrogenetic grid for REE fluorocarbonates and associated minerals. *Am. Mineral.*, **56**, 725–38.

Wolff J. A. (1994) Physical properties of carbonatite magmas inferred from molten salt data, and application to extraction patterns from carbonatite–silicate magma chambers. *Geol. Mag.*, **131**, 145–53.

Wyllie, P. J. (1966) Experimental studies of carbonatite problems: the origin and differentiation of carbonatite magmas. in *The Carbonatites* (eds O. F. Tuttle and J. Gittins), John Wiley and Sons, Inc., pp. 311–52.

Wyllie, P. J. (1977) Mantle fluid compositions buffered by carbonates in peridotite–CO_2–H_2O. *J. Geol.*, **85**, 187–207.

Wyllie, P. J. (1978) Mantle fluid compositions buffered in peridotite–CO_2–H_2O by carbonates, amphibole, and phlogopite. *J. Geol.*, **86**, 687–713.

Wyllie, P. J. (1980) The origin of kimberlites. *J. Geophys. Res.*, **85**, 6902–10.

Wyllie, P. J. (1987) Discussion of recent papers on carbonated peridotite, bearing on mantle metasomatism and magmatism. *Earth Planet. Sci. Lett.*, **82**, 391–7, 401–2.

Wyllie, P. J. (1989) Origin of carbonatites – evidence from phase equilibrium studies. in *Carbonatites: Genesis and Evolution*, (ed. K. Bell) Unwin, Hyman, London, pp. 500–45.

Wyllie, P. J. and Biggar, G. M. (1966) Fractional crystallization in the 'carbonatite systems' $CaO–MgO–CO_2–H_2O$ and $CaO–CaF_2–P_2O_5–CO_2–H_2O$. Papers and Proceedings of the 4th General Meeting, International Mineralogical Association, I. M. A. Volume, Mineralogical Society of India, 92–105.

Wyllie, P. J. and Huang W. L. (1975) Peridotite, kimberlite, and carbonatite explained in the system $CaO–MgO–SiO_2–CO_2$. *Geology*, **3**, 621–4.

Wyllie, P. J. and Huang, W. L. (1976) Carbonation and melting reactions in the system $CaO–MgO–SiO_2–CO_2$ at mantle pressures with geophysical and petrological applications. *Contrib. Mineral Petrol.*, **54**, 79–107.

Wyllie, P. J. and Jones, A. J. (1985) Experimental data bearing on the origin of carbonatites, with particular reference to the Mountain Pass rare earth deposit, in *Appl. Miner. Park.* W. C. et al., Ed., P. 935–949. Am. Institute Min. Metal. Petroleum Eng., Inc., New York.

Wyllie, P. J. and Rutter, M. (1986) Experimental data on the solidus for peridotite–CO_2, with applications to alkaline magmatism and mantle metasomatism. *EOS, Trans. Am. Geophys. Union*, **67**, 390.

REFERENCES

Wyllie, P. J. and Tuttle, O. F. (1960) The system $CaO-CO_2-H_2O$ and origin of carbonatites. *J. Petrol.*, **1**, 1–46.

Wyllie, P. J., Huang, W. L., Otto, J. and Byrnes, A. P. (1983) Carbonation of peridotites and decarbonation of siliceous dolomites represented in the system $CaO-MgO-SiO_2-CO_2$ to 30 kbar. *Tectonophys.*, **100**, 359–88.

Yaxley, G. M., Crawford, A. J. and Green, D. H. (1991) Evidence for carbonatite metasomatism in spinel peridotite xenoliths from western Victoria, Australia. *Earth Planet. Sci. Lett.*, **107**, 305–17.

CHAPTER FIVE
Formation of rare earth minerals in hydrothermal systems
R. Gieré

5.1 Introduction

Historically, the most important economic occurrences of RE minerals are of igneous or sedimentary origin, i.e. carbonatite-hosted deposits (e.g. Mountain Pass, California) and coastal placers (e.g. monazite beach sands, Western Australia; cf. Neary and Highley, 1984). Therefore, the attention of geologists has often been drawn away from hydrothermal processes which, under certain conditions, may lead to the formation of equally important REE deposits (e.g. Bayan Obo, Inner Mongolia, China). A second reason for neglecting REE in hydrothermal processes lies in the fact that for a long time the REE were regarded as generally immobile elements. As pointed out by Mariano (1989a), however, hydrothermal occurrences of RE minerals, and REE skarns and replacement bodies in particular, may soon exceed the carbonatite-hosted deposits in economic importance as a light REE resource.

In this chapter it will be shown that RE minerals occur very frequently in hydrothermal systems. It is also of note that several RE minerals have been described for the first time in hydrothermal environments, e.g. bastnäsite, cerite, lanthanite, törnebohmite (cf. Geijer, 1921), burbankite (Pecora and Kerr, 1953), gasparite (Graeser and Schwander, 1987), huanghoite (Semenov and Chang, 1961), paraniite (Demartin, Gramaccioli and Pilati, 1994), sahamalite (Jaffe, Meyrowitz and Evans, 1953) and stillwellite (McAndrew and Scott, 1955). Furthermore, we should bear in mind that cerium as oxide was discovered in metasomatic rocks (in 1803 by J. J. Berzelius and W. Hisinger at Bastnäs, Sweden).

A hydrothermal system is considered here to be any system in which heated aqueous solutions interact with rocks or melts. According to this general definition, hydrothermal fluids are not restricted to direct or indirect involvement of igneous rocks or melts. Weathering processes will be excluded from the following discussion.

Rare Earth Minerals: Chemistry, origin and ore deposits. Edited by Adrian P. Jones, Frances Wall and C. Terry Williams. Published in 1996 by Chapman & Hall. ISBN 0 412 61030 2

5.2 RE minerals of hydrothermal origin

A list of RE minerals frequently reported from hydrothermal systems is given in Table 5.1. In addition to these minerals listed, there is wide variety of other, less important RE minerals which have been found in hydrothermal environments, e.g. ancylite, brannerite, burbankite, calkinsite, cappelenite, carbocernaite, chernovite, crandallite, davidite, fergusonite, fersmite, florencite, gasparite, huanghoite, ilimaussite, joaquinite, kainosite, lanthanite, paraniite, pyrochlore, rhabdophane, röntgenite, sahamalite, thorite and törnebohmite (Geijer, 1921; Jaffe, Meyrowitz and Evans, 1953; Pecora and Kerr, 1953; Semenov and Chang, 1961; Staatz and Conklin, 1966; Graeser, Schwander and Stalder, 1973; Caruso and Simmons, 1985; Staatz, 1985; Graeser and Schwander, 1987; Bermanec et al., 1988; Mariano, 1989b; Oreskes and Einaudi, 1990; Yuan et al., 1992; Demartin, Gramaccioli and Pilati, 1994; Ngwenya, 1994; for more references see Gramaccioli, 1977; Clark, 1984). In many localities the hydrothermal RE minerals are associated with apatite or fluorite which are both usually rich in REE. Furthermore, there are many common rock-forming minerals that incorporate relatively small, but nevertheless significant, amounts of REE into their structures; examples from hydrothermal systems include anhydrite, baryte, calcite, garnet, laumontite, prehnite, vesuvianite scheelite, tourmaline and wolframite (Crook and Oswald, 1979; Möller et al., 1979; Alderton, Pearce and Potts, 1980; Morgan and Wandless, 1980; Raimbault, 1985; Cressey, 1987; Barrett, Jarvis and Jarvis, 1990; Bau and Möller, 1992; Chen, Halls and Stanley, 1992; Collins and Strong, 1992; Jamtveit and Hervig, 1994).

Hydrothermal RE minerals may be found in a wide variety of geological settings ranging from fillings in fissures and breccias to veins, stockworks, skarns and large-scale metasomatic replacement bodies (Table 5.1). The term skarn, as used in this chapter, describes contact metasomatic replacements only i.e. tactites; metasomatic replacements without direct contacts to igneous rocks are termed replacement bodies. Alpine-type fissures (Alpine clefts) are usually open joints, cavities or vugs which were partially filled with a variety of minerals during metamorphism and uplift of mountain belts (Niggli, 1940); besides containing very common minerals such as quartz, adularia, rutile and chlorite, many of these fissures host a number of RE minerals which are often of museum quality (Plates 1 and 2).

RE minerals may also occur as daughter minerals in fluid inclusions, as demonstrated by Kwak and Abeysinghe (1987) and Salvi and Williams-Jones (1990). The former authors pointed out the difficulties in identifying such minerals within tiny fluid inclusions. It is very likely, therefore, that RE daughter minerals have been overlooked in the past and could be found in many other localities, particularly in high-temperature metasomatic environments.

Some characteristics of selected hydrothermal RE deposits are given in

Table 5.1 List of the most commonly reported RE minerals of hydrothermal origin

Mineral	Simplified formula	Modes of hydrothermal occurrence	References
Aeschynite	(REE,Ca,Fe,Th) $(Ti,Nb)_2 (O,OH)_6$	Alpine-type fissures Veins Replacement bodies	15,21,38 48 50
Allanite	$(Ca,REE)_2$ $(Al,Fe)_3 (SiO_4)_3 OH$	Alpine-type fissures Cavities Veins Skarns Replacement bodies	2,21,52 8 1,5,7,8,16,18,23,24,27,46,58,60 1,4,6,7,12,13,14,29,35,43,51 41
Apatite	$Ca_5 (PO_4)_3$ (OH,F,Cl)	cf. Table 5.2 and text	cf. Table 5.2 and text
Bastnäsite	$REE (CO_3) F$	Alpine-type fissures Breccia fillings Veins Skarns Replacement bodies	21,52 4,9,42 4,9,10,16,18,22,31,32,34,39,59 1,4,7 40,41,50
Britholite	$(REE,Ca)_5$ $(SiO_4,PO_4)_3 (OH,F)$	Veins Replacement bodies	39,55 41
Brockite	(Ca,Th,REE) $PO_4 \cdot H_2O$	Veins	16,28
Cerite	$(REE,Ca)_9 (Mg,Fe)$ $Si_7 (O,OH,F)_{28}$	Skarns	1,4,7
Fluocerite	$REE F_3$	Breccia fillings Veins Skarns	27 32 1,4
Fluorite	$Ca F_2$	cf. Table 5.2 and text	cf. Table 5.2 and text
Gadolinite	$REE_2 Fe Be_2 Si_2 O_{10}$	Alpine-type fissures	3,21,52
Monazite	$REE PO_4$	Alpine-type fissures Breccia fillings Veins Skarns Replacement bodies	21,30,33,44,52 42 10,16,18,19,22,23,36,39,40,46,53,60 4,13,43 40,50
Parisite	$REE_2 Ca (CO_3)_3 F_2$	Fracture fillings Breccia fillings Veins Replacement bodies	26 27 27,34,39,59,60 50

Table 5.1 *Continued*

Mineral	Simplified formula	Modes of hydrothermal occurrence	References
Stillwellite	(REE,Ca) B SiO_5	Skarn	6,35
Synchisite	REE Ca $(CO_3)_2$ F	Alpine-type fissures	2,21,33,54
		Fracture fillings	26
		Veins	31,34,39,46,59,60
Titanite	(Ca,REE) Ti SiO_4 (O,OH,F)	Alpine-type fissures	25,49
		Veins	46,47,48,56,58
		Skarns	29
		Replacement bodies	41
Xenotime	Y PO_4	Alpine-type fissures	17,21,33,45,52
		Breccia fillings	27,42
		Veins	28
Zircon	(Zr,REE) SiO_4	Veins	19,28,48,53,57
		Skarns	43,57
		Replacement bodies	50
Zirconolite	(Ca,REE) Zr Ti_2 O_7	Fracture fillings	51
		Veins	11,48
		Skarns	20,29,37,57

References:
(1) Geijer, 1921, (2) Parker, de Quervain and Weber, 1939, (3) Parker and de Quervain, 1940, (4) Glass and Smalley, 1945, (5) Olson et al., 1954, (6) McAndrew and Scott, 1955, (7) Geijer, 1963, (8) Bromley, 1964, (9) Perhac and Heinrich, 1964, (10) Staatz and Conklin, 1966, (11) Rekharskiy and Rekharskaya, 1969, (12) Rudashevskiy, 1969, (13) Papunen and Lindsjö, 1972, (14) Pavelescu and Pavelescu, 1972, (15) Sommerauer and Weber, 1972, (16) Staatz, Shaw and Wahlberg, 1972, (17) Graeser, Schwander and Stalder, 1973, (18) Watson and Snyman, 1975, (19) von Backström, 1976, (20) Zhuravleva, Berezina and Gulin, 1976, (21) Gramaccioli, 1977, (22) van Wambeke, 1977, (23) Isokangas, 1978, (24) Exley, 1980, (25) Mercolli, Schenker and Stalder, 1984, (26) Caruso and Simmons, 1985, (27) Metz et al., 1985, (28) Staatz, 1985, (29) Gieré, 1986, (30) Mannucci et al., 1986, (31) von Gehlen, Grauert and Nielsen, 1986, (32) Förster, Hunger and Grimm, 1987, (33) Graeser and Schwander, 1987, (34) Haack, Schnorrer-Köhler and Lüders, 1987, (35) Kwak and Abeysinghe, 1987, (36) Bermanec et al., 1988, (37) Williams and Gieré, 1988, (38) Dollinger, 1989, (39) Mariano, 1989b, (40) Drew, Meng and Sun, 1990, (41) Lira and Ripley, 1990, (42) Oreskes and Einaudi, 1990, (43) Pan and Fleet, 1990, (44) Demartin et al., 1991a, (45) Demartin et al., 1991b, (46) Pan and Fleet, 1991, (47) Gieré, 1992, (48) Gieré and Williams, 1992, (49) Stalder, pers. comm. 1992, (50) Yuan et al., 1992, (51) Zakrzewski et al., 1992, (52) Braun, 1993, (53) Kerrich and King (1993), (54) Krzemnicki, 1993, (55) Özgenç, 1993, (56) Pan, Fleet and McRae, 1993, (57) Rubin, Henry and Price, 1993, (58) Banks et al., 1994, (59) Ngwenya, 1994, (60) Pan, Fleet and Barnett, 1994.

Table 5.2 Selected REE mineral deposits of hydrothermal origin

Locality	Principle REE minerals	Associated F-minerals: Apatite	Associated F-minerals: Fluorite	Type of deposit	Most likely source of fluids	References
Bayan Obo (China)	Bastnäsite, Monazite	×	×	Replacement body and veins or stockworks	Alkaline intrusives/carbonatites or mantle (?)	Nakai et al. (1989), Drew, Meng and Sun (1990), Chao, Back and Minkin (1992), Yuan et al. (1992)
Buffelsfontein/Buffalo Mine (South Africa)	Monazite, Allanite, Bastnäsite	×	***	Veins	Granite	Watson and Snyman (1975)
Gallinas Mountains (USA)	Bastnäsite		×	Veins and breccia fillings	Alkalic trachyte	Glass and Smalley (1945), Perhac and Heinrich (1964)
Karonge/Gakara (Burundi)	Bastnäsite, Monazite			Veins and stockwork	Carbonatites or mantle (?)	van Wambeke (1977), Mariano (1989a,b), Lehmann et al. (1994)
Korsnäs (Finland)	Monazite	×		Vein and skarn	Pegmatite or carbonatite (?)	Papunen and Lindsjö (1972), Isokangas (1978)
Mary Kathleen (Australia)	Allanite, Stillwellite	×		Skarn	Granite	Hawkins (1975), Kwak and Abeysinghe (1987), Maas, McCulloch and Campbell (1987)
Olympic Dam (Australia)	Bastnäsite, Florencite, Monazite		×	Breccia fillings	Granite and alkaline mafic/ultramafic dykes	Oreskes and Einaudi (1990, 1992), Johnson and McCulloch (1995)
Rodeo de Los Molles (Argentina)	Britholite, Allanite	×	***	Replacement body	Granitic intrusives	Lira and Ripley (1990)

*** means that fluorite is present, but that it formed later than the REE minerals.

Table 5.2. This list could be extended to include several other deposits because it has been recognized that many of the 'magmatic' RE deposits may in fact have been enriched considerably by alteration of the primary ores due to interaction with late- to post-magmatic hydrothermal fluids (e.g. Olson et al., 1954; Mariano, 1989b, Morteani, 1991). Table 5.2 shows that apatite and fluorite are important gangue minerals in many hydrothermal deposits, where they may have played a key role in the formation of the RE minerals (see below). It is also apparent from this compilation that the source of the mineralizing fluids is not in all cases related to carbonatite intrusions. The Mary Kathleen U-REE skarn in Queensland (Australia), for example, was formed by fluids derived from a nearby granite intrusion (Kwak and Abeysinghe, 1987; Maas, McCulloch and Campbell, 1987). In other cases the fluid source is less apparent: the large RE deposit at Bayan Obo bears characteristics of a mineralization formed from carbonatite-derived fluids and, although carbonatite dykes are known from this deposit, no large carbonatite intrusion has been identified in the vicinity (Drew, Meng and Sun, 1990); judging from available isotopic data, however, the fluids may have been derived directly from the mantle without large involvement of crustal material (Nakai et al., 1989, Yuan et al., 1992).

5.3 Zoning in hydrothermal RE minerals

Several RE minerals formed during hydrothermal processes exhibit a pronounced zoning which is usually only seen in backscattered electron (BSE) or cathodoluminescence images. Extremely complex chemical zonation patterns have been documented for aeschynite, allanite, garnet, titanite and zirconolite (Williams and Gieré, 1988; Pan and Fleet, 1991; Sorensen, 1991; Zakrzewski et al., 1992; Gieré and Williams, 1992; Jamtveit and Hervig, 1994).

An example is shown in Figure 5.1, a BSE image of zirconolite formed during metasomatic alteration of a spinel–calcite marble at the contact with the Bergell Intrusion, Switzerland/Italy (Gieré, 1986). Zirconolite from this locality (containing up to 14.5 wt% RE_2O_3) exhibits three discrete zones, each with different average REE contents (Figure 5.2); the standard deviations, though, are relatively large for several of the REE and, in particular, for U and Th (Williams and Gieré, 1988). This variation is due to a superimposed extremely fine-scale oscillatory zoning (<1 μm; Figure 5.1) which cannot be resolved chemically by electron probe microanalysis. Preliminary analyses performed with an analytical transmission electron microscope (probe diameter of 50 nm), however, show that adjacent zones may differ by several weight per cent in UO_2 and ThO_2 (Lumpkin et al., 1994). It is unlikely that this very fine-scale, oscillatory zoning would result from changes in temperature or pressure during the crystallization of zirconolite in a contact metamorphic environment. Therefore, the zoning is

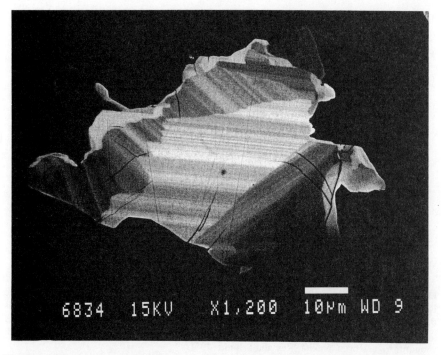

Figure 5.1 Backscattered electron image of metasomatic zirconolite from the Bergell contact aureole. Note fine-scale, oscillatory zoning superimposed on sector zoning. The fine-scale, submicron zoning results mainly from variations in U and Th concentrations.

interpreted as reflecting fluctuations in the composition of the metasomatic fluid. This conclusion is supported by the observation that zirconolite does not appear to fractionate between U, Th and the different REE because of its very flexible structure. Little or no fractionation has also been reported for other RE minerals such as titanite (Staatz, Conklin and Brownfield, 1977). Thus, minerals with this property can be used as monitors of the fluid composition, provided that P and T are constant during their formation (Gieré and Williams, 1992). The zoning patterns in such RE minerals, however, can also be influenced by the competitive crystallization of other RE minerals with strong preferences for certain REE. Simultaneous crystallization of e.g. allanite, apatite, monazite or REE fluocarbonates, minerals that generally exhibit a crystallochemical preference for the lighter REE (Puchelt and Emmermann, 1976; Fleischer, 1978; Clark, 1984; Henderson, 1984; Grauch, 1989; Banks et al., 1994), could impose a heavy REE-dominated composition onto zirconolite. Oscillatory zoning, however, could not be easily explained by competitive crystallization of other RE minerals.

Minerals with strong fractionation trends, on the other hand, are more suitable for quantifying the concentrations of REE in the coexisting fluid,

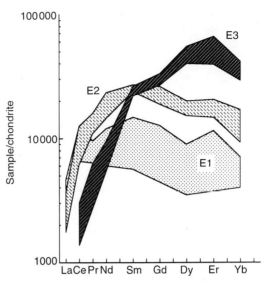

Figure 5.2 Chondrite-normalized REE patterns for the three main zones observed for the Bergell zirconolites. Judging from the geometric alignments of the different zones, zone E3 is the youngest and E1 is the oldest, documenting a progressive increase in HREE with time during the crystallization of zirconolite. (Details and data from Williams and Gieré, 1988; chondrite values from Wakita, Rey and Schmitt, 1971.)

provided that reliable experimental or field-based data are available for the partitioning of REE between the mineral and the fluid. Unfortunately, very few minerals have been studied in this respect; experimental investigations exist for apatite (Ayers and Watson, 1993) and fluorite (Marchand, 1976), and field-based empirical determinations are reported for apatite, scheelite, wolframite and fluorite (Raimbault, 1985, Raimbault et al., 1993). Moreover, Banks et al. (1994) were able to determine directly the distribution coefficients of the REE partitioning between allanite and the hydrothermal solutions, which originated from the Capitan pluton (New Mexico) and were subsequently trapped in fluid inclusions. These coefficients can be used to estimate the concentration of REE in fluids derived from the Bergell intrusion, because the conditions of allanite formation in the Bergell skarn were similar to those in the Capitan pluton. The calculations yield total REE contents of 65–115 ppm in the fluid, with Ce as most abundant REE (30–55 ppm); these data, however, represent approximate concentrations only. The results, displayed in Figure 5.3, emphasize the strong partitioning of the REE into allanite (particularly La and Ce).

Plate 1 RE minerals from Alpine-type fissures. (a) Bastnäsite from Ariège (France); crystal size: 26 mm; collection: Natural History Museum, London; photograph: NHM. (b) Monazite from Alp Moos, Vals (GR, Switzerland); crystal size: approximately 1 mm; collection: Eidgenössische Technische Hochschule Zürich (coll. Koenigsberger); photograph: W. F. Oberholzer.

(c)

(d)

(c) Xenotime (associated with hematite rosettes) from Fibbia, San Gottardo (TI, Switzerland); crystal size 16 mm; collection: Eidgenössische Technische Hochschule Zürich (coll. Wiser; Inv. Nr Wi8966); photograph: W. F. Oberholzer. (d) Kainosite (associated with adularia) from Val Curnera (GR, Switzerland); crystal size: 12 mm; collection: Eidgenössische Technische Hochschule Zürich (Inv. Nr 77031); photograph: W. Baur.

(a)

(b)

Plate 2 RE minerals from Alpine-type fissures. (a) Synchisite from Valais (VS, Switzerland); crystal size 10 mm; collection: Natural History Museum, London; photograph: NHM. (b) Allanite from Tansania; size of entire mineral aggregate: approximately 100 × 70 mm; collection: Natural History Museum, London; photograph: NHM. (c) Gadolinite from Piz Blas, Tavetsch (GR/TI,

(c)

(d)

Switzerland); crystal size: 2.5 mm; collection: Eidgenössische Technische Hochschule Zürich (Inv. Nr 1940.11); photograph: W. F. Oberholzer. (d) Davidite from Selva, Tavetsch (GR, Switzerland); crystal size: 7 mm; collection: Eidgenössische Technische Hochschule Zürich (Inv. Nr 75086); photograph: W. F. Oberholzer.

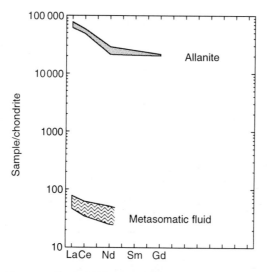

Figure 5.3 Chondrite-normalized REE patterns for metasomatic allanite and coexisting metasomatic fluid in the Bergell contact aureole. (Allanite data from Gieré, 1986; data for fluid calculated from the data given by Banks *et al.*, 1994) for the REE partitioning between hydrothermal allanite and magmatic fluid trapped in fluid inclusions; chondrite values from Wakita, Rey and Schmitt, 1971.)

5.4 REE in hydrothermal fluids

The best evidence for the presence of REE in hydrothermal fluids is provided by direct measurements on hot natural aqueous solutions. Such analyses, however, are only rarely reported in the literature (selected analyses are given in Table 5.3), and are usually restricted to fluids from active hydrothermal systems (Kraynov *et al.*, 1969; Michard and Albarède, 1986; Michard *et al.*, 1983, 1984; Michard, Beaucaire and Michard, 1987; Sanjuan, Michard and Michard, 1988; Michard, 1989). Fluids in ancient hydrothermal systems are accessible for direct chemical analysis only when trapped in fluid inclusions (Norman, Kyle and Baron, 1989; Ghazi *et al.*, 1993; Banks *et al.*, 1994).

The available data show that the concentrations of REE are generally very low compared to other components in hydrothermal waters. As pointed out by Michard (1989), low-pH waters appear to be richer in REE than neutral or alkaline solutions at similar conditions; this is also indicated by preliminary data from the Yellowstone hydrothermal system (USA), where total REE concentrations are as high as 200 ppb in acid waters (Lewis, Palmer and Kemp, 1993). The highest concentrations in active hydrothermal systems have been reported for low-pH, sulphate-rich fluids from Valles caldera ($\Sigma REE = 339$ ppb; Michard, 1989). The total REE content of these

Table 5.3 Concentrations of REE (ppb) in selected submarine and continental hydrothermal fluids, and in fluid inclusions

	Submarine hydrothermal vents		Continental hydrothermal waters					Fluid inclusions			
Location	East Pacific Rise	Mid-Atlantic Ridge	USA New Mexico	Dominica	USA California	Tibet	Bulgaria	USA New Mexico	Copper Flat	USA New Mexico	Capitan pluton
	13°N	23°N	Valles Caldera	Wotten Waven	Salton Sea	Gulu		Copper Flat		Capitan pluton	
Sample #	24G2 vent	HS 88 51	Va 199	DO-13	SS-12	AH9	Bu26	CF4	CF8a	MTE	CPU-2
La	n.d.	n.d.	n.d.	n.d.	n.d.	n.d.	n.d.	2800	71000	72050	334700
Ce	2.19	0.410	200	1.005	0.706	0.063	0.052	2800	46000	83970	582600
Pr	n.d.	n.d.	n.d.	n.d.	n.d.	n.d.	n.d.	n.d.	n.d.	6270	57750
Nd	1.12	0.244	90	0.612	0.226	0.022	0.0214	1500	26000	19270	190500
Sm	0.179	0.060	16	0.161	0.0226	0.0047	0.0053	390	3800	3430	33760
Eu	0.528	0.234	2.7	0.089	0.305	0.0004	0.0006	80	510	590	2750
Gd	0.150	0.053	13.5	0.190	0.025	0.005	0.0032	n.d.	n.d.	2460	32020
Tb	n.d.	n.d.	n.d.	n.d.	n.d.	n.d.	n.d.	50	350	360	4650
Dy	0.088	0.033	9.5	0.172	0.0173	0.0044	0.0028	n.d.	n.d.	1920	25780
Ho	n.d.	n.d.	n.d.	n.d.	n.d.	n.d.	n.d.	n.d.	n.d.	380	4620
Er	n.d.	0.012	3.98	0.108	0.0106	0.003	0.0017	n.d.	n.d.	1080	10500
Tm	n.d.	n.d.	n.d.	n.d.	n.d.	n.d.	n.d.	n.d.	n.d.	190	1560
Yb	0.027	0.008	3.3	0.110	0.00933	0.0034	0.0013	120	850	1220	7780
Lu	n.d.	n.d.	n.d.	n.d.	n.d.	n.d.	n.d.	20	130	180	1150
Σ(REE)	4.28	1.054	339	2.447	1.322	0.106	0.088	7760	148640	193370	1290100
T (°C)	320	330	42.9	97	351	68	47	274	279	400–600	400–600
pH	3.92 (20°C)	4.7 (25°C)	1.33	3.48	5.7	7.7	9.50				
References	Michard et al. (1983, 1984)	Michard (1989)	Michard (1989)	Michard (1989)	Michard (1989)	Michard and Albarède (1986)		Norman, Kyle and Baron (1989)		Banks et al. (1994)	

n.d. = not determined.
T represents the venting temperature (hydrothermal fluids) and the homogenization temperature (fluid inclusions).
pH values measured in the field at temperatures close to the venting temperature, unless stated otherwise.

fluids increases with increasing sulphate concentration and decreasing pH, but does not seem to correlate with the amount of Cl (Figure 5.4), indicating that the REE here are complexed by sulphate; this is also supported by the recent calculations of Haas, Shock and Sassani (in press). In ancient hydrothermal systems the largest amounts of REE (ΣREE = 1290 ppm) have been found in very saline magmatic solutions (Banks *et al.*, 1994), showing that high-temperature saline solutions are very effective in carrying REE and are the most likely candidates for the transport of large amounts of REE.

Indirect evidence for the presence of REE in hydrothermal fluids is provided by the formation of hydrothermal RE minerals in different geological settings (Tables 5.1 and 5.2). Similarly, geochemical studies have shown in numerous examples that REE have been added or removed by hydrothermal fluids during fluid–rock interaction. Recent examples are given by Vander Auwera and Andre (1991), Chen, Halls and Stanley (1992) and Hopf (1993), and more references may be found in Humphris (1984), Cathelineau (1987), Gieré (1986, 1990a) and Grauch (1989). Several studies also revealed that the REE are often transported and deposited together with Zr and Ti, as well as with actinides (McLennan and Taylor, 1979; Gieré, 1986, 1990a; Maas, McCulloch and Campbell, 1987; Fryer and Taylor, 1987; Graeser and Schwander, 1987; Gieré and Williams, 1992;

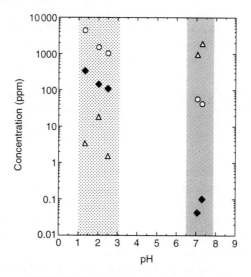

Figure 5.4 Abundance of REE (ΣREE; filled diamonds), sulphate (open circles) and chloride (open triangles) in relation to measured pH in hydrothermal fluids from Valles caldera (New Mexico), for acid, sulphate-rich waters (stippled; $T = 43$–$90°C$) and alkaline, chlorine-rich waters (shaded; $T = 90$–$170°C$). Note the correlation between ΣREE and SO_4^{2-} contents, and the different trend for Cl^-. (Data for analyses given by Michard, 1989.)

Zakrzewski *et al.*, 1992; Braun, 1993; Kerrich and King, 1993; Krzemnicki, 1993; Rubin, Henry and Price, 1993), suggesting that transport and deposition of these elements in hydrothermal systems may be ascribed to similar processes.

The presence and concentration of REE in hydrothermal fluids are mainly controlled by temperature, pressure and the composition of the fluid phase; these variables, combined with crystallochemical factors and reaction kinetics, also control the REE partitioning between the minerals and the solution as well as between the solution and the melt from which the fluids may have been derived (cf. Grauch, 1989; Jamtveit and Hervig, 1994). In ancient hydrothermal systems it is often difficult to determine the source of the REE-bearing fluids. Comparison of the REE patterns of hydrothermal minerals and rocks with those of possible source rocks, for instance, usually does not suffice to establish unequivocally a genetic relationship because multiple fractionation may occur on the way from REE mobilization to deposition. In many cases, therefore, chemical analyses and petrographic observations obtained during field work and microscopic examinations should be complemented by stable and radiogenic isotope data (e.g. Vander Auwera and Andre, 1991).

5.5 Transport of REE

An extensive database is available for the aqueous geochemistry of REE at standard conditions (25°C, 1 bar); general reviews are given by Choppin (1983, 1984), David (1986), Brookins (1989), Wood (1990a, 1993) and Millero (1992). Experimental studies on the speciation of REE in solutions at higher temperature and pressure, however, are currently sparse (e.g. Bilal and Langer, 1987); further investigations are urgently needed in order to understand better the formation of RE minerals and REE deposits under hydrothermal conditions. In the absence of sufficient experimental data, several theoretical attempts have been made to predict the behaviour of REE in hydrothermal solutions (Sverjensky, 1984; Wood, 1990b; Bau, 1991; Haas, Shack and Sassani, 1993 and in press; Haas and Shock, 1994).

These theoretical models predict that the REE have a strong tendency to form aqueous complexes and that they occur as bare cations only in low-temperature solutions. The stability constants of most REE complexes studied generally increase with increasing temperature, but decrease with increasing pressure (Wood, 1990b: up to 350°C and at saturation water vapour pressure; Haas, Shock and Sassani, in press: up to 1000°C and 5 kbars). The models also predict that the divalent oxidation state of some of the REE (Sm, Eu, Yb) becomes increasingly important as the temperature rises. The calculations further indicate that a variety of ligands, including F^-, Cl^-, OH^-, SO_4^{2-} and CO_3^{2-}, are likely candidates for complexing the REE in hydrothermal fluids. The speciation of REE in the fluid, however, does not only depend on temperature and pressure but is also strongly controlled by the relative concentration of the various ligands and by the pH of the solution. The relative abundance of REE-fluoride and REE-carbonate complexes, for example, depends on the total concentration of fluorine relative to that of carbon, which occurs as CO_3^{2-}, HCO_3^- or $CO_{2(aq)}$, depending on the pH of the fluid. Likewise, the pH determines the sulphur distribution which, moreover, is also controlled by the oxidation state of the solution. The combined effects of pH and relative abundance of ligands is exemplified by the calculations of Haas, Shock and Sassani, (in press), revealing that in a general chloride-rich hydrothermal solution (molar Cl/F ratio = 1000) REE are complexed predominantly by chloride, fluoride and hydroxide at acidic, neutral and basic pH conditions, respectively; the calculations also predict that fluoride ions compete effectively with chloride even at F/Cl molar ratios as low as 2×10^{-5} in an acid to neutral fluid at 300°C.

The predictions of Haas, Shock and Sassani (in press) further indicate that light REE and heavy REE can be complexed by different ligands in the same solution, which may lead to fractionation among the REE. These effects were observed primarily in acid to neutral fluids. Basic fluids are not likely to cause significant fractionation because at pH > 5–6 nearly all of each REE cation appears to be complexed by hydroxide at the studied conditions.

The theoretical models allow the calculation of speciation of REE in fluids from active hydrothermal systems where chemical and physical parameters of the solution can be directly determined. Preliminary calculations (Haas, Shock and Sassani, in press) suggest, for example, that the trivalent REE are complexed mainly by chloride (and fluoride, if present) in the submarine hot and acid hydrothermal fluids, whereas sulphate complexes predominate over chloride complexes in the acid waters from Valles caldera (cf. stippled area in Figure 5.4).

In ancient hydrothermal environments, however, it is much more difficult to determine the parameters required for the calculation of REE speciation. Fluid inclusions can provide important information on various chemical and physical parameters, and are commonly used to determine temperature and chloride concentration. In many cases where hydrothermal mobility of REE has been documented, the temperature, pressure and fluid composition have to be estimated by using thermodynamic data applied to mineral solubilities and phase equilibria. The derived parameters should allow the estimation of REE speciation on the basis of the currently available models.

It is also possible to assess empirically the most likely REE complexes, though not the exact speciation, by studying the hydrothermal RE minerals as well as the parageneses in which they occur. Most of the complexes inferred from paragenetic observations have already been incorporated into the theoretical models. Complexes of REE with phosphorus species, however, have not received the attention they deserve, according to petrographic studies. Phosphate minerals such as apatite, monazite and xenotime are common constituents of RE mineral assemblages in various environments (cf. Tables 5.1 and 5.2), suggesting that phosphorus species may be of considerable importance in transporting the REE in certain hydrothermal fluids (e.g. Gieré, 1986, 1990a; Mariano, 1989a; Gieré and Williams, 1992). At low temperatures complexation of REE by phosphate has been recognized as a significant process in natural freshwater systems (Byrne, Lee and Bingler, 1991). At high temperatures REE-phosphate complexes are likely to be even more important, particularly in fluids with elevated phosphate contents. To the author's knowledge, however, only orthophosphate ($H_2PO_4^-$) has been studied in this respect, and an increase in the stability of REE $H_2PO_4^{2+}$ complexes with increasing temperature is predicted (above approximately 150°C; Haas, Shock and Sassani, in press).

Paragenetic relationships indicate further that REE–S^{-II} complexes (i.e. with H_2S, HS^- or S^{2-} as ligands) may exist in certain fluids containing large amounts of reduced sulphur (Gieré, 1993). However, if other potentially strong complex-forming ligands are present, the inferred S^{-II} complexes with REE would most probably not be important because of the strong affinity of the hard REE ions for hard ligands such as F^-, OH^- and PO_4^{3-}, especially at elevated temperatures. (The REE cations are classified as hard ions, i.e. as ions with a small size and high charge. Soft species, on the other hand, are large, highly polarizable ions with a relatively low charge. Hard cations bind preferentially with hard ligands, whereas soft cations tend to form complexes with soft ligands. For details and discussion see Ahrland (1968), Nancollas (1970), Crerar *et al.* (1985) and Brimhall and Crerar (1987).)

The existence of mixed-ligand complexes such as REE-hydroxofluoride, -hydroxocarbonate or -fluocarbonate complexes has also been proposed (for references see Wood, 1990b), but only limited experimental data are currently available (Bilal and Koss, 1982). Furthermore, Bandurkin (1961) suggested that mixed-metal polynuclear complexes such as K–REE–fluoride complexes may be responsible for the REE transport in hydrothermal fluids rich in potassium and sodium (for examples see Gieré, 1990a).

5.6 Formation of RE minerals

Because of the generally low concentrations of REE in most hydrothermal fluids (Table 5.3), mechanisms must exist that effectively lead to an enrichment of REE in specific geological environments. RE minerals can be precipitated from hydrothermal solutions by a variety of mechanisms; important among these are probably (1) changes in temperature or pressure of the solutions, (2) mixing of the REE-bearing solutions with fluids of different chemical composition, (3) interaction of the solutions with the wall rock, and (4) crystallization of gangue minerals.

Determination of the principal causes of REE precipitation is difficult because natural hydrothermal fluids usually possess a very complex chemical composition, and because several of the mentioned mechanisms may be active simultaneously. Despite these difficulties, an attempt is made here in addressing the most likely possibilities.

5.6.1 Changes in temperature and pressure

When hot solutions flow towards the Earth's surface there is usually a decrease in temperature and confining pressure. Simple cooling of the hydrothermal fluids may lead to supersaturation with respect to some RE minerals, if their solubilities decrease with decreasing temperature. Not all RE minerals exhibit lower solubilities at lower temperatures, however; zircon, for example, appears to be more soluble at lower temperatures (Korzhinskaya, 1990). For monazite there is uncertainty with regard to the solubility–temperature relationship: thermodynamic calculations of Wood and Williams-Jones (1994) point to a retrograde solubility (below approximately 250°C; in model seafloor hydrothermal vent fluid), whereas the limited experimental data of Ayers and Watson (1991) indicate a slight increase in solubility with increasing temperature (between 800 and 1000°C at 10 kbar; in pure H_2O).

A decrease in pressure can be relatively fast or even sudden, when hydrothermal solutions encounter open pore space such as fractures, fissures, cavities or brecciated rocks. Tables 5.1 and 5.2 show that many hydrothermal RE minerals and some deposits occur in one of these geological settings. It is therefore possible that the solubility of certain RE minerals decreases as the confining pressure is decreased. This has been confirmed so far only for monazite (Ayers and Watson, 1991). By contrast, the solubility of apatite increases with decreasing pressure (Ayers and Watson, 1991). Unfortunately, experimental investigations on solubilities of other RE minerals are sparse (e.g. Jonasson, Bancroft and Nesbitt, 1985; Aja, Wood and Williams-Jones, 1994). Furthermore, a sudden pressure decrease can lead to a loss of volatiles such as H_2S and CO_2 and a concomitant rise in pH, which in turn could induce the breakdown of REE complexes. Loss of CO_2, for example, could displace the equilibrium

$$REE(CO_3)^+ + 2H^+ \rightleftharpoons REE^{3+} + CO_{2(aq)} + H_2O \qquad (5.1)$$

to the right side and possibly initiate deposition of RE minerals. Large amounts of volatiles are also lost during boiling of hydrothermal solutions; thus boiling may induce precipitation of RE minerals in a way similar to that inferred for the deposition of e.g. Fe or Zn (cf. Brimhall and Crerar, 1987). To the author's knowledge, however, this mechanism has not yet been suggested in order to explain the formation of REE mineralizations, although there is evidence for boiling of REE-rich fluids (Norman, Kyle and Baron, 1989).

5.6.2 Fluid mixing

Mixing of fluids with different chemical compositions can occur in different ways and may involve more than two solutions. For binary fluid mixing there are basically four possibilities: (1) isothermal dilution of a brine, (2) dilution and cooling of a brine, (3) isothermal mixing of two brines, and (4) mixing of two brines and cooling.

An example of fluid mixing as a possible mechanism of REE deposition is provided by active hydrothermal systems: the hot (T ≈ 350°C), acid (pH ≈ 3.5), H_2S- and chloride-rich, REE-bearing submarine vent fluids suddenly encounter dramatically different conditions when they are ejected into the cold (T ≈ 2°C), alkaline (pH ≈ 8) and SO_4^{2-}-rich seawater at the seafloor. This results in the immediate precipitation of sulphide minerals and in the majority of the REE being deposited at the vent sites (e.g. Barrett et al., 1988). However, the deposition of REE at the vent sites (mainly as anhydrite and baryte; Michard and Albarède, 1986; Barrett, Jarvis and Jarvis, 1990) results not only from fluid mixing but is probably also related to sulphide precipitation (see below and Gieré, 1993).

5.6.3 Interaction of fluids with wall rock

RE minerals can also crystallize as a result of interaction between the REE-bearing hydrothermal solutions and the wall rocks. Possible effects of the fluid/wall rock interaction include changes in pH or redox conditions, and/or changes in the composition of the solutions.

A change in pH from acid to neutral, or a change in redox conditions from oxidizing to reducing, for instance, may destabilize REE complexes or lower the solubility of certain RE minerals. Ayers and Watson (1991) have shown, for example, that the solubility of monazite increases with decreasing pH (see also Wood and Williams-Jones, 1994); precipitation of monazite may thus result from neutralization of an initially acid fluid. Similarly, a change in redox conditions due to fluid/wall rock interaction can lead to conversion of e.g. S^{+VI} to S^{-II}, which in turn could affect the speciation of REE (e.g. breakdown of REE SO_4^+ complexes) and lead to precipitation of RE minerals. The effect of changing redox conditions has been inferred as a possible cause for U deposition (e.g. Caruso and Simmons, 1985). Similarly, oxidation of trivalent Ce may lead to the precipitation of primary cerianite (CeO_2) although, to the author's knowledge, this mechanism has not yet been documented in hydrothermal systems. Oxidation of Ce^{3+}, however, is recognized as a relatively important process during alteration (see below) and weathering (e.g. Meintzer and Mitchell, 1988) of RE minerals, and particularly during development of supergene mineralizations (Mariano, 1989a, 1989b; Braun et al., 1990, 1993; Waber, 1992).

A change in fluid composition resulting from interaction with the wall rock can occur, for example, if substantial quantities of a very soluble salt such as $CaCl_2$ are added to a saturated solution containing the less soluble salts NaCl and KCl. In this case the total salinity generally increases and, if excess $CaCl_2$ is added and no dilution by meteoric water occurs, the less soluble salts will precipitate on supersaturation (the 'salting out' effect of Kwak and Abeysinghe, 1987). This mechanism could be of considerable importance for the formation of RE minerals in skarns and carbonate-hosted veins because large amounts of $CaCl_2$ are produced during dissolution of the carbonate minerals (Kwak and Tan, 1981). Carbonate rocks are in fact very common hosts of REE mineralizations (Table 5.2). The distinct U–REE daughter mineral assemblages in fluid inclusions from the Mary Kathleen deposit have been interpreted as being caused by the salting out mechanism (Kwak and Abeysinghe, 1987).

5.6.4 Crystallization of gangue minerals

In order to understand this possible precipitation mechanism it is necessary to study the paragenetic associations of the RE minerals. In some of the economically important hydrothermal ore deposits the RE minerals are closely associated with abundant gangue fluorite (Table 5.2). (The term 'gangue' is used here for fluorite and apatite, but not strictly in an economic sense. In some of the mentioned deposits, fluorite is in fact an important mining product. Moreover, both apatite and fluorite may be very rich in REE and thus are often regarded as actual REE minerals.) The assemblage bastnäsite + fluorite, for instance, is very common (although not ubiquitous), and is typically related to postmagmatic, hydrothermal activity; this is also documented for a number of deposits not listed in Table 5.2 (e.g. Metz et al., 1985; Salvi and Williams-Jones, 1990). Even at Mountain Pass, where much of the REE mineralization is clearly of magmatic origin, the bastnäsite + fluorite assemblage is restricted to late-stage hydrothermal veins (Olson et al., 1954). Table 5.2 further shows that apatite is usually a main constituent (although gangue mineral) in the ores when fluorite is absent. At many other localities apatite or fluorite has also been reported to be associated closely and often intergrown with hydrothermal RE minerals, suggesting simultaneous crystallization (e.g. Gieré, 1990a and references therein; Pan and Fleet, 1990, 1991).

These features could be explained by considering that crystallization of fluorite or apatite reduces the activities of F^- and PO_4^{3-} in the fluid, i.e. of hard ligands that are likely to complex the REE in certain solutions. A decrease in the concentration of F^- and PO_4^{3-} could lead to the stepwise breakdown of initially neutral REE complexes to REE^{3+} ions, as described schematically by the equilibria

$$REE\ F_3°_{(aq)} \rightleftharpoons REE\ F_2^+ + F^- \qquad (5.2)$$

$$REE\ F_2^+ \rightleftharpoons REE\ F^{2+} + F^- \qquad (5.3)$$

$$REE\ F^{2+} \rightleftharpoons REE^{3+} + F^- \qquad (5.4)$$

or

$$REE(PO_4)°_{(aq)} \rightleftharpoons REE^{3+} + PO_4^{3-}. \qquad (5.5)$$

The resulting trivalent REE ions then might be deposited as RE minerals. Thus fluorite and apatite formation should in many cases initiate the precipitation of the RE minerals (Gieré, 1990a). Similarly, Kosterin (1959) and Caruso and Simmons (1985) suggested that crystallization of calcite, dolomite or siderite in hydrothermal systems leads to a decrease in the CO_3^{2-} concentration in the fluid, resulting in a breakdown of the inferred REE–carbonate complexes.

These relationships suggest that knowledge of the mechanisms leading to precipitation of fluorite or apatite would indirectly help in understanding the formation of RE minerals in many hydrothermal systems.

The solubility of apatite has been studied in hydrothermal solutions at high P and T by Ayers and Watson (1991). These authors found that pH is the most important determinant of apatite solubility, in agreement with earlier experiments performed at room temperature (McCann, 1968). The experiments of Ayers and Watson (1991) have shown that the solubility of apatite decreases with increasing pH; furthermore, they indicated that the solubility of apatite is much smaller at $X_{CO_2} = 0.5$ than in pure water. From these experiments it follows that apatite precipitates when acid fluids are neutralized or are diluted by CO_2. This is most likely to be the case when hydrothermal fluids originating from calc-alkaline intrusions encounter carbonate rocks. The experimental results thus provide an explanation for the presence of apatite in many skarns and carbonate-hosted veins (cf. case study below).

The behaviour of fluorite in hydrothermal systems has been studied in detail by various authors (cf. Richardson and Holland, 1979a). Richardson and Holland (1979b) have shown that the most likely mechanisms for fluorite deposition are (1) simple cooling of the hydrothermal solution, (2) dilution of the solution with cool meteoric water, (3) mixing of different brines, and (4) an increase in pH of an acid solution. According to these authors, an increase in the Ca concentration of the fluid due to dissolution of carbonate wall rocks can further lead to precipitation of significant quantities of fluorite, but only if the initial fluid is close to a pure NaCl solution. The REE deposit at Bayan Obo is a good example of a deposit of primary economic importance, where large amounts of dolomite have been replaced by fluorite and RE minerals. Formation of this huge deposit was probably facilitated by the presence of an impermeable shale cap rock forcing the HF-bearing solutions to dissolve large quantities of dolomite (Drew, Meng and Sun, 1990).

As natural hydrothermal solutions tend to have a very complex chemical composition, the deposition mechanisms are most likely to be more complicated than those stated above. In H_2S-rich hydrothermal fluids, for example, the formation of RE minerals may additionally be influenced by the crystallization of sulphide minerals; in several geological environments the REE appear to have been trapped at the site of sulphide precipitation (Gieré, 1993). However, at these sites the REE are not incorporated into sulphide minerals (Morgan and Wandless, 1980), but rather into accessory RE minerals or minerals with a high affinity for REE. These minerals are texturally closely associated with the sulphides, indicating that the deposition of REE from the hydrothermal fluids could be genetically related to the crystallization of base metal sulphides. This suggested genetic relationship is best illustrated by considering chloride-rich hydrothermal fluids in which ore-forming elements such as Fe, Cu, Zn or Pb are carried as chloride complexes (e.g. Barnes, 1979). Sulphide precipitation from these particular fluids can be described by the general reaction

$$MeCl_2^\circ{}_{(aq)} + H_2S_{(aq)} \rightleftharpoons \underset{\text{sulphide mineral}}{MeS} + 2H^+ + 2Cl^-, \tag{5.6}$$

which may be formulated for a variety of sulphide minerals (Eugster 1985, 1986). This conversion of metal chloride solutes to sulphide minerals liberates H^+ (or HCl°) which increases the acidity of the hydrothermal fluid. Deposition of the dissolved REE might then occur in response to such a change in pH, induced by crystallization of the sulphide minerals, rather than as a consequence of reducing the activity of ligands essential for their complexation. An example for this type of mechanism will be discussed below.

5.7 Alteration of hydrothermal RE minerals

Alteration, i.e. corrosion, dissolution or transformation, of RE minerals has been observed in many hydrothermal systems. This is to be expected because fluids of variable composition and temperature may pass through the same system at various stages of its geological history. Therefore, studying both primary and secondary RE minerals helps in unravelling the evolution of a particular hydrothermal system. In some cases alteration causes remobilization of REE and eventually leads to the precipitation of secondary RE minerals at other localities (e.g. Maas, McCulloch and Campbell, 1987). RE minerals of both hydrothermal and magmatic origin can be altered, but the following examples were selected from hydrothermal mineralizations only.

Alteration affects primarily REE fluocarbonate and fluoride minerals. At Karonge/Gakara (Burundi), for instance, hydrothermal bastnäsite has been transformed into monazite, an alteration that Mariano (1989a) ascribed to interaction with late phosphate-bearing solutions. In contrast to that, Watson and Snyman (1975) reported that monazite was replaced by bastnäsite at Buffelsfontein (South Africa). These observations may be represented schematically by the reaction

$$REE(CO_3)F + H_3PO_4^\circ{}_{(aq)} \rightleftharpoons REE\,PO_4 + HF^\circ{}_{(aq)} + H_2O + CO_{2(aq)}. \tag{5.7}$$

bastnäsite   monazite

It is apparent that reaction (5.7) depends on the pH (via dissociation of HF° and $H_3PO_4^\circ$) as well as on the partial pressure of CO_2. Therefore, depending on the prevailing conditions, the reaction proceeds in one or the other direction.

Alteration of fluocerite into secondary bastnäsite is known from various localities (e.g. Geijer, 1921, Förster, Hunger and Grimm, 1987; see also compilation by Perhac and Heinrich, 1964). This transformation can be described by the reaction

$$REE\,F_3 + CO_{2(aq)} + H_2O \rightleftharpoons REE(CO_3)F + 2\,HF^\circ{}_{(aq)} \tag{5.8}$$

fluocerite   bastnäsite

which depends, like reaction (5.7), on pH and P_{CO_2}.

In their description of the first natural occurrence of gasparite ($REEAsO_4$), Graeser and Schwander (1987) noted that this was formed by pseudomorphic replacement of synchysite in an Alpine-type fissure. Thus primary REE fluocarbonate minerals may be altered into arsenates provided that the hydrothermal fluids contain arsenic.

Transformations are also known to take place among minerals belonging to the same group, e.g. fluocarbonates. For example, in the Tundulu carbonatite complex (Malawi) some of the parisite occurring in the hydrothermal mineralization was produced by alteration of primary synchysite (Ngwenya 1994). Williams-Jones and Wood (1992) have pointed out in their theoretical study that, at a given temperature and pressure transformations among fluocarbonates are very likely to result from variations in either Ca^{2+} or CO_3^{2-} activities in the fluid. This may readily be seen, for instance, for the bastnäsite–parisite transformation, which could be described by the following schematic reaction:

$$2\text{REE}(\text{CO}_3)\text{F} + \text{CO}_3^{2-} + \text{Ca}^{2+} \rightleftharpoons \text{REE}_2\text{Ca}(\text{CO}_3)_3\text{F}_2. \qquad (5.9)$$
bastnäsite   parisite

The activities of Ca^{2+} or CO_3^{2-} in the fluid could be changed, for example, by precipitation of calcite. Jaffe, Meyrowitz and Evans (1953) reported a similar transformation, but observed sahamalite as an additional alteration product associated with parisite.

Although alteration has been documented most often for REE fluocarbonates and fluorides, a number of other hydrothermal RE minerals can also be affected. For instance, zircon may be corroded and dissolved in hydrothermal solutions at high temperatures and pressures (Sinha, Wayne and Hewitt, 1992). Monazite is sometimes replaced by bastnäsite (see above), and in other cases by parisite + apatite (Maruéjol, Cuney and Turpin, 1990). Allanite of both hydrothermal and magmatic origin is quite commonly attacked by fluids and transformed into REE fluocarbonate minerals, most often into bastnäsite (e.g. Mineyev, Makarochkin and Zhabin, 1962; Perhac and Heinrich, 1964; Mineyev *et al.* 1973; Caruso and Simmons, 1985; Littlejohn, 1981; Lira and Ripley, 1990; Pan, Fleet and Barnett, 1994). This bastnäsitization of allanite can be described schematically as

$$\text{allanite} + \text{fluid} \rightleftharpoons \text{bastnäsite} + \text{clay mineral} + \text{thorite} \pm \text{fluorite} \pm \text{magnetite}, \qquad (5.10)$$

where the clay mineral is represented by kaolinite, montmorillonite or illite. Formation of thorite as a reaction product is due to the elevated Th contents commonly observed in allanite. Bastnäsite is also known to form by alteration of various other hydrothermal RE minerals, including britholite (Lira and Ripley, 1990), cerite (Geijer, 1921), titanite (Pan, Fleet and MacRae, 1993), and fluorite (Haack, Schnorrer-Köhler and Lüders, 1987).

Corrosion and reprecipitation have been described for REE-bearing zirconolite in high-temperature hydrothermal fluids (Gieré and Williams, 1992); this is in contrast to the results of many experimental investigations which have shown that zirconolite is extremely resistant to hydrothermal attack at lower temperatures (for review see Smith and Lumpkin, 1993). The corrosive nature of these high-temperature fluids is probably due to large concentrations of phosphorus and fluorine in solution (see below).

Alteration of many RE minerals may further be enhanced by damage to the crystal structure caused by the radioactive decay of alpha-emitting actinides (e.g. Ewing, Haaker and Lutze, 1982; Ewing *et al.* 1987). Radiation damage and eventually a crystalline to aperiodic transformation would be expected, particularly for minerals such as allanite, monazite, thorite, zircon and zirconolite, and would be most apparent in very old specimens rich in U and Th. Figure 5.5 shows an altered zirconolite grain from Phalaborwa, South Africa; the 2060 Ma crystal is almost completely aperiodic (metamict)

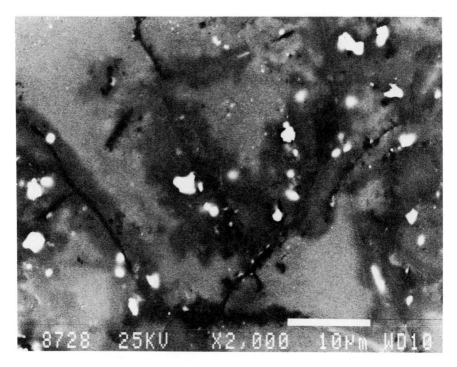

Figure 5.5 Backscattered electron image of 2060 Ma metamict (aperiodic) zirconolite from Phalaborwa (South Africa). Note dark areas representing alteration zones resulting from interaction with a sulphur- and silica-bearing fluid which penetrated along cracks and microfractures. Bright spots are galena crystals containing part of the radiogenic lead (see Gieré et al., 1994). Photograph: SEM Laboratory, University of Basel.

due to relatively high contents of UO_2 ($\approx 2\,wt\%$) and ThO_2 ($\approx 5\,wt\%$). The zirconolite has been altered by a sulphur- and SiO_2-bearing fluid which penetrated along fractures into the mineral (Lumpkin, Smith and Gieré, 1994; Gieré et al., 1994). Because the principal effects of the crystalline to aperiodic transformation are volume expansion and microfracturing, the fractures visible in Figure 5.5 are probably a direct consequence of the radioactive decay of Th and U. Alteration of primary hydrothermal RE minerals may also be accompanied by oxidation. For example, Styles and Young (1983) found that both fluocerite and bastnäsite have been transformed into an assemblage consisting of bastnäsite-(La) and cerianite. These transformations both involve oxidation of Ce^{3+} to Ce^{4+} and a concomitant separation of La and Ce into two different phases. Oxidation during alteration thus may lead to the formation of peculiar mineral associations, e.g. bastnäsite-(Ce) in contact with bastnäsite-(La), which cannot readily be explained by simultaneous crystallization.

Similarly, the occurrence of allanite-(La) together with allanite-(Ce) in hydrothermal veins from the Hemlo gold deposit, Canada, was not interpreted as a paragenesis, but rather as an assemblage in which the two minerals were formed at different times (Pan and Fleet, 1991). This is also the case for Alpine-type fissure fillings occurring in the Val Formazza–Monte Giove area, Northern Italy (Graeser and Schwander, 1987; Braun, 1993), which contain both monazite-(Ce) and monazite-(Nd).

5.8 Case study

The results of a study on carbonate-hosted veins from Northern Italy will be summarized here, as an example for the formation of RE minerals in hydrothermal systems (for details see Gieré, 1990b, 1992, 1993; Gieré and Williams, 1992).

Titanium-rich veins containing several RE minerals were discovered in an old marble quarry at the southeastern margin of the Adamello batholith, the largest igneous complex of Tertiary age in the Alps. The veins occur in Triassic graphite-bearing, pure dolomite marbles at the contact with a major tonalite intrusion. Field relationships combined with petrographic observations indicate that the veins were formed contemporaneously with, and outlasted, a syn-intrusive deformation event caused by the tonalite emplacement.

5.8.1 Petrographic features

The Ti-rich veins are characterized by four distinct mineral zones, consisting of the following assemblages (Figure 5.6):

1. forsterite + calcite;
2. pargasite + calcite + titanite + sulphides;
3. phlogopite + calcite + titanite + sulphides;
4. titanian clinohumite + spinel + calcite + sulphides.

The RE minerals (zirconolite, aeschynite, titanite and fluorapatite) are present in the three central vein zones only, which are further characterized by containing abundant sulphides (mainly pyrrhotite with minor chalcopyrite). Accessory pyrite occurs exclusively as a secondary mineral in cracks and along grain boundaries of the other sulphides. Textural relationships indicate that pyrrhotite was formed contemporaneously with most other minerals. Fluorapatite is abundant throughout the three Ti-rich central vein zones, and is closely associated with the other RE minerals (Figure 5.7) as well as with the sulphides, titanian clinohumite and phlogopite. Furthermore, all hydrous minerals, including titanite, contain appreciable amounts of fluorine and thus demonstrate that, in addition to sulphur and phosphorus, fluorine was a significant constituent of the metasomatic fluid. Graphite is another important accessory phase, which occurs in the three central vein zones, but

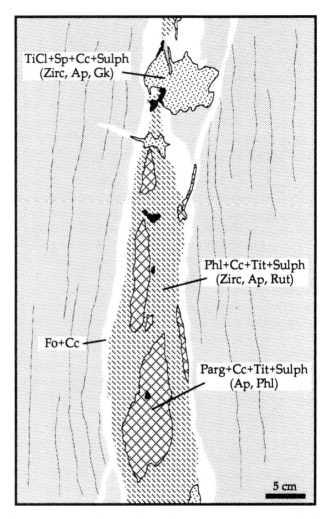

Figure 5.6 Mineralogical map of a hydrothermal vein in contact metamorphic dolomite marble (shaded) from the Adamello contact aureole. Ap = apatite, Cc = calcite, Fo = forsterite, Gk = geikielite, Parg = pargasite, Phl = phlogopite, Rut = rutile, Sp = spinel, Sulph = sulphides, TiCl = titanian clinohumite, Tit = titanite, Zirc = zirconolite. Black areas represent large aggregates of sulphides (mainly pyrrhotite with minor chalcopyrite).

was not found in the forsterite + calcite zone; graphite is also absent in a 1–2 cm wide zone of bleached dolomite marble occurring along the interface between the wall rock and the outermost vein zone. Other accessory minerals present in the central vein zones include rutile, geikielite, zircon, thorianite and galena.

CASE STUDY

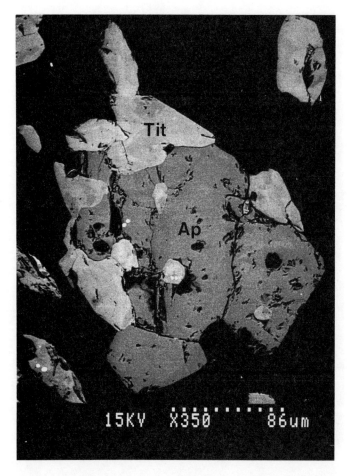

Figure 5.7 Backscattered electron image of titanite (Tit) coexisting with fluor-apatite (Ap) in the phlogopite zone, Adamello. Dark areas are phlogopite and calcite. The two small bright spots (high-Z areas) within the large apatite crystal are zircons; all other bright grains are sulphides.

5.8.2 Characterization of the RE minerals

Zirconolite is quite abundant but it was found in two vein zones only: in the phlogopite zone it is invariably anhedral, often corroded, and exhibits complex chemical zonation patterns (Figure 5.8a). In the titanian clinohumite zone zirconolite is also characterized by a pronounced discontinuous chemical zoning, but it is generally idiomorphic and shows no evidence of corrosion (Figure 5.8b). The considerable compositional variation observed for zirconolite (in wt%: $\Sigma(REE_2O_3)$ = 0.77–16.8, UO_2 = 0.59–24.0, ThO_2 =

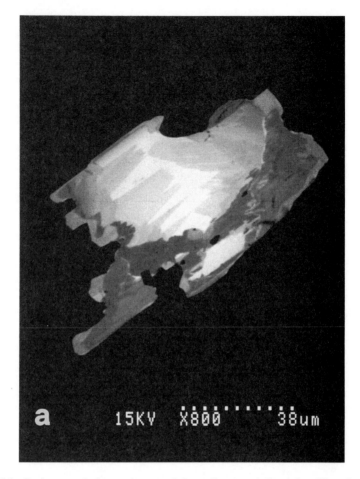

Figure 5.8 Backscattered electron images of zirconolite crystals from the phlogopite zone (a) and from the titanian clinohumite zone (b), Adamello veins. Note contrasting appearance of the crystals in the two vein zones. (For details see Gieré and Williams, 1992.)

0.67–17.1) can be attributed to four major substitutions described by the exchange vectors (a) $(Th,U)(Mg,Fe^{2+})\,Ca_{-1}Ti_{-1}$, (b) $REE\,Al\,Ca_{-1}Ti_{-1}$, (c) $REE\,Fe^{2+}(Nb,Ta)Ca_{-1}Ti_{-2}$ and (d) $HfZr_{-1}$. Exchange vector (b) is effective at total REE_2O_3 contents up to approximately 5 wt%, whereas vector (c) operates at higher concentrations. Consideration of the geometric relationships between the different zones in zirconolite from the two vein assemblages allows the derivation of the compositional evolution of zirconolite with time (Figure 5.9), the chondrite-normalized REE patterns being either Ce- or Nd-dominated for individual zones (Figures 5.9 and 5.10). Because of the flexibility of the zirconolite structure and because the

CASE STUDY

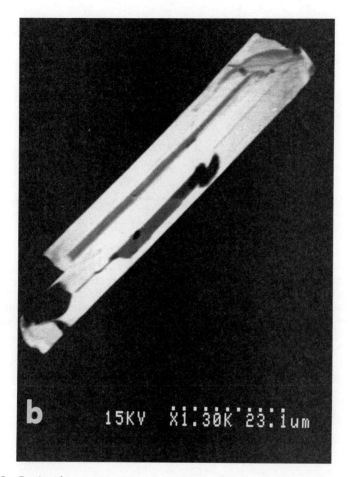

Figure 5.8 *Continued*

different zones appear to have been formed isothermally, this compositional evolution most probably reflects changes in fluid composition during crystallization of zirconolite. Corrosion and reprecipitation of zirconolite have led to the release of Th and U into the hydrothermal fluid (Figure 5.11).

Accessory aeschynite-(Ce) was observed in the phlogopite zone. It contains high amounts of REE ($\Sigma(REE_2O_3) \approx 39\,wt\%$) and exhibits a strong enrichment in light REE relative to heavy REE (Figure 5.12). Aeschynite-(Ce) is further characterized by high contents of titanium ($TiO_2 = 32\,wt\%$) and tungsten ($WO_3 = 3.9\text{--}4.8\,wt\%$). Like zirconolite, it possesses a very complex compositional zoning with narrow bands of alternating high-Z and low-Z areas, and also appears to be corroded.

Titanite is a very abundant mineral in the pargasite and phlogopite zones,

FORMATION OF RE MINERALS IN HYDROTHERMAL SYSTEMS

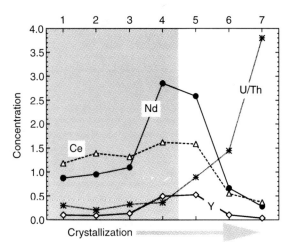

Figure 5.9 Compositional evolution of zirconolite during its crystallization. Analyses 1–4 are from phlogopite zone (shaded area), analyses 5 to 7 are from titanian clinohumite zone. The diagram was deduced from a combination of electron microprobe data, geometric analysis of backscattered electron images and textural observations in thin section. Concentration axis is dimensionless, and represents for Ce, Nd and Y the chondrite-normalized values divided by 10^4 (chondrite values from Wakita, Rey and Schmitt, 1971); for details, data and errors, see Gieré and Williams, 1992.)

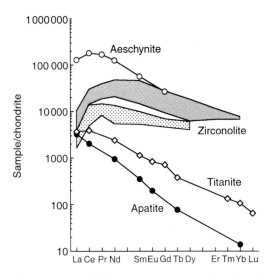

Figure 5.10 Chondrite-normalized REE patterns for the RE minerals occurring in the Adamello veins. Two compositional ranges are given for zirconolite, i.e. for low-REE (stippled; $\Sigma(REE_2O_3) < 5\,wt\%$) and for high-REE varieties (shaded). (Data from Gieré and Williams, 1992; chondrite values from Wakita, Rey and Schmitt, 1971.)

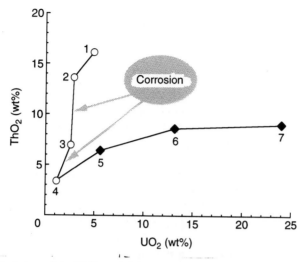

Figure 5.11 Variation of the ThO_2 and UO_2 concentrations in zirconolite during its crystallization. Analyses 1–4 are from the phlogopite zone (open symbols), analyses 5–7 are from the titanian clinohumite zone (cf. Figure 5.9). The diagram was deduced from a combination of electron microprobe data, geometric analysis of backscattered electron images and textural observations in thin section. Note the loss in actinides resulting from corrosion, and the subsequent progressive enrichment of zirconolite in UO_2 and ThO_2 with time. (For details, data and errors see Gieré and Williams, 1992.)

where it commonly occurs as euhedral crystals with a pronounced zoning. It has an average total REE concentration of approximately 6870 ppm and exhibits strongly light REE-dominated chondrite-normalized patterns (Figure 5.10). The Adamello titanites always contain fluorine (up to 0.59 wt% F); the fluorine concentration is correlated positively with that of aluminium.

Fluorapatite is almost always idiomorphic and closely associated with the other RE minerals (Figure 5.7), suggesting simultaneous formation. It is the only chlorine-bearing phase in the veins (Cl ≈ 0.25 wt%, F ≈ 2.8 wt%). Fluorapatite is also the only RE mineral for which no zoning can be observed in BSE images. Furthermore, the composition of fluorapatite is almost identical with respect to major elements and REE (Σ(REE) ≈ 3600 ppm) in all three vein zones. The chondrite-normalized REE patterns exhibit a strong light REE enrichment (Fig. 5.10).

5.8.3 Whole-rock chemical composition of veins

Whole-rock analyses of the individual zones demonstrate that the REE are present in concentrations up to several hundred ppm (Σ(REE)$_{max}$ ≈ 800 ppm) in the three central vein zones whereas they commonly are undetectable (light REE < 0.5 ppm; heavy REE < 0.06 ppm; Gieré, 1990b,

1993) in the dolomite marble host rock. This pronounced REE enrichment of the veins relative to the host rock is clearly emphasized by the chondrite-normalized REE patterns (Figure 5.12) which reveal a minimum enrichment of approximately two orders of magnitude for most REE. Figure 5.12 further shows that the Ti-rich central vein zones have REE patterns very similar to those found in the nearby tonalite intrusion. A negative Eu anomaly is observed in both tonalite and individual vein samples.

The central vein zones are rich in Ti, Zr and actinides (several wt% TiO_2; up to several hundred ppm Zr, U and Th), whereas these elements are not present or only in very small concentrations in the dolomite marble ($TiO_2 <$ 0.08 wt%, Zr < 16 ppm, U ≈ 6 ppm, Th < 0.1 ppm). Furthermore, the high sulphur, fluorine and phosphorus contents of the Ti-rich central vein zones (S up to 7 wt%, F ≈ 4000 ppm, P_2O_5 up to 2.3 wt%) are in sharp contrast to the extremely low concentrations found in the dolomite marble (S $<$ 10 ppm, F = 50–150 ppm, P_2O_5 = 0.02–0.04 wt%), and thus document the presence of these components in the metasomatic fluid.

These data, together with whole-rock stable and radiogenic isotope data (Gieré, Oberli and Meier, 1988; Gieré 1990b; Gieré and Hoering, 1990), indicate a magmatic origin of the fluid, which was then diluted by components from the dolomite host rock.

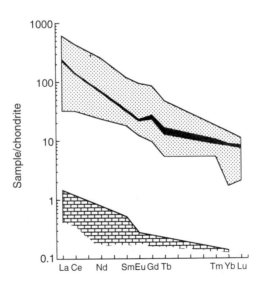

Figure. 5.12 Chondrite-normalized REE patterns displaying the compositional ranges for the Ti-rich veins (stippled), for the tonalite intrusion (black) and for the dolomite marble. The data for the dolomite marble represent for Sm and Nd the actual concentrations (isotope dilution analysis), for all others the detection limits (instrumental neutron activation analysis). (Data from Gieré, 1990b; chondrite values from Wakita, Rey and Schmitt, 1971.)

CASE STUDY

5.8.4 Conditions of vein formation

Thermodynamic analysis of the phase assemblages indicates that the vein minerals crystallized in a relatively water-rich ($X_{CO_2} \approx 0.2$) environment at a temperature of 500–600°C and an estimated total pressure of 2000 bars. The mole fraction of CO_2 was constrained mainly by the titanium phases in the phlogopite zone (rutile and titanite; no perovskite), and thus is theoretically valid for that zone only. The temperatures derived from phase relationships are in good agreement with temperatures calculated from stable isotopic data for coexisting minerals.

Pyrrhotite is the main iron-bearing phase in all vein assemblages containing the RE minerals, and is well suited to study the relationship between the fugacities of oxygen and various sulphur species in the fluid. Figure 5.14 displays a phase diagram for the system Fe–Cu–Pb–Ti–C–O–H–S, where $\log f_{H_2S}$ is plotted against $\Delta \log f_{O_2}$ (i.e. f_{O_2} relative to the fayalite–

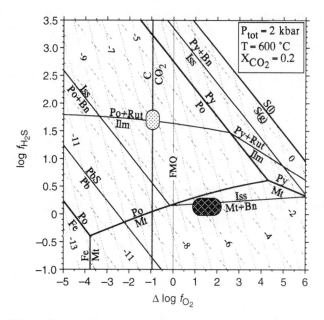

Figure 5.13 Phase diagram for the system Fe–Cu–Pb–Ti–C–O–H–S calculated for the P–T–X conditions of vein formation at Adamello (P_{tot} = 2 kbar, T = 600°C, X_{CO_2} = 0.2). S(g)–S(l) line represents the sulphur condensation. Fine dashed lines give contours for $\log f_{S_2}$; fine dotted lines give contours for $\log f_{SO_2}$; dark cross-hatched field represents estimated conditions for the original magmatic fluid; stippled field shows approximate conditions of vein formation (see text for discussion). Bn = bornite, FMQ = fayalite–magnetite–quartz buffer, Ilm = ilmenite (a_{FeTiO_3} = 0.5), Iss = intermediate solid solution ($CuFeS_{1.8}$), Mt = magnetite, PbS = galena, Po = pyrrhotite, Py = pyrite, Rut = rutile. (Thermodynamic data from Robie, Hemingway and Fisher, 1979; Barton and Skinner, 1979; Barker and Parks 1986; Shi, 1992.)

magnetite–quartz [FMQ] buffer) for several reactions of interest. Because graphite occurs in the three central vein zones, the reaction

$$C + O_2 \rightleftharpoons CO_2 \quad (5.11)$$
Graphite

has been drawn into the diagram (thick solid vertical line for $X_{CO_2} = 0.2$, corresponding to the mole fraction of CO_2 inferred from the iron-free silicate–oxide assemblages in the phlogopite zone). The equilibria involving ilmenite were calculated for $a_{FeTiO_3} = 0.5$ because magnesian ilmenite occurs in the phlogopite zone which also contains rutile. As shown in Figure 5.13, reaction (5.11) intersects the ilmenite–rutile reaction at log $f_{H_2S} \approx$ 1.7, i.e. within the stability fields of both galena and intermediate solid solution ($\approx$ 'chalcopyrite' above 557°C), consistent with the petrographic observations. These results, combined with thermodynamic data for equilibria involving H_2O, SO_2, H_2, O_2 and S_2, and with experimental data on the partitioning of F and Cl between apatite and fluid (Korzhinskiy, 1981), allowed the fugacities of several species in the metasomatic fluid to be calculated. These fugacities are listed in Table 5.4, which shows that the fluid was relatively rich in $HCl°_{(aq)}$ and $HF°_{(aq)}$. Figure 5.13 and Table 5.4 also reveal that H_2S was the dominant sulphur species in the fluid, implying a rather low pH (for details of thermodynamic analysis, see Gieré, 1990b).

5.8.5 Transport of REE by the hydrothermal fluid

The results of the thermodynamic analysis together with the evidence from the whole-rock chemical composition of the veins demonstrate that the hydrothermal fluid contained high concentrations of fluorine, chlorine, phosphorus and sulphur. Because of the mentioned strong affinity of the hard REE cations for hard ligands at high temperatures, fluoride and phosphate are the most likely ligands for complexing the REE in the Adamello fluid. This conclusion is supported by petrographic evidence, notably by the abundance of idiomorphic, REE-rich fluorapatite as well as by the observed textural relations.

At 600°C and 2000 bars, REE–chloride complexes appear to be important only in acid fluids with a high molar Cl/F ratio (Haas, Shock and Sassani, in press). From these predictions and from the large abundance of fluorine relative to chlorine in the fluid (Table 5.4), it is likely that chloride complexes played a relatively minor role in the transport of the REE.

For several reasons, carbonate complexes were probably not important for the REE transport in this hydrothermal system First, the fluid had a rather low pH, and carbonate complexes of REE only form in near-neutral to basic pH conditions (Wood, 1990b). Furthermore, REE–carbonate complexes appear to be generally of little importance in high-temperature

Table 5.4 Fugacities (bars) of selected species in the vein-forming fluid at Adamello. Calculated for $T = 600\,°C$, $P_{tot} = 2000$ bars, and $X_{CO_2} = 0.2$ (corresponding to the mole fraction of CO_2 inferred from the mineral assemblage in the phlogopite zone)

$\log f_{H_2O}$	3.0
$\log f_{CO_2}$	3.1
$\log f_{O_2}$	-20.7
$\log f_{H_2}$	1.4
$\log f_{S_2}$	-5.1
$\log f_{SO_2}$	-5.1
$\log f_{H_2S}$	1.7
$\log f_{HF°}$	-2.1
$\log f_{HCl°}$	-1.1

fluids, as suggested by the calculations of Haas, Shock and Sassani (in press). Finally, large amounts of CO_2 were probably produced only at the site of vein formation and not during separation of the fluid from the tonalitic melt.

From the presence of a negative Eu anomaly in the whole-rock vein samples, it appears most likely that Eu was transported in its bivalent oxidation state, in agreement with the predictions for high-temperature fluids (Wood, 1990a; Sverjensky, 1984).

In the absence of fluid inclusions suitable for direct analysis of trace elements, the concentrations of three REE have been estimated by using experimental data for the partition of REE between apatite and fluid. The results, listed in Table 5.5, indicate that the REE were present in concentrations of several hundred ppm (ΣREE). These concentrations are within the same order of magnitude as those found in fluid inclusion liquids from the Capitan Pluton (Banks et al., 1994). The results must be regarded as rough estimates only, as large uncertainties result from the extrapolation of experimental data obtained in simple systems to a complex natural fluid at significantly different conditions (cf. Table 5.5).

5.8.6 Deposition of REE

Based on the petrographic observations and the available analytical data, the following model has been derived to explain formation of the Adamello veins and of the RE minerals in particular.

As mentioned above, the hydrothermal fluids originated from the nearby tonalite intrusion, representing fluids which probably separated from the tonalitic melt. Such fluids are expected to be initially acid and relatively oxidizing with a probable oxygen fugacity of 1–2 log units above FMQ (cf.

Table 5.5 Estimated concentration of REE (ppm) in the vein-forming fluid at Adamello (inferred conditions: $T \approx 600°C$, $P_{tot} \approx 2000$ bars, $X_{CO_2} \approx 0.2$). Calculated with the experimentally determined K_D values given by Ayers and Watson (1993) for the partition of REE between apatite and pure water, and apatite and water + dissolved albite (experimental conditions: $T = 1000°C$, $P_{tot} = 10$ kbar). Ce and Yb content of Adamello apatite from Gieré and Williams (1992); Gd content interpolated from chondrite-normalized REE pattern shown in Figure 5.10. The given range includes uncertainties in both K_D values and REE concentrations in apatite. Please note that the calculated REE concentrations are rough estimates only!

	Apatite/H$_2$O		Apatite/H$_2$O + albite	
	Minimum	Maximum	Minimum	Maximum
Ce	34.1	261.3	163.3	172.6
Gd	0.6	2.6	2.6	2.6
Yb	0.1	0.6	0.2	0.4
Total	34.8	264.4	166	175.5

Figure 5.13; judging from the silicate assemblage in the tonalite; Frost and Lindsley, 1992).

This tonalitic fluid interacted with the dolomite marble, causing oxidation of the wall rock graphite (according to reaction [5.11]) in the immediate vicinity of the veins, and produced the graphite-free assemblage forsterite + calcite (Figure 5.6). Formation of this vein zone may be expressed as

$$2CaMg(CO_3)_2 + SiO_{2(aq)} \rightleftharpoons 2CaCO_3 + Mg_2SiO_4 + 2CO_{2(aq)}. \quad (5.12)$$
dolomite　　　　　　　　　　calcite　　forsterite

This decarbonation reaction leads to a dilution of the metasomatic fluid (consistent with the stable isotopic data), decreasing the activity of ligand-forming anions (e.g. Cl$^-$) and inducing precipitation of sulphide minerals (Crerar and Barnes, 1976; Barnes, 1979). This conclusion is drawn from reactions similar to reaction (5.6), which is considered to be representative for precipitation of the sulphides in the Adamello veins. Reaction (5.6) assumes that the ore-forming metals are transported as chloride complexes; this is feasible for the Adamello example because of the relatively high fugacity of HCl° in the fluid (Table 5.4).

The general sulphide-forming reaction (5.6), probably initiated by fluid dilution due to progression of reaction (5.12), would lead to an increase in the fluid acidity. In carbonate rocks this acidity is generally neutralized by calcite dissolution according to the schematic reaction

CASE STUDY

$$CaCO_3 + 2H^+ + 2Cl^- \rightleftharpoons CaCl_2°_{(aq)} + CO_{2(aq)} + H_2O \quad (5.13)$$
calcite

which produces a $CaCl_2°$-rich brine (Kwak and Tan, 1981; Eugster, 1985).

In high-temperature fluids copper is probably transported mainly as $CuCl°$ (Crerar and Barnes, 1976), while iron commonly forms $FeCl_2°$ complexes (Chou and Eugster, 1977; Crerar et al., 1985). Combining this information with reactions (5.6) and (5.13) leads to a chalcopyrite-forming model reaction

$$3CaCO_3 + 4H_2S_{(aq)} + 2CuCl°_{(aq)} + 2FeCl_2°_{(aq)} \rightleftharpoons$$
calcite

$$2CuFeS_2 + 3H_2O + 3CaCl_2°_{(aq)} + 3CO_{2(aq)} + H_{2(aq)} \quad (5.14)$$
chalcopyrite

or, expressed in a different way,

$$6CaCO_3 + 8H_2S_{(aq)} + 4CuCl°_{(aq)} + 4FeCl_2°_{(aq)} \rightleftharpoons$$
calcite

$$4CuFeS_2 + 8H_2O + 6CaCl_2°_{(aq)} + 5CO_{2(aq)} + C. \quad (5.15)$$
chalcopyrite \qquad\qquad\qquad\qquad Graphite

The presence of both graphite (C) and chalcopyrite in the three Ti-rich central assemblages can be explained by reaction (5.15). Reactions (5.14) and (5.15) could eventually shift the oxygen fugacity of the fluid to values lower than those of the original fluid derived from the intrusion, permitting the precipitation of pyrrhotite (cf. Figure 5.13). Pyrrhotite formation can readily be described by reaction (5.6); the resulting decrease in pH could then induce the breakdown of REE complexes which may be expressed, for example, as

$$REE(PO_4)°_{(aq)} + 3H^+ \rightleftharpoons REE^{3+} + H_3PO_4°_{(aq)} \quad (5.16)$$

or

$$REEF_3°_{(aq)} + 3H^+ \rightleftharpoons REE^{3+} + 3HF°_{(aq)} \quad (5.17)$$

for the case of a complete, three-step breakdown of the suggested phosphate or fluoride complexes to trivalent REE cations. (In high-temperature fluids metal ions tend to form neutral complexes: Seward, 1981; Crerar et al., 1985). Moreover, crystallization of fluorapatite, formulated schematically as

$$5CaCO_3 + 3H_3PO_4°_{(aq)} + HF°_{(aq)} \rightleftharpoons$$
calcite

$$Ca_5(PO_4)_3F + 5CO_{2(aq)} + 5H_2O \quad (5.18)$$
apatite

would further assist in destabilizing the REE complexes and eventually in depositing the REE. Reaction (5.18) also explains the observed close association of fluorapatite with the other RE minerals (Figure 5.7).

According to the model presented above, the calcite crystals present in each of the sulphide-bearing assemblages (Figure 5.6) would have been formed only after neutralization of the acidity produced by sulphide precipitation and after crystallization of apatite. An early crystallization of apatite within the sulphide-bearing, Ti-, Zr-, REE- and actinide-rich central vein zones is also indicated by the textural relationships.

This model explains the mineralogical zoning with respect to some of the minerals present in the Adamello veins. The model is further consistent with the observed textural relationships and the geochemical data which suggest that the formation of RE minerals is genetically linked to the formation of fluorapatite and to the precipitation of base metal sulphides. In this case the REE appear to be trapped efficiently at the site of fluorapatite and pyrrhotite formation. Thus, the Adamello veins provide a good example for the complex interdependence between crystallization of sulphides, changes in fluid composition and the formation of RE minerals.

5.9 Conclusions

RE minerals of hydrothermal origin are very common and, in addition to being very interesting for mineralogical and petrological studies, constitute an important economic resource. These RE minerals are often zoned and altered, reflecting changes in the physical conditions or in the fluid composition during the evolution of a hydrothermal system. The available data suggest that large quantities of REE (several hundred ppm) can probably only be transported by high-temperature magmatic fluids with relatively high salinities. Speciation of the REE cations in hydrothermal fluids is strongly dependent on the relative concentrations of possible ligands and on the pH of the solution. Precipitation of RE minerals from hydrothermal solutions may take place in response to fluid mixing, fluid/wall rock interaction, crystallization of gangue minerals, and as a result of changes in temperature and pressure; it is often difficult to determine the principal cause of deposition, because of the complex composition of natural fluids. Carbonate rocks appear to be particularly efficient in trapping REE, especially when they occur in contact with igneous rocks.

I hope that this chapter encourages experimental geologists to perform careful studies on various aspects of REE geochemistry, but particularly on solubility and speciation at higher temperatures and pressures. Such investigations are urgently needed in order to understand better the complex mechanisms leading to the formation of RE minerals in hydrothermal systems.

Acknowledgements

I am very grateful to Adrian Jones, Frances Wall and Terry Williams for their careful reviews and constructive criticism, and for their patience in editing this book. I would like to express my special gratitude to Terry Williams for his encouragement and valuable suggestions, and for providing photographs of exceptional RE minerals from the collections of the Natural History Museum, London. I am particularly grateful also to Walter F. Oberholzer and Peter Brack for allowing me to include their pictures of outstanding fissure minerals from the collections of ETH Zürich. I would also like to thank David Banks, Peter Berlepsch, Matthias Braun, Ron Frost, Urs Gerber, Stefan Graeser, Richard Guggenheim, Johnson Haas, Michael Krzemnicki, Vala Ragnarsdottir, Arnold Stahel, Bruce Watson and Bruce Yardley for their contributions towards completing this manuscript.

This article was written during my visit at the Geophysical Laboratory, Carnegie Institution of Washington. The help of Carnegie librarians Shaun Hardy and Merri Wolf has been invaluable. I would like to thank the Schweizerischer Nationalfonds (Grants No. 21-33830.92 and 8220-037190) and the Freiwillige Akademische Gesellschaft Basel (colour plates) for their generous financial support.

References

Ahrland, S. (1968) Thermodynamics of complex formation between hard and soft acceptors and donors. *Struct. Bond.*, **5**, 118–149.

Aja, S. U., Wood, S. A. and Williams-Jones, A. E. (1994) An experimental determination of the solubility of neodymium hydroxybastnaesite at 25°C. *EOS, Trans. Am Geophys. Union*. **75/16** (suppl.), 139.

Alderton, D. H. M., Pearce, J. A. and Potts, P. J. (1980) Rare earth element mobility during granite alteration: evidence from southwest England. *Earth. Planet. Sci. Lett.*, **49**, 149–65.

Ayers, J. C. and Watson, E. B. (1991) Solubility of apatite, monazite, zircon, and rutile in supercritical aqueous fluids with implications for subduction zone geochemistry. *Phil. Trans. R. Soc. Lond. A*, **335**: 365–75.

Ayers, J. C. and Watson, E. B. (1993) Apatite/fluid partitioning of rare-earth elements and strontium: experimental results at 1.0 GPa and 1000°C and application to models of fluid–rock interaction. *Chem. Geol.*, **110**, 299–314.

Bandurkin, G. A. (1961) Behavior of the rare earths in fluorine-bearing media. *Geochem. Int.*, **2**, 159–67.

Banks, D. A., Yardley, B. W. D., Campbell, A. R. and Jarvis, K. E. (1994) REE composition of an aqueous magmatic fluid: a fluid inclusion study from the Capitan Pluton, New Mexico. *Chem. Geol.*, **113**, 259–72.

Barker, W. W. and Parks, T. C. (1986) The thermodynamic properties of pyrrhotite and pyrite: a re-evaluation. *Geochim. Cosmochim. Acta*. **50**, 2185–94.

Barnes, H. L. (1979) Solubility of ore minerals, in *Geochemistry of hydrothermal ore deposits* (ed. H. L. Barnes), Wiley & Sons, New York, pp. 404–60.
Barrett, T. J., Jarvis, I. and Jarvis, K. E. (1990) Rare earth element geochemistry of massive sulfides–sulfates and gossans on the Southern Explorer Ridge. *Geology*, **18**, 583–6.
Barrett, T. J., Jarvis, I., Longstaffe, F. J. and Farquhar, R. (1988) Geochemical aspects of hydrothermal sediments in the Eastern Pacific Ocean: an update. *Can. Mineral.*, **26**, 841–58.
Barton, P. B. and Skinner, B. J. (1979) Sulfide mineral stabilities, in *Geochemistry of hydrothermal ore deposits* (ed. H. L. Barnes), Wiley & Sons, New York, pp. 278–403.
Bau, M. (1991) Rare-earth element mobility during hydrothermal and metamorphic fluid–rock interaction and their significance of the oxidation state of europium. *Chem. Geol.*, **93**, 219–30.
Bau, M. and Möller, P. (1992) Rare earth element fractionation in metamorphogenic hydrothermal calcite, magnesite and siderite. *Mineral. Petrol.*, **45**, 231–46.
Bermanec, V., Tibljas, D., Gessner, M. and Kniewald, G. (1988) Monazite in hydrothermal veins from Alinci, Yugoslavia. *Mineral. Petrol.*, **38**, 139–50.
Bilal, B. A. and Koss, V. (1982) Complex formation of trace elements in geochemical systems, VI. Study on the formation of hydroxo fluoro mixed ligand complexes of the lanthanide elements in fluorite bearing systems. *Polyhedron*, **1**, 239–41.
Bilal. B. A. and Langer, P. (1987) Complex formation of trace elements in geochemical systems: stability constants of fluoro complexes of the lanthanides in a fluorite bearing model system up to 200°C and 1000 bar. *Inorg. Chim. Acta*, **140**, 297–8.
Braun, M. C. (1993) Untersuchungen von REE-Mineralien aus dem Monte Giove – Gebiet. Unpublished Master's thesis, Universität Basel, Switzerland.
Braun, J.-J., Pagel, M., Herbillon, A. and Rosin, C. (1993) Mobilization and redistribution of REEs and thorium in a syenitic lateritic profile: a mass balance study. *Geochim. Cosmochim. Acta*, **57**, 4419–34.
Braun, J.-J., Pagel, M., Muller, J.-P. *et al.* (1990) Cerium anomalies in lateritic profiles. *Geochim. Cosmochim. Acta*, **54**, 781–95.
Brimhall, G. H. and Crerar, D. A. (1987) Ore fluids: magmatic to supergene, in *Thermodynamic Modeling of Geological Materials: Minerals, Fluids and Melts* (eds I. S. E. Carmichael and H. P. Eugster), *Reviews in Mineralogy*, **17**, Mineralogical Society of America, Washington, DC, 235–321.
Bromley A. V. (1964) Allanite in the Tan-Y-Grisiau microgranite, Merionethshire, North Wales. *Am. Mineral.*, **49**, 1747–52.
Brookins, D. G. (1989) Aqueous geochemistry of rare earth elements, in *Geochemistry and Mineralogy of the Rare Earth Elements*, (eds B. R. Lipin and G. A. McKay), *Reviews in Mineralogy*, **21**, Mineralogical Society of America, Washington, DC, 201–25.
Byrne, R. H., Lee, J. H. and Bingler, L. S. (1991) Rare earth element complexation by PO_4^{3-} ions in aqueous solution. *Geochim. Cosmochim. Acta*, **55**, 2729–35.

Caruso, L. and Simmons, G. (1985) Uranium and microcracks in a 1000-meter core, Redstone, New Hampshire. *Contrib. Mineral. Petrol.*, **90**, 1–17.
Cathelineau, M. (1987) U–Th–REE mobility during albitization and quartz dissolution in granitoids: evidence from southeast French Massif Central. *Bull. Mineral.*, **110**, 249–59.
Chao, E. C. T., Back, J. M. and Minkin, J. A. (1992) Host-rock controlled epigenetic, hydrothermal metasomatic origin of the Bayan Obo REE–Fe–Nb ore deposit, Inner Mongolia, P.R.C. *Appl. Geochem.*, **7**, 443–58.
Chen, J., Halls, C. and Stanley, C. J. (1992) Rare earth element contents and patterns in major skarn minerals from Shizhuyuan W, Sn, Bi and Mo deposit, South China. *Geochem. J.*, **26**, 147–58.
Choppin, G. R. (1983) Comparison of the solution chemistry of the actinides and lanthanides. *J. Less-Common Metals*, **93**, 323–30.

REFERENCES

Choppin, G. R. (1984) Lanthanide complexation in aqueous solutions. *J. Less-Common Metals*, **100**, 141–51.
Chou, I.-M. and Eugster, H. P. (1977) Solubility of magnetite in supercritical chloride solutions. *Am. J. Sci.* **277**, 1296–14.
Clark, A. M. (1984) Mineralogy of the rare earth elements, in *Rare Earth Element Geochemistry* (ed. P. Henderson), *Developments in Geochemistry*, **2**, Elsevier, Amsterdam, pp. 33–61.
Collins, C. J. and Strong, D. F. (1992) The distribution of REE in hydrothermal vein fluorite and calcite from the St. Lawrence fluorspar district, NFLD. Geological Association of Canada/Mineralogical Association of Canada, Programme with abstracts, **15**, pp. A19–20.
Crerar, D. A. and Barnes H. L. (1976) Ore solution chemistry V. Solubilities of chalcopyrite and chalcocite assemblages in hydrothermal solution at 200° to 350°C. *Econ. Geol.*, **71**, 772–94.
Crerar, D. A., Wood, S. A., Brantley, S. L. and Bocarsly, A. (1985) Chemical controls on the solubility of ore-forming minerals in hydrothermal solutions. *Can. Mineral.*, **23**, 333–52.
Cressey, G. (1987) Skarn formation between metachalk and agglomerate in the Central Ring Complex, Isle of Arran, Scotland. *Mineral. Mag.*, **51**, 231–46.
Crook, W. W., Oswald, S. G. (1979) New data on cerian vesuvianite from San Benito County, California. *Am. Mineral.*, **64**, 367–8.

David, F. (1986) Thermodynamic properties of lanthanide and actinide ions in aqueous solution. *J. Less-Common Metals*, **121**, 27–42.
Demartin, F., Gramaccioli, C. M. and Pilati, T. (1994) Paraniite-(Y), a new tungsten arsenate mineral from Alpine fissures. *Schweiz Mineral. Petrogr. Mitt.*, **74/2**, 155–60.
Demartin, F., Pilati, T., Diella, V. *et al.* (1991a) Alpine monazite: further data. *Can. Mineral.*, **29**, 61–7.
Demartin, F., Pilati, T., Diella, V. *et al.* (1991b) The chemical composition of xenotime from fissures and pegmatites in the Alps. *Can. Mineral.*, **29**, 669–75.
Dollinger, H. (1989) Petrographische und geochemische Untersuchungen des Altkristallins zwischen Nägelisgrätli und Oberaarjoch – Hydrothermale Veränderungen granitischer Gesteine in der Grimselregion. Unpublished Master's Thesis, Universität Bern, Switzerland.
Drew, L. J., Meng, Q. and Sun, W. (1990) The Bayan Obo iron–rare–earth–niobium deposits, Inner Mongolia, China. *Lithos*, **26**, 43–65.

Eugster, H. P. (1985) Granites and hydrothermal ore deposits: a geochemical framework. *Mineral. Mag.*, **49**, 7–23.
Eugster, H. P. (1986) Minerals in hot water. *Am. Mineral.*, **71**, 655–73.
Ewing R. C., Haaker R. F. and Lutze, W. (1982) Leachability of zircon as a function of alpha dose. *Scientific basis for radioactive waste management V*. (ed. W. Lutze) North-Holland, New York, pp. 389–397.
Ewing, R. C., Chakoumakos, B. C., Lumpkin, G. R. and Murakami, T. (1987) The metamict state. *Mater. Res. Soc. Bull.*, May/June, 58–66.
Exley, R. A. (1980) Microprobe studies of REE-rich accessory minerals: implications for Skye granite petrogenesis and REE mobility in hydrothermal systems. *Earth. Planet. Sci. Lett.*, **48**, 97–110.

Fleischer, M. (1978) Relative proportions of the lanthanides in the minerals of the bastnäsite group. *Can. Mineral.*, **16**, 361–3.
Förster, H.-J., Hunger, H.-J. and Grimm, L. (1987) Elektronenstrahlmikroanalytische Untersuchungen von Erzmineralen aus der Zinn-Lagerstätte Altenberg (Erzgebirge, DDR). *Chem. Erde*, **46**, 73–9.
Frost, B. R. and Lindsley, D. H. (1992) Equilibria among Fe–Ti oxides, pyroxenes, olivine, and quartz: Part II. Application. *Am. Mineral.*, **77**, 1004–20.

Fryer, B. J. and Taylor, R. P. (1987) Rare earth element distributions in uraninites: implications for ore genesis. *Chem. Geol.*, **63**, 101–8.

Geijer, P. (1921) The cerium minerals of Bastnäs at Riddarhyttan. *Sver. Geol. Unders. Ser. C*, No. **304**, Årsbok **14/6**, 5–24.

Geijer, P. (1963) The geological significance of the cerium mineral occurrences of the Bastnäs type in Central Sweden. *Arkiv Mineral. Geol.*, **3/4**, 99–105.

Ghai, A. M., Vanko, D. A., Roedder, E. and Seeley, R. C. (1993) Determination of rare earth elements in fluid inclusions by inductively coupled plasma–mass spectrometry (ICP-MS). *Geochim. Cosmochim. Acta*, **57**, 4513–6.

Gieré, R. (1986) Zirconolite, allanite and hoegbomite in a marble skarn from the Bergell contact aureole: implications for mobility of Ti, Zr and REE. *Contrib. Mineral. Petrol.*, **93**, 459–70.

Gieré, R. (1990a) Hydrothermal mobility of Ti, Zr and REE: examples from the Bergell and Adamello contact aureoles (Italy). *Terra Nova*, **2**, 60–7.

Gieré, R. (1990b) Quantification of element mobility at a tonalite/dolomite contact (Adamello Massif, Provincia di Trento, Italy). Unpublished PhD Thesis No. 9141, Eidgenössische Technische Hochschule, Zürich, Switzerland.

Gieré, R. (1992) Compositional variation of metasomatic titanite from Adamello (Italy). *Schweiz Mineral. Petrogr. Mitt.*, **72**, 167–77.

Gieré, R. (1993) Transport and deposition of REE in H_2S-rich fluids: evidence from accessory mineral assemblages. *Chem. Geol.*, **110**, 251–68.

Gieré, R. and Hoering, T. C. (1990) Isotopic composition of sulfides in Ti-rich veins from the Adamello contact aureole (Italy). Carnegie Institution of Washington, DC, Annual Report of Geophysical Laboratory, (1989–1990), 2200, pp. 33–7.

Gieré, R. and Williams, C. T. (1992) REE-bearing minerals in a Ti-rich vein from the Adamello contact aureole (Italy). *Contrib. Mineral. Petrol.*, **112**, 83–100.

Gieré, R., Oberli, F. and Meier, M. (1988) Mobility of Ti, Zr and REE: mineralogical, geochemical and isotopic evidence from the Adamello contact aureole (Italy). *Chem. Geol.*, **70**, 161.

Gieré, R., Guggenheim, R., Düggelin, M. et al. (1994) Retention of actinides during alteration of aperiodic zirconolite. Electron Microscopy 94: Proceedings of the 13th International Congress on Electron Microscopy, Applications in Materials Sciences 2B, 1269–70.

Glass, J. J. and Smalley, R. G. (1945) Bastnäsite. *Am. Mineral.*, **30**, 601–15.

Graeser, S. and Schwander, H. (1987) Gasparite-(Ce) and monazite-(Nd): two new minerals to the monazite group from the Alps. *Schweiz Mineral. Petrogr. Mitt.*, **67**, 103–13.

Graeser, S., Schwander, H. and Stalder, H. A. (1973) A solid solution series between xenotime (YPO_4) and chernovite ($YAsO_4$). *Mineral. Mag.*, **39**, 145–51.

Gramaccioli, C. M. (1977) Rare earth minerals in the Alpine and Subalpine region. *Mineral. Rec.*, **8**, 287–93.

Grauch, R. I. (1989) Rare earth elements in metamorphic rocks, in *Geochemistry and Mineralogy of the Rare Earth Elements* (eds B. R. Lipin and G. A. McKay), *Reviews in Mineralogy*, **21**, 147–67, Mineralogical Society of America, Washington, DC.

Haack, U., Schnorrer-Köhler, G. and Lüders, V. (1987) Seltenerd-Minerale aus hydrothermalen Gängen des Harzes, *Chem. Erde*, **47**, 41–5.

Haas, J. R. and Shock, E. L. (1994) Predictions rare earth element (REE) speciation in a simulated hydrothermal solution as a function of pressure, temperature and fluid pH. *EOS, Trans. Am. Geophys. Union*, **75/16** (suppl.), 361.

Haas, J. R., Shock, E. L. and Sassani, D. A. (1993) Predictions of high temperature stability constants for aqueous complexes of the rare earth elements. *Geol. Soc. Am. Abstracts*, **25/6**, A-437.

REFERENCES

Haas, J. R., Shock, E. L. and Sassani, D. A. (in press) Rare earth elements in hydrothermal systems: estimates of standard partial molal thermodynamic properties of aqueous complexes of the REE at high pressures and temperatures. *Geochim. Cosmochim. Acta.*

Hawkins, B. W. (1975) Mary Kathleen uranium deposit, in *Economic Geology of Australia and Papua New Guinea. 1. Metals* (ed. C. L. Knight), Australasian Institute of Mining and Metallurgy, Parkville, Victoria, pp. 398–402.

Henderson, P. (1984) General geochemical properties and abundances of the rare earth elements, in *Rare Earth Element Geochemistry* (ed. P. Henderson), *Developments in Geochemistry*, **2**, Elsevier, Amsterdam, pp. 1–32.

Hopf, S. (1993) Behaviour of rare earth elements in geothermal systems of New Zealand. *J. Geochem. Explor.*, **47**, 333–57.

Humphris, S. E. (1984) The mobility of rare earth elements in the crust, in *Rare Earth Element Geochemistry* (ed. P. Henderson), *Developments in Geochemistry*, **2**, Elsevier, Amsterdam, pp. 317–42.

Isokangas, P. (1978) Finland, in *Mineral Deposits of Europe, 1, Northwest Europe* (eds S. H. U. Bowle, A. Kvalheim and A. W. Haslam), Institution of Mining and Metallurgy and Mineralogical Society, London, pp. 39–92.

Jaffe, H. W., Meyrowitz, R. and Evans, H. T. (1953) Sahamalite, a new rare earth carbonate mineral. *Am. Mineral.*, **38**, 741–54.

Jamtveit, B. and Hervig, R. L. (1994) Constraints on transport and kinetics in hydrothermal systems from zoned garnet crystals. *Science*, **263**, 505–8.

Johnson, J. P. and McCulloch, M. T. (1995) Sources of mineralising fluids for the Olympic Dam deposit (South Australia): Sm-Nd isotopic constraints. *Chem. Geol.*, **121**, 177–99.

Jonasson, R. G., Bancroft, G. M. and Nesbitt, H. W. (1985) Solubilities of some hydrous REE phosphates with implications for diagenesis and seawater concentrations. *Geochim. Cosmochim. Acta*, **49**, 2133–9.

Kerrich, R. and King, R. (1993) Hydrothermal zircon and baddeleyite in Val-d'Or Archean mesothermal gold deposits: characteristics, compositions, and fluid-inclusion properties, with implications for timing of primary gold mineralization. *Can. J. Earth Sci.*, **30**, 2334–51.

Korzhinskaya, V. S. (1990) The stability of zircon in the system $ZrO_2-SiO_2-H_2O-H_2O-HCl$ from 400 to 600°C and to 2 kbar, in *Experiment-89; informative volume* (ed. V. A. Zharikov), Akad. Nauk SSSR, Inst. Eksp. Mineral., Chernogolovka, pp. 84–6.

Korzhinskiy, M. A. (1981) Apatite solid solutions as indicators of the fugacity of HCl° and HF° in hydrothermal fluids. *Geochem. Int.*, **18**, 44–60.

Kosterin, A. V. (1959) The possible modes of transport of the rare earths by hydrothermal solutions. *Geochem. Int.*, **4**, 381–7.

Kraynov, S. R., Mer'kov, A. N., Petrova, N. G. *et al.* (1969) Highly alkaline (pH 12) fluosilicate waters in the deeper zones of the Lovozero massif. *Geochem. Int.*, **6/4**, 635–40.

Krzemnicki, M. (1993) Rare earth element-zoning in cafarsite from the Binnatal region (Switzerland). Rare Earth Mineral Conference, London, Abstract volume, p. 63.

Kwak, T. A. and Abeysinghe, P. B. (1987) Rare earth and uranium minerals present as daughter crystals in fluid inclusions, Mary Kathleen U–REE skarn, Queensland, Australia. *Mineral. Mag.*, **51**, 665–70.

Kwak, T. A. P. and Tan, T. H. (1981) The importance of $CaCl_2$ in fluid composition trends – evidence from the King Island (Dolphin) skarn deposit. *Econ. Geol.*, **76**, 955–60.

Lehmann, B., Nakai, S., Höhndorf, A. *et al.* (1994) REE mineralisation at Gakara, Burundi: evidence for anomalous upper mantle in the western Rift Valley. *Geochim. Cosmochim. Acta*, **58**, 985–92.

Lewis, A. J., Palmer, M. R. and Kemp, A. J. (1993) Rare earth element behaviour in the Yellowstone hydrothermal system. Rare Earth Mineral Conference, London, Abstract volume, pp. 69–71.

Lira, R. and Ripley, E. M. (1990) Fluid inclusion studies of the Rodeo de Los Molles REE and Th deposit, Las Chacras Batholith, Central Argentina. *Geochim. Cosmochim. Acta*, **54**, 663–71.

Littlejohn, A. L. (1981) Alteration products of accessory allanite in radioactive granites from the Canadian shield. *Geol. Surv. Can. Pap.*, **81–1B**, 95–104.

Lumpkin, G. R., Smith, K. L. and Gieré, R. (1994) The crystalline to aperiodic transition in zirconolite induced by alpha-recoil damage. 13th Biennial Conference of the Australian Society for Electron Microscopy, Abstracts, p. 124.

Lumpkin, G. R., Blackford, M. G., Smith, K. L. *et al.* (1994) Determination of 25 elements in the complex oxide mineral zirconolite by analytical electron microscopy. *Micron*, **25**, 581–7.

Maas, R., McCulloch, M. T. and Campbell, I. H. (1987) Sm-Nd isotope systematics in uranium–rare earth element mineralization at the Mary Kathleen uranium mine, Queensland. *Econ. Geol.*, **82**, 1805–26.

Mannucci, G., Diella, V., Gramaccioli, C. M. and Pilati, T. (1986) A comparative study of some pegmatitic and fissure monazite from the Alps. *Can. Mineral.*, **24**, 469–74.

Marchand, L. (1976) Contribution à l'étude de la distribution des lathanides dans la fluorine. Unpublished PhD Thesis, Université d'Orléans, France.

Mariano, A. N. (1989a) Economic geology of rare earth elements, in *Geochemistry and Mineralogy of the Rare Earth Elements* (eds B. R. Lipin and G. A. McKay), *Reviews in Mineralogy*, **21**, Mineralogical Society of America, Washington, DC, 309–37.

Mariano, A. N. (1989b) Nature of economic mineralization in carbonatites and related rocks, in *Carbonatites – Genesis and evolution* (ed. K. Bell), Unwin Hyman, London, pp. 149–76.

Maruéjol, P., Cuney, M. and Turpin, L. (1990) Magmatic and hydrothermal REE fractionation in the Xihuashan granites (SE China). *Contrib. Mineral. Petrol.*, **104**, 668–80.

McAndrew, J. and Scott, T. R. (1955) Stillwellite, a new rare-earth mineral from Queensland. *Nature*, **176**, 509–10.

McCann, H. G. (1968) The solubility of fluorapatite and its relationship to that of calcium fluoride. *Archs Oral Biol.*, **13**, 987–1001.

McLennan, S. M. and Taylor, S. R. (1979) Rare earth element mobility associated with uranium mineralisation. *Nature*, **282**, 247–50.

Meintzer, R. E. and Mitchell, R. S. (1988) The epigene alteration of allanite. *Can. Mineral.*, **26**, 945–55.

Mercolli, I., Schenker, F. and Stalder, H. A. (1984) Geochemie der Veränderung von Granit durch hydrothermale Lösungen. *Schweiz Mineral. Petrogr. Mitt.*, **64**, 67–82.

Metz, M. C., Brookins, D. G., Rosenberg, P. E. and Zartman R. E. (1985) Geology and geochemistry of the Snowbird deposit, Mineral County, Montana. *Econ. Geol.*, **80**, 394–409.

Michard, A. (1989) Rare earth element systematics in hydrothermal fluids. *Geochim. Cosmochim. Acta*, **53**, 745–50.

Michard, A. and Albarède, F. (1986) The REE content of some hydrothermal fluids. *Chem. Geol.*, **55**, 51–60.

Michard, A., Beaucaire, C. and Michard, G. (1987) Uranium and rare earth elements in CO_2-rich waters from Vals-les-Bains (France). *Geochim. Cosmochim. Acta*, **51**, 901–10.

Michard, A., Albarède, F., Michard, G. *et al.* (1983) Rare earth elements and uranium in high-temperature solutions from East Pacific Rise hydrothermal vent field (13°N). *Nature*, **303**, 795–97.

Michard, G., Albarède, F., Michard, A. *et al.* (1984) Chemistry of solutions from the 13°N East Pacific Rise hydrothermal site. *Earth Planet. Sci. Lett.*, **67**, 297–307.

Millero, F. J. (1992) Stability constants for the formation of rare earth inorganic complexes as a function of ionic strength. *Geochim. Cosmochim. Acta*, **56**, 3123–32.

REFERENCES

Mineyev, D. A., Makarochkin, B. A. and Zhabin, A. G. (1962) On the behavior of lanthanides during alteration of rare earth minerals. *Geochem.*, **7**, 684–93.

Mineyev, D. A., Rozanov, K. I., Smirnova, N. V. and Matrosova, T. I. (1973) Bastnäsitization products of accessory orthite. *Dokl. Akad. Nauk SSSR Earth Sci. Section*, **210**, 149–52.

Möller, P., Morteani, G., Hoefs, J. and Parekh, P. P. (1979) The origin of ore-bearing solution in the Pb–Zn veins of the Western Harz, Germany, as deduced from rare earth element and isotope distributions in calcites. *Chem. Geol.*, **26**, 197–215.

Morgan, J. W. and Wandless, G. A. (1980) Rare earth element distribution in some hydrothermal minerals: evidence for crystallographic control. *Geochim. Cosmochim. Acta*, **44**, 973–80.

Morteani, G. (1991) The rare earths: their minerals, production and technical use. *Eur. J. Mineral.*, **3**, 641–50.

Nakai, S., Masuda, A., Shimizu, H. and Qi, L. (1989) La–Ba dating and Nd and Sr isotope studies on the Baiyun Obo rare earth element ore deposits, Inner Mongolia, China. *Econ. Geol.*, **84**, 2296–9.

Nancollas, G. H. (1970) The thermodynamics of metal complex and ion-pair formation. *Coord. Chem. Rev.*, **5**, 379–415.

Neary, C. R. and Highly, D. E. (1984) The economic importance of the rare earth elements, in *Rare earth element geochemistry* (ed. P. Henderson), *Developments in Geochemistry*, **2**, Elsevier, Amsterdam, pp. 423–66.

Ngwenya, B. T. (1994) Hydrothermal rare earth mineralisation in carbonatites of the Tundulu complex, Malawi: processes at the fluid/rock interface. *Geochim. Cosmochim. Acta*, **58**, 2061–72.

Niggli, P. (1940) *Mineralien der Schweizer-Alpen, Band I, II.* Wepf, Basel, 300 pp.

Norman, D. I., Kyle, P. R. and Baron, C. (1989) Analysis of trace elements including rare earth elements in fluid inclusion liquids. *Econ. Geol.*, **84**, 162–66.

Olson, J. C., Shawe, D. R., Pray, L. C. and Sharp, W. N. (1954) Rare earth mineral deposits of the Mountain Pass District, San Bernardino County, California. *US Geol. Surv. Prof. Pap.*, **261**, 1–75.

Oreskes, N. and Einaudi, M. T. (1990) Origin of rare earth element-enriched hematite breccias at the Olympic Dam Cu–U–Au–Ag deposit, Roxby Downs, South Australia. *Econ. Geol.*, **85**, 1–28.

Oreskes, N. and Einaudi, M. T. (1992) Origin of hydrothermal fluids at Olympic Dam: preliminary results from fluid inclusions and stable isotopes. *Econ. Geol.*, **87**, 64–90.

Özgenç, I. (1993) Geology and chemistry of britholite veins. Rare Earth Mineral Conference, London, Abstract volume (suppl).

Pan, Y. and Fleet, M. E. (1990) Halogen-bearing allanite from the White River gold occurrence, Hemlo area, Ontario. *Can. Mineral.*, **28**, 67–75.

Pan, Y. and Fleet, M. E. (1991) Vanadian allanite-(La) and vanadian allanite-(Ce) from the Hemlo gold deposit, Ontario, Canada. *Mineral. Mag.*, **55**, 497–507.

Pan, Y., Fleet, M. E. and MacRae, N. D. (1993) Late alteration in titanite ($CaTiSiO_5$): redistribution and remobilization of rare earth elements and implications for U/Pb and Th/Pb geochronology and nuclear waste disposal. *Geochim. Cosmochim. Acta*, **57**, 355–67.

Pan, Y., Fleet, M. E. and Barnett, R. L. (1994) Rare earth mineralogy and geochemistry of the Mattagami Lake volcanogenic massive sulfide deposit, Quebec. *Can. Mineral.*, **32**, 133–47.

Papunen, H. and Lindsjö, O. (1972) Apatite, monazite and allanite; three rare earth minerals from Korsnäs, Finland. *Bull. Geol. Soc. Finland*, **44**, 123–9.

Parker, R. L. and de Quervain, F. (1940) Gadolinit aus den Schweizeralpen. *Schweiz Mineral. Petrogr. Mitt.*, **20**, 11–6.

Parker, R. L., de Quervain, F. and Weber, F. (1939) Über einige seltene Mineralien der Schweizeralpen. *Schweiz Mineral. Petrogr. Mitt.*, **19**, 293–306.

Pavelescu, L. and Pavelescu, M. (1972) Study of some allanites and monazites from the South Carpathians (Romania). *Tscherm. Mineral. Petrogr. Mitt.*, **17**, 208–14.

Pecora, W. T. and Kerr, J. H. (1953) Burbankite and calkinsite, two new carbonate minerals from Montana. *Am. Mineral.*, **38**, 1169–83.

Perhac, R. M. and Heinrich, E. W. M. (1964) Fluorite–bastnäsite deposits of the Gallinas Mountains, New Mexico and bastnäsite paragenesis. *Econ. Geol.*, **59**, 226–39.

Puchelt, H. and Emmermann, R. (1976) Bearing of rare earth patterns of apatites from igneous and metamorphic rocks. *Earth Planet. Sci. Lett.*, **31**, 279–86.

Raimbault, L. (1985) Utilisation des spectres de terres rares des minéraux hydrothermaux (apatite, fluorine, scheelite, wolframite) pour la caractérisation des fluides minéralisateurs et l'identification des magmas sources et des processus évolutifs. *Bull. Mineral.*, **108**, 737–44.

Raimbault, L., Baumer, A., Dubru, M. *et al.* (1993) REE fractionation between scheelite and apatite in hydrothermal conditions. *Am. Mineral.*, **78**, 1275–85.

Rekharskiy, V. I. and Rekharskaya, V. M. (1969) The new zirkelite–jordisite mineral paragenesis. *Dokl. Akad. Nauk SSSR*, **184**, 144–7.

Richardson, C. K. and Holland, H. D. (1979a) The solubility of fluorite in hydrothermal solutions – an experimental study. *Geochim. Cosmochim. Acta*, **43**, 1313–25.

Richardson, C. K. and Holland, H. D. (1979b) Fluorite deposition in hydrothermal systems. *Geochim. Cosmochim. Acta*, **43**, 1327–35.

Robie, R. A., Hemingway, B. S. and Fisher, J. R. (1979) Thermodynamic properties of minerals and related substances at 298.15 K and 1 Bar (105 Pascals) pressure and at higher temperatures. *US Geol. Surv. Bull.*, **1452**, 456 pp.

Rubin, J. N., Henry, C. D. and Price, J. G. (1993) The mobility of zirconium and other 'immobile' elements during hydrothermal alteration. *Chem. Geol.*, **110**, 29–47.

Rudashevskiy, N. S. (1969) Epidote–orthite from metasomatites of southern Siberia. *Vses. Mineral. Obshchest. Zap.*, **98/6**, 739–49.

Salvi, S. and Williams-Jones, A. E. (1990) The role of hydrothermal processes in the granite-hosted Zr, Y, REE deposit at Strange Lake, Quebec/Labrador: evidence from fluid inclusions. *Geochim. Cosmochim. Acta*, **54**, 2403–18.

Sanjuan, B., Michard, A. and Michard, G. (1988) Influence of the temperature of CO_2-rich springs on their Al and REE content. *Chem. Geol.*, **68**, 57–68.

Semenov, E. I. and Chang, P.-S. (1961) Huanghoite – a new rare earth mineral. *Sci. Sin.*, **10**, 1007–11.

Seward, T. M. (1981) Metal complex formation in aqueous solutions at elevated temperatures and pressures, in *Physics and Chemistry of the Earth* (eds D. T. Rickard and F. E. Wickman), Vol. **13/14**. Pergamon Press, Oxford, pp. 113–32.

Shi, P. (1992) Fluid fugacities and phase equilibria in the Fe–Si–O–H–S system. *Am. Mineral.*, **77**, 1050–66.

Sinha, A. K., Wayne, D. M., Hewitt, D. A. (1992) The hydrothermal stability of zircon: preliminary experimental and isotopic studies. *Geochim. Cosmochim. Acta*, **56**, 3551–60.

Smith, K. L. and Lumpkin, G. R. (1993) Structural features of zirconolite, hollandite and perovskite, the major waste-bearing phases in Synroc, in *Defects and Processes in the Solid State: Geoscience applications*. The McLaren Volume (eds J. N. Boland and J. D. Fitzgerald), Elsevier, Amsterdam, pp. 401–22.

Sommerauer, J. and Weber, L. (1972) Aeschynit-(Y, Gd, Dy, Er), ein neues Zerrkluftmineral der zentralen Schweizer Alpen. *Schweiz Mineral Petrogr Mitt*, **52**, 75–91.

Sorensen, S. S. (1991) Petrogenetic significance of zoned allanite in garnet amphibolite from a paleosubduction zone: Catalina Schist, southern California. *Am. Mineral.*, **76**, 589–601.

REFERENCES

Staatz, M. H. (1985) Geology and description of the thorium and rare-earth veins in the Laughlin Peak Area, Colfax County, New Mexico. *US Geol. Surv. Prof. Pap.*, **1049-E**, 1–32.

Staatz, M. H. and Conklin, N. M. (1966) Rare earth thorium carbonate veins of the Road Gulch area, northern Wet Mountains, Colorado. *US Geol. Surv. Prof. Pap.*, **550-B**, 130–4.

Staatz, M. H., Conklin, N. M. and Brownfield, I. K. (1977) Rare earths, thorium, and other minor elements in sphene from some plutonic rocks in west-central Alaska. *J. Res. US Geol. Surv.*, **5/5**, 623–8.

Staatz, M. H., Shaw, E. and Wahlberg, J. S. (1972) Occurrence and distribution of rare earths in the Lemhi Pass thorium veins, Idaho and Montana. *Econ. Geol.*, **67**, 72–82.

Styles, M. T. and Young, B. R. (1983) Fluocerite and its alteration products from the Afu Hills, Nigeria. *Mineral. Mag.*, **47**, 41–6.

Sverjensky, D. A. (1984) Europium redox equilibria in aqueous solution. *Earth Planet. Sci. Lett.*, **67**, 70–8.

Vander Auwera, J. and Andre L. (1991) Trace elements (REE) and isotopes (O, C, Sr) to characterize the metasomatic fluid sources: evidence from the skarn deposit (Fe, W, Cu) of Traversella (Ivrea, Italy). *Contrib. Mineral. Petrol.*, **106**, 325–39.

van Wambeke, L. (1977) The Karonge rare earth deposit, Republic of Burundi: new mineralogical–geochemical data and origin of the mineralization. *Mineral. Deposita*, **12**, 373–80.

von Backström, J. W. (1976) Thorium, in *Mineral Resources of the Republic of South Africa* (ed. C. B. Coetzee), pp. 209–12.

von Gehlen, K., Grauert, B. and Nielsen, H. (1986) REE minerals in southern Schwarzwald veins and isotope studies on gypsum from the central Schwarzwald, F. R. Germany. *N. Jb. Mineral. Mhonats.*, **9**, 393–9.

Waber, N. (1992) The supergene thorium and rare earth element deposit at Morro do Ferro, Poços de Caldas, Minas Gerais, Brazil. *J. Geochem. Explor.*, **45**, 113–57.

Wakita, H., Rey, P. and Schmitt, R. A. (1971) Abundances of the 14 rare earth elements and 12 other elements in Apollo 12 samples: five igneous and one breccia rocks and four soils. Proceedings of the Second Lunar Science Conference, *Geochim Cosmochim Acta*, Suppl. 2 (2), 1319–29.

Watson, M. D. and Snyman, C. P. (1975) The geology and the mineralogy of the fluorite deposits at the Buffalo fluor-spar mine on Buffelsfontein, 347KR, Naboomspruit District. *Trans. Geol. Soc. S. Afr.*, **78**, 137–51.

Williams, C. T. and Gieré, R. (1988) Metasomatic zonation of REE in zirconolite from a marble skarn at the Bergell contact aureole (Switzerland/Italy). *Schweiz Mineral. Petrogr. Mitt.*, **68**, 133–140.

Williams-Jones, A. E. and Wood, S. A. (1992) A preliminary petrogenetic grid for REE fluorocarbonates and associated minerals. *Geochim. Cosmochim. Acta*, **56**, 725–738.

Wood, S. A. (1990a) The aqueous geochemistry of rare-earth elements and yttrium 1. Review of available low-temperature data for inorganic complexes and the inorganic REE speciation of natural waters. *Chem. Geol.*, **82**, 159–86.

Wood, S. A. (1990b) The aqueous geochemistry of rare-earth elements and yttrium 2. Theoretical predictions of speciation in hydrothermal solutions to 350°C at saturation water vapor pressure. *Chem. Geol.*, **89**, 99–125.

Wood, S. A. (1993) The aqueous geochemistry of rare-earth elements: critical stability constants for complexes with simple carboxylic acids at 25°C and 1 bar and their application to nuclear waste management. *Eng. Geol.*, **34**, 229–59.

Wood, S. A. and Williams-Jones, A. E. (1994) The aqueous geochemistry of the rare earth elements and ytrrium 4. Monazite solubility and REE mobility in exhalative massive sulfide-depositing environments. *Chem. Geol.*, **115**, 47–60.

Yuan, Z., Bai, G., Wu, C. et al. (1992) Geological features and genesis of the Bayan Obo REE ore deposit, Inner Mongolia, China. *Appl. Geochem.*, **7**, 429–42.

Zakrzewski, M. A., Lustenhouwer, W. J., Nugteren, H. J. and Williams, C. T. (1992) Rare-earth minerals zirconolite-(Y) and allanite-(Ce) and associated minerals from Koberg mine, Bergslagen, Sweden. *Mineral. Mag.*, **56**, 27–35.

Zhuravleva, L. N., Berezina, L. A. and Gulin, Y. E. N. (1976) Geochemistry of rare and radioactive elements in apatite–magnetite ores in alkali–ultrabasic complexes. *Geochem. Int.*, **13/5**, 147–66.

CHAPTER SIX
Rare earth minerals from the syenite pegmatites in the Oslo Region, Norway

A. O. Larsen

6.1 Introduction

The first descriptions of minerals from the nepheline syenite pegmatites in the Langesundsfjord–Larvik area dates back to 1801, and in 1824 the first new mineral, a rare earth-bearing Ca, Ti, Zr oxide, was described by the famous Swedish chemist, J. J. Berzelius, and named polymignite (Berzelius, 1824). This name has recently been discredited in favour of the mineral name zirconolite (Bayliss *et al.*, 1989). Pyrochlore, also a rare earth-bearing mineral, was the next to be described as a new species from Stavern, by F. Wöhler in 1826. Mineralogists then became more aware of the Langesundsfjord area, mainly because the vicar of Brevik, Hans Morten Thrane Esmark, a skilled mineralogist, sent mineral samples to mineralogical societies in Norway, Sweden and Germany. As a result of his interest in minerals and his professional relations with J. J. Berzelius, the latter discovered a new element in samples sent to him by Esmark. The element was named thorium and the mineral named thorite, after the Norse god Thor, by Berzelius in 1829. The source of the mineral was a nepheline syenite pegmatite at Løvøya in the Langesundsfjord. In the following years a number of new minerals, some rare earth and rare earth-bearing (in italics), were described: aegirine (1834), *wöhlerite* (1843), *leucophanite* (1840), *mosandrite* (1841), *tritomite* (1849), *catapleiite* (1849), *meliphanite* (1852), *astrophyllite* (1854) and *homilite* (1876). For a complete list of older literature, see Brøgger (1890).

During the years 1883–1890 W. C. Brøgger worked on his monograph on the syenite pegmatite minerals in the Oslo Region (Brøgger, 1890). During that time he identified several minerals new to the district, and described eight new mineral species from the Langesundsfjord district: *cappelenite* (1885), *låvenite* (1885), nordenskiöldine (1887), *melanocerite* (1887), *rosenbuschite* (1887), eudidymite (1887), *hiortdahlite* (1888) and hambergite (1890). In addition he described three new minerals which later have been

Rare Earth Minerals: Chemistry, origin and ore deposits. Edited by Adrian P. Jones, Frances Wall and C. Terry Williams. Published in 1996 by Chapman & Hall. ISBN 0 412 61030 2

found to be identical with other species: barkevikite (1887) = ferro-edenitic hornblende; johnstrupite (1887) = mosandrite; and caryocerite (1890) = melanocerite. He also supplemented knowledge of the chemistry, crystallography and physical and optical properties of the already known minerals. At the turn of the twentieth century the list of minerals from the syenite pegmatites in the Oslo Region amounted to approximately 75 species. Very little was added to this list until about 1960, when the syenite pegmatites in the Oslo Region again gained interest among a few Norwegian mineralogists. Mineral samples collected 50–75 years ago were re-examined and re-analysed, and new and interesting localities were visited. Around 1970 a group of amateur mineralogists started systematically collecting syenite pegmatite minerals. Because of this enthusiastic interest and a close connection with the geological and mineralogical institutions in Norway, by 1980 the list of minerals had grown to about 120 species (Raade et al., 1980). During the last decade more than 50 species have been added, and for the first time in over 90 years new minerals were described from the Langesundsfjord area (Raade et al., 1983; Larsen et al., 1992). A list of rare earth and rare earth-bearing minerals are presented in Tables 6.1 and 6.2. The syenite pegmatites

Table 6.1 Rare earth minerals from the syenite pegmatites in the Oslo Region, Norway. The chemical formulae are mainly according to Fleischer and Mandarino (1991) and Clark (1993)

Mineral	Chemical formula
Loparite	$(Ce,La,Na,Ca,Sr)(Ti,Nb)O_3$
Fergusonite	$YNbO_4$
Bastnäsite-(Ce)	$(Ce,La)(CO_3)F$
Parisite-(Ce)	$Ca(Ce,La)_2(CO_3)_3F_2$
Calcio-ancylite-(Ce)	$(Ca,Sr)Ce(CO_3)_2(OH) \cdot H_2O$
Ancylite	$SrCe(CO_3)_2(OH) \cdot H_2O$
Xenotime	YPO_4
Monazite	$(Ce,La,Nd,Th)PO_4$
Cerite	$(Ce,Ca)_9(Mg,Fe^{+2})Si_7(O,OH,F)_{28}$
Britholite	$(Ce,Ca)_5(SiO_4,PO_4)_3(OH,F)$
Tritomite	$(Ce,La,Y,Th)_5(Si,B)_3(O,OH,F)_{13}$
Melanocerite	$(Ce,Ca)_5,(Si,B)_3O_{12}(OH,F) \cdot nH_2O$
Gadolinite-(Ce)	$(Ce,La,Nd,Y)_2Fe^{+2}Be_2Si_2O_{10}$
Hingganite	$(Y,Yb,Er)BeSiO_4(OH)$
Tadzhikite	$Ca_3(Y,Ce)_2(Ti,Al,Fe^{+3})B_4Si_4O_{22}$
Mosandrite	$(Ca,Na,Ce)_{12}(Ti,Zr)_2Si_7O_{31}H_6F_4$
Perrierite-(Ce)	$(Ca,Ce,Th)_4(Mg,Fe^{+2})_2(Ti,Fe^{+3})_3Si_4O_{22}$
Chevkinite-(Ce)	$(Ca,Ce,Th)_4(Mg,Fe^{+2})_2(Ti,Fe^{+3})_3Si_4O_{22}$
Allanite	$(Ce,Ca,Y)_2(Al,Fe^{+3})_3(SiO_4)_3(OH)$
Cappelenite-(Y)	$Ba(Y,Ce)_6Si_3B_6O_{24}F_2$
Stillwellite-(Ce)	$(Ce,La,Ca)BSiO_5$
Kainosite	$Ca_2(Y,Ce)_2Si_4O_{12}CO_3 \cdot H_2O$

Table 6.2 Minerals with REE as minor components from the syenite pegmatites in the Oslo Region

Mineral	RE_2O_3 (wt%)	Y_2O_3 (wt%)	References
Senaite		1.5	Larsen (1989)
Zirconolite	7.4–11.3	2.3–2.5	Brøgger (1890); Mazzi and Munno (1983)
Pyrochlore	5–7.3		Brøgger (1890)
Thorite	0.4–1.5	1.6	Brøgger (1890); Farges and Calas (1991)
Titanite	2.0–2.6	0.5–0.6	Brøgger (1890)
Homilite	0.2		Brøgger (1890)
Wöhlerite	0.3–0.7	0.3–0.6	Brøgger (1890); Mariano and Roeder (1989); Mellini and Merlino (1979)
Rosenbuschite	1.9	1.9	Neumann (1962)
Eudialyte	1.3–5.2	0.3–0.7	Brøgger (1890); Bollingberg et al. (1983)
Catapleiite	0.08	0.3	Bollingberg et al. (1983)
Leucophanite	0.6	0.1	This work
Meliphanite	0.3		This work
Apatite	≤6	≤0.4	This work
Hiortdahlite	1.2	0.8	This work
Götzenite	6.5	4.7	This work
Elpidite	0.3	0.5	This work

of the Oslo Region were among the first to be investigated thoroughly by professional mineralogists, and today rank among the most mineral-rich localities in the world, comparable with the more recently discovered and more intensively investigated alkaline and hyperalkaline intrusions such as Mont St Hilaire in Canada, Ilimaussaq in Greenland, and Lovozero and Khibina in the Kola Peninsula, Russia (see also Chapter 12).

6.2 Geological setting

The Oslo Region, a Permian rift system, stretches 200 km from the Langesund–Larvik area in the south to the Mjøsa area in the north. The southernmost part of the region is made up of larvikite (monzonite) and related rocks which occupy about 1000 km² (Figure 6.1). The geology of this area has been described by Oftedahl (1960), Oftedal, Bergstøl and Svinndal (1960) and Oftedahl and Peterson (1978).

The main body of larvikite consists of a rather uniform rock type. However, as the western boundary is approached, essentially on the islands of the Langesundsfjord, the larvikite is extensively penetrated by veins and bodies of a medium-grained and often schistose nepheline syenite. In this area nepheline syenite pegmatite dykes are very abundant. The majority of these occur as small irregular veins, often not particularly coarse grained. Some of the pegmatites, however, form flattened bodies up to 2 m in thick-

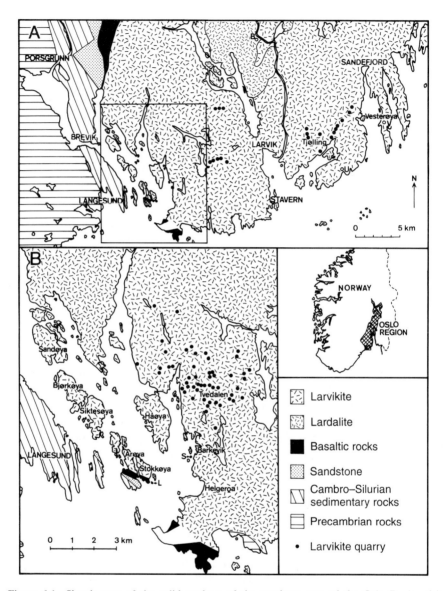

Figure 6.1 Sketch map of the solid geology of the southern part of the Oslo Region (a) and the Langesundsfjord–Tvedalen area (b). At the following localities, mineral collecting is prohibited and the areas are protected by law: L = Låven island, S = Skutesundskjær islet, U = the Ula pegmatite, B = the Bratthagen pegmatite, and V = Vøra pegmatite.

ness. The pegmatites on the islands in the Langesundsfjord have a rather complex mineralogy. The abundance of Zr-, Ti-, Nb- REE- and Be-minerals is conspicuous. At the southern point of Stokkøya and on the neighbouring island Låven, the pegmatites form extensive flattened masses up to several metres in thickness. The island of Låven and the islet of Skutesundskjær near Helgeroa are protected by law because of their interesting mineralogy. This is also the case with the syenite pegmatites at Bratthagen in Lågendalen, at Ula, 9 km southeast of Larvik, and at Vøra on Vesterøya, Sandefjord.

On the mainland east of the Langesundsfjord the number of pegmatite dykes diminishes. In contrast to the pegmatitic veins and lenses in the fjord area, the pegmatites here occur solely as fissure-filling dykes ranging in size from a few centimetres to several metres in thickness. In general, the complexity of minerals is limited, although some pegmatites show a wide variety of mineral species. For example, the huge pegmatite which occurs in the Sage I larvikite quarry at Mørje near Tvedalen has yielded approximately 70 different mineral species. Pegmatite veins and dykes occur over the whole larvikite area stretching east towards Larvik, Sandefjord and Tønsberg.

6.3 The rare earth minerals

There are 22 RE minerals in the syenite pegmatites in the Langesundsfjord district, and a further 10–15 minerals where the rare earths are significant, but minor, elements. Most of the RE minerals are very restricted in occurrence. The RE fluocarbonates, however, are a quite widespread although they always occur in small quantities. Zirconolite, pyrochlore, apatite, wöhlerite and eudialyte, with REE concentrations typically from <1% up to several per cent RE_2O_3, are abundant in the syenite pegmatites on the islands of the Langesundsfjord (pyrochlore, wöhlerite and eudialyte) and on the mainland immediately east of the Langesundsfjord (zirconolite, pyrochlore, apatite and wöhlerite). The following is a brief description of the RE minerals which have been found in the syenite pegmatites in the Oslo Region.

6.3.1 Loparite

Loparite has been found in sparse amounts in the Bratthagen syenite pegmatites in Lågendalen. It is an early-crystallized mineral occurring together with a large suite of accessory minerals (Larsen, Raade and Sæbø, 1993). The mineral occurs as dark brown interpenetration twins up to 10 mm across. Loparite from Bratthagen is metamict. The chemical composition is not yet known.

6.3.2 Fergusonite

Fergusonite has been found in sparse amounts in syenite pegmatite material on the dump from a water tunnel at Buer in Bjørkedalen. Gadolinite-(Ce), described by Segalstad and Larsen (1978a), was discovered in the same dump, although not from the same pegmatite. Fergusonite occurs as millimetre-sized tabular crystals in small vugs together with bavenite and helvite, and is obviously a late-crystallized mineral.

6.3.3 Bastnäsite

Two minerals of the bastnäsite group have been found in the syenite pegmatites in the Oslo Region: bastnäsite-(Ce) and parisite-(Ce). Initially reported from Lille Arøy in the Langesundsfjord by Sverdrup, Bryn and Sæbø (1959), bastnäsite-(Ce) has since been found in several other localities, although mostly in very small amounts. Parisite-(Ce) has been identified at a few localities, sometimes in syntaxic intergrowth with bastnäsite-(Ce). To what extent the two bastnäsite group minerals occur in syntaxy in the various Langesundsfjord district localities remains to be investigated. A few occurrences, however, have been examined (Larsen and Williams, 1993). Bastnäsite-(Ce) is a late-stage, hydrothermal mineral and has been found as well-developed crystals either intergrown with analcime or as isolated crystals in vugs. Crystals of bastnäsite-(Ce) usually have a tabular morphology with bipyramid faces terminated by a pinacoid. Crystals 10–20 mm in size are found in syenite pegmatites in both Tuften and Bjørndalen larvikite quarries in Tvedalen. Aggregates and rosettes of tiny, platy crystals are also observed in Heia larvikite quarry in Tvedalen. In a syenite pegmatite dyke in Treschow quarry in Tvedalen, well-developed bastnäsite-(Ce) crystals have been found lining small vugs in analcime. These vugs are actually negative crystals of a primary, probably gadolinite-group, mineral. This gadolinite mineral is the precursor of bastnäsite-(Ce) which is sometimes intergrown with parisite-(Ce) (Larsen and Williams, 1993). Bastnäsite-(Ce) is reported as an alteration product of chevkinite-(Ce) from Bugården at Sandefjord (Segalstad and Larsen, 1978b), and as an alteration product of apatite group minerals at several localities in the Sandefjord district (S. A. Berge, personal communication, 1993), and at Vøra, Vesterøya, Sandefjord (Raade and Larsen, 1980). The chemical composition of bastnäsite-(Ce) from syenite pegmatites in the Tvedalen area varies within the ranges given by the formula $(La_{0.3-0.5}Ce_{0.4-0.5}Nd_{0.05-0.1})CO_3F$. The other REE, mainly Pr, Sm and Gd, are below 0.05 atomic proportions (Larsen and Williams, in preparation).

6.3.4 Parisite

Brøgger (1890) described parisite in association with a mineral which he called 'weibyeite'. This mineral assemblage was re-examined by Sæbø (1963), and shown to be ancylite-(Ce) and bastnäsite-(Ce).

Parisite-(Ce) has been identified from several localities in the Sandefjord area (Kariåsen, Buer and Vøra at Vesterøya and Nordre Bergan at Råstad), at Sandøy, Hvasser, at Hoftøya, Færder, at several localities in the Tvedalen area (the larvikite quarries at Treschow, Bjørndalen, Tuften, Mørje, Vevja and Sage I), in a syenite pegmatite in a roadcut on highway E-18 at Rønningen in Porsgrunn, and in the southern part of Siktesøya in the Langesundsfjord. Parisite-(Ce) from Bjørndalen and Treschow occurs intergrown with bastnäsite-(Ce), while parisite-(Ce) at Vevja occurs as yellow, hexagonal tabular crystals displaying bipyramid faces terminated by a pinacoid.

The chemical composition of parisite-(Ce) from syenite pegmatites in the Tvedalen area (Treschow, Bjørndalen and Vevja quarries) varies within the range given by the formula $Ca_{0.9-1.0}(La_{0.4-0.7}Ce_{1.0-1.1}Pr_{0.1}Nd_{0.2-0.4}Sm_{0.1-0.2})_{\Sigma 2}(CO_3)_3F_2$ (Larsen and Williams, in preparation).

6.3.5 Calcio-ancylite

Calcio-ancylite-(Ce) was found as a brown, pulverulent material occurring as a decomposition product from an unidentified Ca–REE mineral in a syenite pegmatite at Tuften larvikite quarry in Tvedalen. Identification was confirmed by X-ray diffraction and SEM equipped with EDS. Ca, La, and Ce were the main elements present. The second occurrence of calcio-ancylite-(Ce) was in a syenite pegmatite in the Treschow larvikite quarry in Tvedalen. Here, the mineral occurs as well-developed, pale rose-coloured crystals up to 2 mm in diameter, in vugs associated with epididymite, chiavennite, gadolinite-(Ce), chlorite and zeolites.

6.3.6 Ancylite

'Weibyeite' was described by Brøgger (1890) as a new mineral from a syenite pegmatite on the island Lille Arøy in the Langesundsfjord. Re-examination of Brøgger's original material by Sæbø (1963) showed that 'weibyeite' is ancylite patchily replaced by bastnäsite. Ancylite has been found in a syenite pegmatite at Bratthagen in Lågendalen, at Stålaker larvikite quarry in Tjølling, at Saga I larvikite quarry at Mørje, and at Bjørndalen and Askedalsbotn larvikite quarries in Tvedalen. The mineral has also been found at Vardeåsen and Lysebo. Like the other RE fluocarbonates, ancylite occurs as a very late-crystallized mineral in vugs. The chemical composition is not yet known.

6.3.7 Xenotime

Xenotime, comprising two greenish-brown coloured crystals, was identified by Brøgger (1890) in samples from the Arøy skerries in the Langesundsfjord. No further material has since been found.

6.3.8 Monazite

Monazite occurs in syenite pegmatites at several localities at Vøra near Sandefjord on Vesterøya (where crystals may attain several centimetres in length), Kariåsen, Buer, Bentserød and Husebyåsen. This mineral has also been found in small amounts in larvikite quarries at Bjørndalen, Vevja and Heia in Tvedalen. At Heia monazite occurs as small, round aggregates embedded in analcime together with bastnäsite-(Ce). Sæbø (1966) reports that monazite is relatively abundant in the nepheline syenite pegmatite at Bratthagen in Lågendalen. In the epididymite–eudidymite-bearing nepheline syenite pegmatite on the east side of Vesle Arøy in the Langesundsfjord, monazite occurs as millimetre-sized, well-developed yellow crystals in vugs together with the aforementioned minerals plus albite, polylithionite and analcime. In all localities, monazite is a late-stage hydrothermal mineral. No information on the chemical composition of monazite from the nepheline syenite pegmatites in the Oslo Region is available.

6.3.9 Cerite

Cerite was found at the Saga I larvikite quarry near Tvedalen in 1993. The mineral occurs as tiny crystals in small vugs in analcime associated with calcite, epididymite and second-generation aegirine, biotite and zircon. Cerite has obviously crystallized here as a very late-stage mineral.

Cerite occurs as millimetre-sized, rose- to brownish-red, translucent, blocky crystals exhibiting mainly bipyramidal faces and terminated by a pinacoid. A qualitative analysis using an analytical SEM showed Ce, La, Ca and Si to be the main elements present.

6.3.10 Britholite

Britholite has been found in several syenite pegmatites in the Tvedalen area, at Vesterøya near Sandefjord and at Hedrum quarry near Larvik. Britholite is one of the first minerals to crystallize during the magmatic phase of the pegmatite formation, and is found intergrown with feldspars, hornblende and nepheline. It usually occurs as dark brown hexagonal prisms, but the crystal outlines are not always present.

At some localities the crystals have a core of yellow to yellowish-green coloured fluoapatite. This unusual association was previously noted by

J. F. L. Hausman in 1852 (cited by Brøgger, 1890), and a semiquantitative analysis showed that the mineral was a calcium cerium silicate phosphate. More than 50 years later britholite was described as a new mineral species from south Greenland. Brøgger (1890) made a brief examination of the mineral from Langesundsfjord, and reported it to be more or less isotropic, later confirmed to be metamict. Due to the lack of material, he was unable to do more research on this Ce-rich phosphate mineal, and even today no quantitative chemical data are available on britholite from the syenite pegmatites in the Oslo Region.

6.3.11 Tritomite

Much confusion exists regarding the rear earth-rich borosilicates minerals tritomite, melanocerite and caryocerite. They are very similar to each other, and have been considered as being either identical, or separate species, by different authors. For a brief discussion and review see Miyawaki and Nakai (1987) and Burt (1989). Neumann (1985) and Fleischer and Mandarino (1991) list caryocerite-(Ce) as a thorian melanocerite-(Ce). There is now general agreement that tritomite and melanocerite are boron-substituted members of the apatite group. The minerals are metamict. Recrystallization experiments by heating tritomite and melanocerite in air or under water vapour pressure at 600°–1000°C give X-ray diffraction patterns of either an apatite phase or a CeO_2 phase, or a combination of these (Neumann, 1985). It is highly probable that both tritomite, melanocerite and caryocerite have a common crystalline precursor, a boron-rich apatite group mineral with considerable substitution within the general formula $A_5C_3O_{12}X$, where A is (Ca, Na, REE, Th,); C is (P, Si, B, C), and X is (F, OH, O). A complicating factor for present-day identification is the degree and extent of metamictization, during which elemental leaching and replacement may have taken place.

Chemical analyses of tritomite have been given by Brøgger (1890) and Neumann (1985). The latter author recalculated one of the analyses by Brøgger, together with one of his own analyses, to produce an ideal formula of $(Ce,REE)(Si,B)(O,F)_3$. A third analysis gave a formula close to that of apatite. Neumann (1985) therefore questioned whether tritomite has a unique structure, or whether the material is a mixture of two or more mineral species.

First found in 1849 at the island Låven in the Langesundsfjord tritomite occurs as dark brown tetrahedral crystals in leucophanite or analcime. Masses without crystalline outlines have been found at Stokkøya, Arøya, the Arøy skerries, the Barkevik skerries, Kjeøya and in the larvikite quarries at Håkestad in Tjølling and at Saga I near Tvedalen. Neumann (1985) also mentioned a 'tritomite-like' mineral from one of the larvikite quarries in Tvedalen area, and also from Hjertnesåsen near Sandefjord.

6.3.12 Melanocerite

As discussed above, melanocerite and tritomite are very similar to each other, both chemically and structurally. Like tritomite, melanocerite is probably a Ca-RE-borosilicate of the apatite group. It is metamict, and is dark brown to black in colour. From the original locality at the island Kjeøya near Barkevik on the eastern side of Langesundsfjord, melanocerite-(Ce) occurs as thick tabular crystals with a rhombohedral habit and terminated by a pinacoid. According to Brøgger (1890) melanocerite-(Ce) has a total content of RE_2O_3 of 54.05 wt%. Andersen and Neumann (1985) re-examined the type specimen of 'freyalite', a dubious mineral whose identity has been debated, and found this to be an inhomogeneous thorian melanocerite which has been altered due to metamictization with subsequent leaching of Na, Ca and F.

6.3.13 Gadolinite

Type gadolinite-(Ce) was described from a syenite pegmatite at Buer in Bjørkdalen, east of Porsgrunn, by Segalstad and Larsen (1978a). The syenite pegmatite material was found on the dump from a water tunnel in basaltic rocks close to the margin of the larvikite.

Gadolinite-(Ce) is black, vitreous and completely metamict. Recrystallization by heating the mineral in air gave an X-ray diffraction pattern of a gadolinite phase plus CeO_2. Recrystallization in nitrogen and under hydrothermal conditions at 700°–1000°C produced an X-ray diffraction pattern of a pure gadolinite phase. Gadolinite-(Ce) from Buer is zoned with respect to the REE, with a rim richer in Ce and La, and poorer in Y (25 wt% Ce_2O_3, 16 wt% La_2O_3, 2 wt% Y_2O_3), relative to the core (15 wt% Ce_2O_3, 9 wt% La_2O_3, 14 wt% Y_2O_3).

Black vitreous minerals of the gadolinite group have been found at several syenite pegmatite localities in the Oslo Region (the larvikite quarries at Klåstad, Stålaker and Håkestad in Tjølling, Bratthagen in Lågendalen, the larvikite quarries at Vevja, Treschow, Tuften, Bjørndalen, Auen and Saga I in the Tvedalen area, and at Blåfjell in Langangen). As these minerals have not been analysed, it is not possible to characterize them as being either gadolinite-(Ce) or gadolinite-(Y). However, due to the tendency of the gadolinites to reflect the REE environment in which they crystallized (Segalstad and Larsen, 1978a), gadolinites from the syenite pegmatites in the Oslo Region are most likely to be gadolinite-(Ce).

6.3.14 Hingganite

Hingganite was found in 1991 in a nepheline syenite pegmatite at the Tuften larvikite quarry in Tvedalen. The mineral occurs as millimetre-sized, beige prismatic crystals in small vugs in analcime, and is considered to be a very late-stage mineral. The occurrence of hingganite is locally restricted to a zeolitized part of the pegmatite dyke.

6.3.15 Tadzhikite

Tadzhikite occurs in several nepheline syenite pegmatites in the Barkevik area (the Barkevik skerries, Skutesundskjær, and on the shore on the mainland near Barkevik). The mineral has also been found at Låven and Vesle Arøya in the Langesundsfjord and at the Saga I larvikite quarry near Tvedalen. Tadzhikite usually occurs as thin plates crystallizing in stellate rosettes up to 10 mm across. Tadzhikite is brown in colour, with a pronounced pearly lustre. The chemical composition is not known.

6.3.16 Mosandrite

Mosandrite was described as a new mineral species from the Låven island in the Langesundsfjord in 1840. A similar mineral called johnstrupite was described by Brøgger (1890). Later the related minerals rinkite, rinkolite and lovchorrite have been described from Ilimaussaq, Greenland and Khibina, Kola Peninsula, Russia. The status of all these minerals has been discussed by Petersen, Rønsbo and Leonardsen (1989).

Mosandrite has been found in many nepheline syenite pegmatites in the Langesundsfjord area, but not in localities east of Tvedalen. Notable localities are Låven island, Lille Arøya, the Arøy skerries, Siktesøya, the Barkevik skerries and the Saga I larvikite quarry near Tvedalen. At the latter locality, mosandrite is medium- to dark-brown in colour, and occurs as tabular shaped crystals up to 50 cm long. Mosandrite crystallized early in the nepheline syenite pegmatites in the Langesundsfjord district.

In most of the localities mosandrite has been subjected to hydrothermal alteration, leaving only traces of the original mineral. Alteration products are pyrophanite, magnetite, pyrochlore, fluorite, calcite, aegirine, titanite and lorenzenite (Raade, 1967). Mosandrite from the Langesundsfjord district contains 13.5–16.5 wt% RE_2O_3 and 1.1–3.5 wt% Y_2O_3. Most mosandrites are metamict, due to a small amount of Th in the structure. Petersen, Rønsbo and Leonardsen (1989) noted that the original mosandrite sample from Låven appears to be an intergrowth of two or more minerals with slightly different compositions, and that this material needs re-examination.

6.3.17 Chevkinite and perrierite

Chevkinite-(Ce) and its dimorph, perrierite-(Ce), are found at a few syenite pegmatite localities in the Oslo Region. Segalstad and Larsen (1978b) described chevkinite-(Ce) from three localities (Buer in Bjørkedalen, Stokkøya in the Langesundsfjord, and Bugården at Sandefjord), and perrierite-(Ce) from the Buer in Bjørkedalen. Later, chevkinite was identified from Bentserød and Vøra near Sandefjord, from Valleråsen and Ramsåskollen near Porsgrunn, from Holtehedde in Lågendalen, from Heia larvikite quarry in Tvedalen and from Jahren in Brunlanes.

Both chevkinite-(Ce) and perrierite-(Ce) are black in colour with a vitreous lustre and are completely metamict. On heating at 1000°C in nitrogen, the minerals recrystallize to give their characteristic X-ray diffraction patterns. On heating the minerals in air, however, CeO_2 is expelled as a separate phase (Segalstad and Larsen, 1978b).

Chevkinite-(Ce) and perrierite-(Ce) from the syenite pegmatites in the Oslo Region have concentrations of 47.7–48.0 wt% RE_2O_3 and 0.1–0.4 wt% Y_2O_3. Chevkinite-(Ce) from Bugården at Sandefjord has a La_2O_3/Ce_2O_3 ratio of 0.87, which is lower than for chevkinite-(Ce) and perrierite-(Ce) from other localities (La_2O_3/Ce_2O_3 ratios of 1.15–1.23).

6.3.18 Allanite

Allanite has been found at several syenite pegmatite localities near Sandefjord, at Tuften larvikite quarry at Tvedalen, at Torstein larvikite quarry in Brunlanes, at Buer in Bjørkedalen, and at a road cut on highway E-18 at Slevolden, Eidanger. It occurs as a very late-stage mineral in vugs in feldspars or in zeolites, where the crystals are developed as brown to black coloured tabular prisms, and may attain 5 mm in length. The chemical composition is not accurately known.

6.3.19 Cappelenite

Cappelenite-(Y) was described by Brøgger (1890). Only a few specimens of this mineral have been found in a small nepheline syenite pegmatite dyke on the eastern side of the island of Vesle Arøy in the Langesundsfjord. Cappelenite-(Y) from this locality is yellow-brown in colour, one crystal shows well-developed hexagonal prism faces, modified by bipyramids and terminated by a small pinacoid. Cappelenite-(Y) is completely metamict, but recrystallization above 700°C produces the characteristic X-ray diffraction pattern. In this way Shen and Moore (1984) were able to obtain a single crystal on which they performed a crystal structure analysis. Based on the analysis by Brøgger (1890), the chemical formula for cappelenite-(Y) is $BaY_6Si_3B_6O_{24}F_2$ (Shen and Moore, 1984). Cappelenite-(Y) is one of the very few yttrian minerals from the nepheline syenite pegmatites in the Oslo Region.

6.3.20 Stillwellite

Stillwellite-(Ce) was described from the Langesundsfjord area by Neumann, Bergstøl and Nilssen (1966). This mineral was identified in two samples originally acquired by the Geological Museum in Oslo in 1898 and 1906. The locality of these samples was given as the Barkevik skerries at the eastern side of the Langesundsfjord; however no further samples have been found.

Stillwellite-(Ce) occurs as brownish-yellow trigonal prisms with a maximum size of 20 mm by 3 mm, intergrown with feldspar. Associated minerals include feldspar, nepheline, zircon, fluorite, aegirine, pyrophanite and pyrochlore. Stillwellite-(Ce) gives an X-ray diffraction pattern with slightly diffuse lines, indicating incipient metamictization. A semiquantitative X-ray fluorescence analysis of the REE showed 25 wt% La_2O_3 and 30 wt% Ce_2O_3 and minor amounts of Pr, Nd and Sm, together with 0.5 wt% ThO_2.

6.3.21 Kainosite

Kainosite has been found only once in a syenite pegmatite material at the Bjørndalen larvikite quarry in Tvedalen. The mineral occurs as beige-coloured radiating masses up to 1 mm across embedded in a calcite matrix. Associated minerals include lorenzenite and pyrophanite. The pegmatite comprises microcline, nepheline and aegirine as the main minerals, plus abundant mosandrite (partly decomposed), pyrochlore and fluorite. Kainosite probably crystallized as a result of hydrothermal decomposition of mosandrite.

6.3.22 Other rare earth-bearing minerals

The REE play a subordinant role in many minerals from the nepheline syenite pegmatite minerals in the Oslo Region, as shown in Table 6.2. However, very little is known about the distribution of individual REE in the different minerals. Complete analyses have been carried out on only a few of the listed minerals, namely rosenbuschite (Neumann, 1962), eudialyte and catapleiite (Bollingberg *et al.*, 1983).

6.4 Concluding remarks

The nepheline syenite pegmatites in the Oslo Region host a variety of RE minerals and rare earth-bearing minerals. Unfortunately, very little recent work has been done on these minerals using modern analytical equipment and techniques. Therefore their chemical compositions are often not accurately known. Even less work has been done on the systematic geochemical and petrological contexts of the RE minerals from the classical Langesundfjord localities. It is hoped that these shortcomings will be addressed; suggestions for future studies on minerals from the nepheline syenite pegmatites in the Oslo Region are.

1. chemical analyses of the more common minerals such as apatite, britholite, pyrochlore, eudialyte and wöhlerite, together with a geographical and geochemical correlation with the magmatic evolution of the larvikite complex as suggested by Petersen (1978);
2. chemical analyses and X-ray diffraction studies on melanocerites, tritomites and 'caryocerites' in order to establish their compositions and relationships to the apatite group minerals;
3. chemical analyses and X-ray diffraction studies on minerals of the gadolinite group which often are found in association with meliphanite;
4. chemical analyses of mosandrites and a re-examination of the original mosandrite material, as suggested by Petersen, Rønsbo and Leonardsen (1989), and establishment of a nomenclature for the mosandrite/rinkite minerals;
5. confirmation of the chemical composition of the RE minerals which have been found during the last two or three decades, such as loparite, fergusonite, ancylite, monazite, cerite, tadzhikite, allanite and kainosite.

Acknowledgements

I thank S. A. Berge for valuable information on the mineral localities. Thanks are also due to A. Jenkins for correcting the English text.

References

Andersen, T. and Neumann, H. (1985) Identity of 'freyalite', an alleged rare earth-rich variety of thorite, and its pre-metamict composition. *Am. Mineral.*, **70**, 1059–64.

Bayliss, P., Mazzi, F., Munno, R. and White, T. J. (1989) Mineral nomenclature: zirconolite. *Mineral. Mag.*, **53**, 565–9.

Berzelius, J. J. (1824) Undersökning af några mineralier. *Kon. Vetenskaps Akad. Stockh. Handl.* 334–58.

Bollingberg, H. J., Ure, A. M., Sorensen, I. and Leonardsen, E. S. (1983) Geochemistry of some eudialyte–eucolite specimens and coexisting catapleiite from Langesund, Norway. *Tschermaks Min. Petr. Mitt.*, **32**, 153–69.

Brøgger, W. C. (1890) Die Mineralien der Syenitpegmatitgänge der südnorwegischen Augit- und Nephelinsyenite. *Z. Krist.*, **16**, 1–898.

Burt, D. M. (1989) Compositional and phase relations among rare earth element minerals in *Geochemistry and Mineralogy of Rare Earth Elements* (eds B. R. Lipin and G. A. McKay, *Reviews in Mineralogy*, **21**, Mineralogical Society of America, pp. 259–307.

Clark, A.M. (1993) *Hey's Mineral Index*, Natural History Museum Publications, Chapman and Hall, London, 851 pp.

Farges, F. and Calas, G. (1991) Structural analysis of radiation damage in zircon and thorite: an X-ray absorption spectroscopic study. *Am. Mineral.*, **76**, 60–73.

REFERENCES

Fleischer, M. and Mandarino, J. A. (1991) *Glossary of Mineral Species*, Mineral. Rec. Inc.

Larsen, A. O. (1989) Senaite from syenite pegmatite at Tvedalen in the southern part of the Oslo Region, Norway. *Nor. Geol. Tidsskr.*, **69**, 235–8.

Larsen, A. O. and Williams, C. T. (1993) Co-existing bastnäsite-(Ce) and parisite-(Ce) from the Langesundfjord district, Norway. *Rare Earth Minerals: Chemistry, Origin and Ore Deposits*, Mineralogical Society Spring Meeting, Natural History Museum. London, U.K. Conference Extended Abstracts, pp. 64–5.

Larsen, A. O., Raade, G. and Sæbø, P. C. (1993) Lorenzenite from the Bratthagen nepheline syenite pegmatites, Lågendalen, Oslo Region, Norway. *Nor. Geol. Tidsskr.*, **72**, 381–4.

Larsen, A. O., Åsheim, A., Raade, G. and Taftø, J. (1992) Tvedalite, $(Ca,Mn)_4Be_3Si_6O_{17}(OH)_4 \cdot 3H_2O$, a new mineral from syenite pegmatite in the Oslo Region, Norway. *Am. Mineral.*, **77**, 438–43.

Mariano, A. N. and Roeder, P. L. (1989) Wöhlerite: chemical composition, cathodoluminescence and environment of crystallization. *Can. Mineral.*, **27**, 709–20.

Mazzi, F. and Munno, R. (1983) Calciobetafite (new mineral of the pyrochlore group) and related minerals from Campi Flegrei, Italy; crystal structures of polymignite and zirkelite: comparison with pyrochlore and zirconolite. *Am. Mineral.*, **68**, 262–76.

Mellini, M. and Merlino, S. (1979) Refinement of the crystal structure of wöhlerite. *Tschermaks Min. Petr. Mitt.*, **26**, 109–23.

Miyawaki, R. and Nakai, I. (1987) Crystal structures of rare-earth minerals. *Rare Earths*, **11**, 1–133.

Neumann, H. (1962) Rosenbuschite and its relation to götzenite. Contributions to the mineralogy of Norway, No. 13. *Nor. Geol. Tidsskr.*, **42**, 179–86.

Neumann, H. (1985) Norges mineraler. *Norges Geologiske Undersøkelse, skrifter*, **68**, 1–278.

Neumann, H., Bergstøl, S. and Nilssen, B. (1966) Stillwellite in the Langesundfiord nepheline syenite pegmatite dykes. Contributions to the mineralogy of Norway, No. 34. *Nor. Geol. Tidsskr.*, **46**, 327–34.

Oftedahl, C. (1960) Permian igneous rocks of the Oslo Graben, Norway. *XXI Int. Geol. Congess, Norden. Guidebook* **i**, 1–23.

Oftedahl, C. and Peterson, J. S. (1978) Southern part of the Oslo rift. *Norges Geologiske Undersøkelse*, **337**, 163–82.

Oftedal, I. W., Bergstøl, S. and Svinndal, S. (1960) The Larvik–Langesund and the Fen areas. *XXI Int. Geol. Congress, Norden. Guidebook*, **k**, 1–17.

Petersen, J. S. (1978) Structure of the larvikite–lardalite complex, Oslo-region, Norway, and its evolution. *Geol. Rundschau*, **67**, 330–42.

Petersen, O. V., Rønsbo, J. G. and Leonardsen, E. S. (1989) Nacareniobsite-(Ce), a new mineral species from the Ilimaussaq alkaline complex, South Greenland, and its relation to mosandrite and the rinkite series. *N. Jb. Miner. Monatsh.*, **1989**, 84–96.

Raade, G. (1967) Ramsayite as an alteration product of mosandrite. *Nor. Geol. Tidsskr.*, **47**, 249–50.

Raade, G. and Larsen, A. O. (1980) Polylithionite from syenite pegmatite at Vøra, Sandefjord, Oslo Region, Norway. Contributions to the mineralogy of Norway, No. 65. *Nor. Geol. Tidsskr.*, **60**, 117–24.

Raade, G. Haug, J., Kristiansen, R. and Larsen, A. O. (1980) Langesundsfjord. *Lapis*, **5**, 22–8.

Raade, G., Åmli, R., Mladeck, M. H. et al. (1983) Chiavennite from syenite pegmatites in the Oslo Region, Norway. *Am. Mineral.*, **68**, 628–33.

Sæbø, P. C. (1963) The identity of weibyeite. Contributions to the mineralogy of Norway, No. 20. *Nor. Geol. Tidsskr.*, **43**, 441–3.

Sæbø, P. C. (1966) The first occurrence of the rare mineral barylite, $Be_2BaSi_2O_7$, in Norway. Contributions to the mineralogy of Norway, No. 35. *Nor. Geol. Tidsskr.*, **46**, 335–48.

Segalstad, T. V. and Larsen, A. O. (1978a) Gadolinite-(Ce) from Skien, southwestern Oslo region, Norway. *Am. Mineral.*, **63**, 188–95.

Segalstad, T. V. and Larsen, A. O. (1978b) Chevkinite and perrierite from the Oslo region, Norway. *Am. Mineral.*, **63**, 499–505.

Shen, J. and Moore, P. B. (1984) Crystal structure of cappelenite, $Ba(Y,RE)_6[Si_3B_6O_{24}]F_2$: a silicoborate sheet structure. *Am. Mineral.*, **69**, 190–5.

Sverdrup, T. L., Bryn, K. Ø. and Sæbø, P. C. (1959) Bastnäsite, a new mineral for Norway. Contributions to the mineralogy of Norway, No. 2. *Nor. Geol. Tidsskr.*, **39**, 237–47.

CHAPTER SEVEN
Rare earth element mineralization in peralkaline systems: the T-Zone REE–Y–Be deposit, Thor Lake, Northwest Territories, Canada

R. P. Taylor and P. J. Pollard

7.1 Introduction

The Thor Lake rare-metal (Be, Y, REE, Nb, Ta, Zr, Ga) deposits are located approximately 100 km southeast of Yellowknife, Northwest Territories and about 5 km north of the Hearne Channel of Great Slave Lake (Figure 7.1). The rare-metal deposits at Thor Lake are located within hydrothermally altered and brecciated peralkaline granite and syenite. Together, the peralkaline granite and syenite comprise the youngest known intrusive phase of the Blatchford Lake igneous complex (Davidson, 1978), which is Early Proterozoic (Aphebian) in age and is interpreted as having formed during the development of the Authapuscow aulacogen that underlies the eastern arm of Great Slave Lake (Davidson, 1982; Hoffman et al., 1977). Five zones of rare-metal mineralization have been identified: the Fluorite, Lake, R-, S- and T-Zones (Trueman et al., 1988). The T-Zone deposit, which contains economically significant grades of REE, Y and Be, forms the focus of this study.

Although the igneous rocks of the Blatchford Lake complex that host the Thor Lake rare-metal deposits have been the subject of intensive study (Davidson, 1978, 1982), very little is known about the origin of the rare-metal deposits and their associated hydrothermal alteration. The geological and geochemical data summarized here form a part of an integrated study designed to identify the origin of the ore-forming fluid responsible for rare earth element (REE) and Be mineralization in the T-Zone deposit, and to determine the conditions of ore deposition and hydrothermal alteration.

Rare Earth Minerals: Chemistry, origin and ore deposits. Edited by Adrian P. Jones, Frances Wall and C. Terry Williams. Published in 1996 by Chapman & Hall. ISBN 0 412 61030 2

7.2 Geological setting of the Blatchford Lake igneous complex

The Blatchford Lake igneous complex comprises a multiphase, subcircular ring complex approximately 23 km in diameter, emplaced into a terrane of Archean metasedimentary and plutonic rocks in the southern part of the Slave Province (Figure 7.1). Radiometric dating studies show that the igneous complex is Early Proterozoic in age and was intruded at circa 2150 Ma (see later section on geochronology for details), and such an age is corroborated by the regional field relationships which suggest that it is older than the Aphebian Great Slave Supergroup that underlies the eastern arm of Great Slave Lake directly to the south (Davidson, 1978). A simplified geological map (Figure 7.1), which summarizes information contained in a series of 1:50 000 scale maps (Davidson, 1981), illustrates the setting.

The igneous complex itself is made up of several distinctive and successively intruded plutonic phases, which are from oldest to youngest: (1) Caribou Lake Gabbro, a marginal unit that is confined to the western margin of the complex; (2) Whiteman Lake Quartz Syenite, which ranges from syenite to granite in composition; (3) Hearne Channel Granite and Mad Lake Granite, which are probably comagmatic, and are aluminous hornblende–biotite granites; (4A) Grace Lake Granite, which is the largest unit of the igneous complex, underlying an area of 155 km² north of Hearne Channel and is a hypersolvus arfvedsonite–alkali feldspar granite of per-

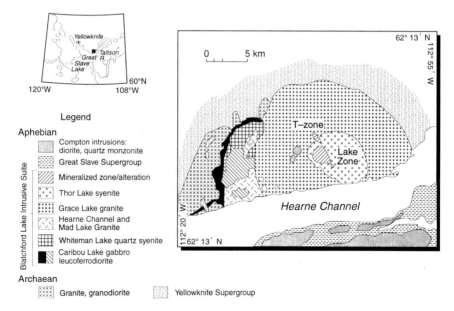

Figure 7.1 Geological map of the Blatchford Lake igneous complex showing the location of the T-Zone and Lake Zone rare-metal deposits (modified from Davidson, 1982).

alkaline composition; (4B) Thor Lake Syenite which, like the Grace Lake Granite, has a peralkaline composition. The Thor Lake Syenite occupies a roughly oval area of 30 km^2 in the centre of the Grace Lake Granite intrusion. The contact between these two peralkaline phases is gradational across a few metres, and is represented simply by a decrease in the quartz content from granite to syenite, with no accompanying change in grain size, nor any evidence of discordance.

Both the peralkaline granite and syenite are cut by ENE-trending diabase dykes. Davidson (1982, pp. 72–3) considered the Blatchford Lake igneous complex as consisting of two parts 'the earlier, western part including the Caribou Lake Gabbro, Whiteman Lake Quartz Syenite, Hearne Channel Granite and Mad Lake Granite; and the later, dominating Grace Lake Granite with its Thor Lake Syenite core'. It is within the latter two units that the Thor Lake rare-metal deposits are located (Figure 7.1).

Although the geochronological database provided in Wanless et al. (1979) places some constraints upon the timing of magmatic events in the Blatchford Lake complex, very little is known about the age of rare-metal mineralization. Two U–Pb zircon ages, for the Hearne Channel Granite (2175 ± 7 Ma) and the Thor Lake Syenite (2094 ± 10 Ma), published by Bowring, Van Schmus and Hoffman (1984) provide some support for the results of the K–Ar study. However, the specimen of Thor Lake Syenite from which zircons were separated is an altered sample from the Lake Zone (coordinates 62°6′N, 112°35′W) comprising '... aggregates of radiating albite crystals with cross-cutting veins of carbonate and zircon' (Bowring, Van Schmus and Hoffman, 1984, p. 1320). It is therefore evident that the 2094 Ma age records the timing of a hydrothermal, rather than a magmatic, event. This latter event involved the late-stage deposition of carbonate and zircon in the vicinity of the Lake Zone rare-metal deposit. Recently, U–Pb zircon and monazite analyses (Sinclair, Hunt and Birkett, 1994) have yielded an age of 2176.2 ± 1.3 Ma for the crystallization of the Grace Lake Granite. This age is indistinguishable from the U–Pb zircon age of 2175 Ma for the Hearne Channel Granite, but significantly older (c. 80 Ma) than the U–Pb age for hydrothermal zircon in cross-cutting veins from the Lake Zone.

Both the Grace Lake Granite and Thor Lake Syenite have agpaitic indices >1.0 and acmite in their CIPW norms. They are characterized by having high contents of F, Zr, Y and Nb (Davidson, 1982; Trueman et al., 1988). Their peralkalinity distinguishes them from the other intrusive rocks in the Blatchford Lake complex which are all aluminous in nature. Peralkaline granites and syenites typically are developed in rift or proto-rift (i.e. intracratonic) settings such as Nigeria (Bowden, 1985), the Oslo Graben (Neumann, 1978) and parts of the Arabian Shield (Stoeser, 1986). It is in such environments that the more evolved peralkaline units often host large-tonnage, disseminated deposits of Nb, Ta, Sn, Y, Th, U, Zr and REE (cf., Kinnaird, 1985; Jackson, 1986).

7.3 The Thor Lake rare-metal deposits

All of the rare-metal deposits, with the exception of the T-Zone, occur within the western part of the Thor Lake Syenite. The T-Zone deposit is unique in that it crosscuts the northwestern contact of the Thor Lake syenite and extends into the Grace Lake Granite (Figure 7.1), a relationship that has been interpreted as supporting an age postdating that of the emplacement of both of the peralkaline intrusive phases. Such an age relationship probably applies to all the zones of rare-metal mineralization in the area (Trueman, Pedersen and de St Jorre, 1984).

The T-Zone rare-metal deposit, which is the focus of the present study, trends NNW from the northern margin of the Lake Zone. It is an irregular feature, with a dyke-like configuration, which extends for 1 km and varies in width up to 250 m (Figure 7.2). The vertical extent of mineralization in the T-Zone has been verified to a depth of 150 m. Contacts between the rare-metal deposit and the host intrusive rocks are sharp. The T-Zone has been subdivided by Trueman, Pedersen and de St Jorre (1984) into two separate areas of rare-metal mineralization (Figure 7.2): the subcircular body at the northern end ('North T-Zone deposit'), and an irregular body which crops

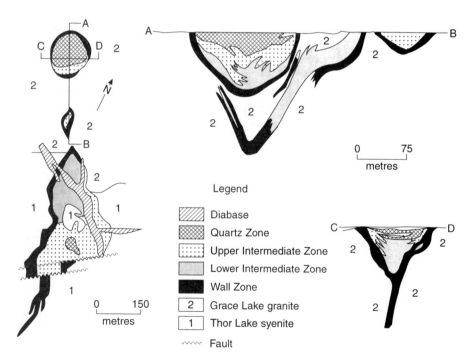

Figure 7.2 Geological map of the T-Zone deposit (left); cross-sections AB and CD (right) are from the North T-Zone REE–Y–Be deposit (after Trueman *et al.*, 1988).

out at the southern end ('South T-Zone deposit'). Both rare-metal deposits appear to be connected at depth. Alteration in the T-Zone rare-metal deposit is strongly zonal and has been subdivided by de St Jorre (1986) into four distinct lithologies designated the Wall, Lower Intermediate, Upper Intermediate and Quartz zones. The distribution of these alteration zones is shown in Figure 7.2.

Rare-metal mineralization in the two parts of the T-Zone (Highwood Resources Ltd., Annual Report, 1986) comprises 0.51 million tonnes at 1.11% BeO, 0.17% Y_2O_3, 0.28% RE_2O_3, and 0.58% Nb_2O_5 (North T-Zone); and 1.25 million tonnes at 0.62% BeO, 0.1% Y_2O_3, 0.2% RE_2O_3, and 0.46% Nb_2O_5 (South T-Zone).

7.4 The north T-Zone REE–Y–Be deposit

The outermost part of the T-Zone deposit, termed the Wall Zone (Figure 7.2), consists of a series of units dominated by microcline, albite and quartz. A series of banded aplites occur together with a massive albitite, and a so-called 'breccia' comprising very coarse-grained microcline crystals set in a quartz matrix. Columbite is a common accessory mineral in the Wall Zone. High gallium values (up to 4000 ppm) are associated with the feldspars, in particular the albite (de St Jorre and Smith, 1988).

The rocks of the Lower Intermediate Zone are in sharp contact with the rocks of the underlying Wall Zone, but are in gradational contact and interlayered with the rocks of the overlying Upper Intermediate Zone (Figure 7.2). The Lower Intermediate Zone is distinguished from the other alteration zones by the presence of partially altered xenoliths of the peralkaline granite and syenite host rocks, and by the mafic character of its rocks; this due to the abundance of Fe-rich minerals. Quartz, biotite, chlorite, feldspar and magnetite are the predominant minerals in the Lower Intermediate Zone. Phenakite and columbite are the principal minerals of economic interest in the Lower Intermediate Zone, whilst REE-fluocarbonates, Ga-bearing albite and zircon are locally present in lesser amounts (de St Jorre, 1986).

Typically, the rocks of the Upper Intermediate Zone have gradational boundaries and are intercalated with rocks of the underlying Lower Intermediate Zone and overlying Quartz Zone (Figure 7.2). Rocks of the Upper Intermediate Zone are the most diverse, both texturally and mineralogically, of any of the alteration zones in the T-Zone rare-metal deposit. The major minerals that are present are aegirine, K-feldspar, albite, REE-fluocarbonates, quartz, polylithionite, biotite, chlorite, magnetite, fluorite, thorite and phenakite. Important accessory minerals include xenotime, gadolinite, zircon, columbite, monazite, carbonates, Fe–Ti oxides and various sulphides.

The Quartz Zone occupies the core of the T-Zone rare-metal deposit. Generally, this zone of massive, vuggy quartz appears to have a gradational contact with the Upper Intermediate Zone. REE-fluocarbonates, fluorite, sulphides (sphalerite and pyrite) and carbonate minerals (ankerite and siderite) are present in minor amounts.

Many of the rocks from the Lower and Upper Intermediate Zones show evidence of the partial to complete replacement of pre-existing minerals or mineral assemblages. However, these zones (together with the rocks of the Wall Zone and the Quartz Zone) appear to have formed mainly by mineral precipitation into open space, together with alteration/replacement of some of the earliest minerals (e.g. aegirine and/or albite) during deposition of later mineral phases. At the deposit scale, the timing relationships indicate that the Wall Zone formed at a early stage and was followed successively by the development of the Lower Intermediate Zone, the Upper Intermediate Zone and the Quartz Zone.

The most abundant REE-bearing minerals in the North T-Zone are the fluocarbonates of the bastnäsite group (Plate 3a,b) which comprise more than 90% of the light REE (La–Sm) enrichment in the T-Zone rare-metal deposit (de St Jorre, 1986). Although the REE-fluocarbonates occur as accessory minerals in all the major lithologies in the Upper Intermediate Zone, they are most abundant in the quartz–bastnäsite unit defined by de St Jorre (1986). This latter unit is located at the upper boundary of the Upper Intermediate Zone, where it is intercalated with rocks of the Quartz Zone. Textural relationships indicate that the REE-fluocarbonates were amongst the earliest of the hydrothermal minerals to have been deposited in the Upper Intermediate Zone.

Phenakite (Be_2SiO_4) is the most abundant beryllium mineral in the T-Zone deposit, and xenotime (YPO_4) is the major Y-bearing mineral of economic interest. The distribution of phenakite and xenotime, like that of the principal REE-bearing minerals, the fluocarbonates, is irregular and results from the wide diversity of rock types present in the Upper Intermediate Zone. Regardless of the nature of its host lithology in the Upper Intermediate Zone, phenakite is always amongst the very last of the hydrothermal minerals to have formed. Phenakite commonly occurs together with fluorite, and in places they enclose, or partially replace, earlier mineral phases including the REE-fluocarbonates (Plate 3c). Locally, other hydrothermal assemblages (e.g. ankerite + sulphides) also overprint and partially replace the REE-fluocarbonate mineralization (Plate 3d).

In economic terms, the Upper Intermediate Zone contains the most significant enrichments of REE, Y and Be in the North T-Zone deposit (de St Jorre, 1986). For example, a rare earth fluocarbonate-rich subzone of the Upper Intermediate Zone contains about 60 000 tonnes grading 8% RE_2O_3 (Sinclair, Jambor and Birkett, 1992). Representative partial chemical analyses of the most important rock types in the Upper Intermediate Zone are presented in Table 7.1, and chondrite-normalized REE plots of adjacent REE-fluocarbonate and mica-rich assemblages are presented in Figure 7.3. Inspection of the chemical data in Table 7.1 shows that mica-rich lithologies

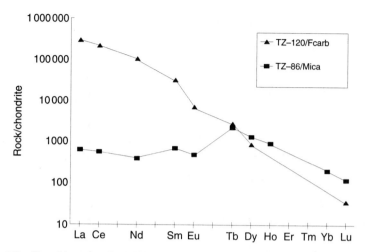

Figure 7.3 Chondrite-normalized REE patterns for adjacent REE-fluocarbonate (TZ-120) and polylithionite mica (TZ-86) assemblages from the Upper Intermediate Zone of the North T-Zone rare-metal deposit. (The data for whole-rock samples are from Table 7.1; the REE data are normalized to the chondritic values of Taylor and McLennan, 1985.)

in the Upper Intermediate Zone (e.g. sample TZ-86) are enriched in Li and Th because of the abundance of polylithionite and thorite, respectively, but have relatively low contents of REE (Fig. 7.3). In contrast, rocks containing the REE-fluocarbonate mineralization (e.g. TZ-120) have lower concentrations of most elements, with the obvious exception of the light REE.

Minerals of the REE-fluocarbonate group, identified in this study by XRD and SEM-EDS (scanning electron microscope equipped with an energy dispersive spectrometer), are in order of decreasing abundance: bastnäsite $(REE)(CO_3)F$, parisite $(REE)_2Ca(CO_3)_3F_2$ and synchysite $(REE)Ca(CO_3)_2F$. In hand specimen the REE-fluocarbonates, which typically occur in massive form or as mineral pseudomorphs, have a distinctive brick-red colour that permits their easy identification (Plate 3b). In thin section the REE-fluocarbonates can be seen to occur typically as complex intergrowths (Figure 7.4a). Grain size is extremely variable, from $<10\,\mu m$ to $>1\,mm$, although single crystals occur only rarely. Most grains are in fact 'polycrystals', syntactic intergrowths of two or perhaps three species (Donnay and Donnay, 1953), with the most common combination being that of bastnäsite and parisite (Figure 7.4b). The syntactic intergrowths display a variety of crystal habits and textures. The most common habits are as tabular or acicular grains with subhedral form. The polycrystals can occur singly, or in massive aggregates that often display radial or plumose textures (Figure 7.4c). Typically, any combination of the latter can occur as components of mineral pseudomorphs whose outlines suggest the replacement of pre-existing alkali pyroxene or amphibole (Figure 7.4a and d).

Table 7.1 Partial chemical analyses of whole-rock samples from the North T-Zone rare-metal deposit

Lithology	UIZ	UIZ	UIZ	WZ	LIZ	QZ
Mineralogy	RF, qtz fl, ank	RF, qtz fl, ank	poly, kfs qtz, thor fl	kfs, ab qtz, col	qtz, bt, kfs, mag ph, col, zr	qtz, RF fl, ank sp, py
Sample	TZ-120	TZ-121	TZ-86	#7	#17	#16
F (wt%)	6.7	7.0	13.6	0.03	1.48	0.18
Li (ppm)	72	128	4800	5	220	18
Be (ppm)	n.a.	n.a.	n.a.	14	800	135
Cs (ppm)	<10	<10	2.9	n.a.	n.a.	n.a.
Ga (ppm)	n.a.	n.a.	n.a.	290	25	1
Y (ppm)	n.a.	n.a.	n.a.	70	1130	133
Zr (ppm)	n.a.	n.a.	n.a.	20	1103	20
Nb (ppm)	n.a.	n.a.	n.a.	579	4890	20
Ta (ppm)	<6	<6	31	3	22	2
Th (ppm)	776	611	1451	31	1702	37
U (ppm)	<25	<25	25	8	130	1
La (wt%)	11.5	8.8	0.026	0.004	0.037	0.001
Ce (wt%)	22.8	19.0	0.062	0.006	0.109	0.005
Nd (wt%)	7.8	6.0	0.030	n.a.	n.a.	n.a.
Sm (wt%)	0.73	0.59	0.016	n.a.	n.a.	n.a.
Eu (ppm)	609	456	42	n.a.	n.a.	n.a.
Tb (ppm)	154	117	123	n.a.	n.a.	n.a.
Dy (ppm)	309	229	469	n.a.	n.a.	n.a.
Ho (ppm)	<150	<150	84	n.a.	n.a.	n.a.
Yb (ppm)	<20	<35	51	n.a.	n.a.	n.a.
Lu (ppm)	1.4	1.2	4.7	n.a.	n.a.	n.a.

Abbreviations: UIZ = Upper Intermediate Zone; WZ = Wall Zone; LIZ = Lower Intermediate Zone; QZ = Quartz Zone; RF = REE-fluocarbonate; qtz = quartz; fl = fluorite; ank = ankerite; poly = polylithionite; kfs = K-feldspar; thor = thorite; ab = albite; col = columbite; bt = biotite; mag = magnetite; ph = phenakite; zr = zircon; sp = sphalerite; py = pyrite; n.a. = not analyzed. Analyses TZ-120, TZ-121, and TZ-86 (this study); #7, #17, and #16 (Trueman *et al.*, 1988). F by specific ion electrode; Li and Be by atomic absorption spectrometry; Ga, Y, Zr and Nb by X-ray fluorescence spectrometry; and Cs, Ta, Th, U and REE by instrumental neutron activation analysis.

Chemical analyses of the REE-fluocarbonate minerals present in the rocks of the Upper Intermediate Zone are presented in Table 7.2. A JEOL JSM-6400 SEM equipped with a LINK SYSTEMS eXL energy dispersive microanalysis system (EDS) was used for the quantitative analysis of the REE-fluocarbonates. Analytical data, determined during the course of this study, for REE-bearing standards are summarized in Table 7.3.

Because of the absence of calcium layers in its structure, bastnäsite has the highest REE contents of any of the fluocarbonate species. In bastnäsite from the Upper Intermediate Zone ΣRE_2O_3 contents of 71–74 wt% are the norm, with the light REE (La–Sm) typically being the only ones that occur

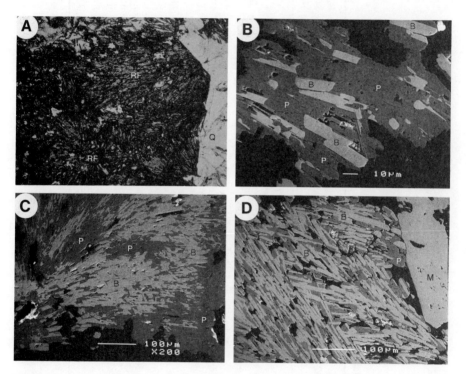

Figure 7.4 (a) Photomicrograph of REE-fluocarbonate polycrystals in mineral pseudomorph (Plane polarized light, field of view = 5 mm). (b–d) Backscattered electron images that illustrate the variety of textures displayed by the REE-fluocarbonate polycrystals in mineralized rocks of the Upper Intermediate Zone. Abbreviations are the same as in Table 7.2.

Table 7.2 Compositional data for REE-fluocarbonate minerals from the Upper Intermediate Zone

	BAS ($n=5$) EU Mean (1σ)	BAS ($n=5$) AN/INC (PY) Mean (1σ)	BAS ($n=6$) AN/PS Mean (1σ)	BAS ($n=7$) SUB/INC (PA) Mean (1σ)	BAS ($n=8$) EU Mean (1σ)	BAS ($n=12$) SUB/PS/PL Mean (1σ)	BAS ($n=7$) SUB/PS Mean (1σ)	BAS ($n=14$) SUB/PS Mean (1σ)
				Concentrations (wt%)				
CaO	0.61 (0.35)	0.34 (0.19)	0.20 (0.15)	0.05 (0.05)	0.18 (0.15)	0.16 (0.16)	0.24 (0.27)	0.18 (0.03)
SrO	1.20 (0.32)	0.63 (0.17)	0.29 (0.21)	0.21 (0.19)	0.48 (0.51)	0.20 (0.16)	0.23 (0.16)	0.41 (0.15)
FeO	n.d.	0.13 (0.11)	n.d.	n.d.	n.d.	n.d.	n.d.	n.d.
La_2O_3	18.29 (0.75)	18.63 (1.64)	19.18 (0.24)	19.36 (0.96)	17.06 (1.11)	18.06 (1.56)	19.44 (1.01)	19.01 (0.26)
Ce_2O_3	36.21 (0.33)	35.61 (0.71)	36.54 (0.37)	36.94 (0.49)	37.02 (0.54)	36.11 (0.75)	37.53 (0.35)	36.60 (0.30)
Pr_2O_3	3.58 (0.22)	3.65 (0.15)	3.58 (0.20)	3.72 (0.28)	3.94 (0.27)	3.74 (0.36)	3.68 (0.37)	3.63 (0.35)
Nd_2O_3	11.98 (0.61)	12.82 (1.16)	12.37 (0.18)	12.81 (0.77)	14.36 (0.88)	13.77 (0.92)	11.65 (0.85)	12.46 (0.31)
Sm_2O_3	1.30 (0.15)	1.50 (0.32)	1.19 (0.15)	1.32 (0.30)	1.56 (0.36)	1.52 (0.34)	1.01 (0.25)	1.23 (0.23)
Gd_2O_3	0.30 (0.26)	—	—	—	—	0.24 (0.25)	—	—
Dy_2O_3	n.d.	n.d.	n.d.	n.d.	n.d.	n.d.	n.d.	n.d.
Y_2O_3	n.d.	n.d.	n.d.	n.d.	n.d.	n.d.	n.d.	n.d.
ThO_2	0.27 (0.24)	n.d.	n.d.	n.d.	n.d.	n.d.	n.d.	n.d.
Cl	n.d.	n.d.	n.d.	n.d.	n.d.	n.d.	n.d.	n.d.
Total	73.74	73.31	73.35	74.41	74.60	73.80	73.78	73.52

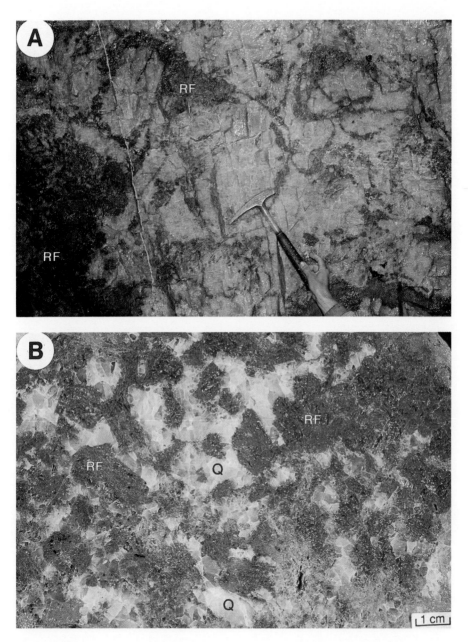

Plate 3 (a) REE-fluocarbonate–quartz mineralization in the decline of the North T-Zone; note the hammer for scale. (b) REE-fluocarbonate (RF) + quartz (Q) assemblage from the Upper Intermediate Zone.

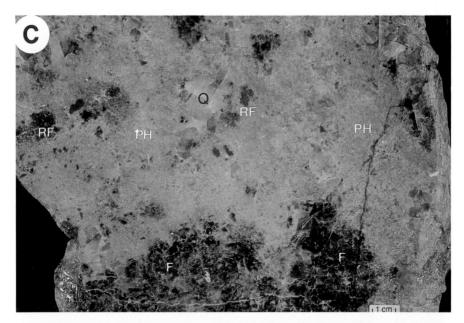

(c) Early REE-fluocarbonate (RF) mineralization from the Upper Intermediate Zone, overprinted and replaced by later phenakite (PH) + fluorite (F) assemblage. (d) Early REE-fluocarbonate (RF) mineralization from the Upper Intermediate Zone, overprinted and replaced by later ankerite (AK) + pyrite (PY) assemblage. (Courtesy of W. D. Sinclair, Geological Survey of Canada.)

Table 7.2 *Continued*

	PAR (n = 4) SUB/INC (B) Mean (1σ)	PAR (n = 5) AN/PS Mean (1σ)	PAR (n = 3) AN/PS Mean (1σ)	PAR (n = 5) SUB/PS Mean (1σ)	PAR (n = 8) AN/INT Mean (1σ)	PAR (n = 4) SUB/INC (B) Mean (1σ)	PAR (n = 10) SUB/PS/PL Mean (1σ)	PAR (n = 8) SUB/PS Mean (1σ)
				Concentrations (wt%)				
CaO	9.70 (0.33)	10.16 (0.12)	9.75 (0.91)	10.17 (0.20)	10.70 (0.26)	10.35 (0.11)	10.15 (0.43)	10.18 (0.21)
SrO	0.23 (0.15)	0.09 (0.11)	0.16 (0.11)	n.d. –	0.24 (0.11)	0.12 (0.12)	n.d. –	n.d. –
FeO	6.53 (1.80)	2.05 (1.17)	0.82 (0.71)	2.74 (2.48)	3.80 (1.67)	0.92 (1.17)	2.34 (1.95)	1.72 (1.60)
La_2O_3	11.57 (0.62)	13.25 (0.98)	14.98 (1.03)	14.35 (0.47)	11.43 (0.65)	13.02 (0.51)	13.54 (0.92)	13.89 (0.42)
Ce_2O_3	25.96 (0.23)	30.01 (0.67)	30.30 (0.89)	28.78 (0.78)	26.68 (1.21)	29.62 (0.85)	29.45 (0.80)	30.08 (0.75)
Pr_2O_3	3.29 (0.07)	3.09 (0.08)	2.94 (0.11)	3.10 (0.20)	3.32 (0.22)	3.34 (0.29)	3.23 (0.22)	3.44 (0.30)
Nd_2O_3	11.31 (0.48)	10.65 (0.16)	10.87 (0.52)	11.04 (0.67)	12.21 (0.48)	12.49 (0.68)	11.29 (0.41)	11.49 (0.40)
Sm_2O_3	1.58 (0.26)	1.21 (0.07)	1.18 (0.12)	1.18 (0.25)	1.81 (0.25)	1.57 (0.22)	1.39 (0.09)	1.23 (0.19)
Gd_2O_3	0.68 (0.10)	n.d. –	n.d. –	n.d. –	0.84 (0.11)	0.28 (0.28)	n.d. –	n.d. –
Dy_2O_3	n.d. –	n.d. –	n.d. –	n.d. –	n.d. –	n.d. –	n.d. –	n.d. –
Y_2O_3	0.37 (0.09)	0.25 (0.13)	0.21 (0.16)	n.d. –	0.52 (0.10)	0.30 (0.22)	0.14 (0.15)	n.d. –
ThO_2	0.79 (0.44)	n.d. –	n.d. –	n.d. –	1.21 (0.62)	n.d. –	n.d. –	n.d. –
Cl	0.17 (0.05)	0.25 (0.08)	0.23 (0.12)	0.22 (0.04)	0.15 (0.03)	0.22 (0.07)	0.18 (0.12)	0.20 (0.06)
Total	72.18	71.01	71.44	71.58	72.91	72.23	71.71	72.23

All analyses by SEM-EDS. Typical operating conditions were an accelerating voltage of 20 kV; a beam current of 0.80 nA; Be-window mode; counting times of 100–300 s. Data reduction utilized the ZAF4-FLS program; n = number of analyses; 1σ = one standard deviation. Abbreviations: BAS = bastnäsite; PAR = parisite; EU = euhedral; SUB = subhedral; AN = anhedral; INC = inclusion in (PA) parisite, (B) bastnäsite or (PY) pyrite; PS = mineral pseudomorph; PL = plumose intergrowth; n.d. = not detected.

Table 7.3 Compositional data for REE-bearing standards

	REE3 ($n = 9$)		Glass S236 ($n = 9$)		Glass S254 ($n = 5$)	
	Mean (1σ)	Pub[1]	Mean (1σ)	Pub[2]	Mean (1σ)	Pub[2]
	Concentrations (wt%)					
SiO_2	26.21 (0.20)	27.15	34.12 (0.30)	33.56	31.70 (0.21)	32.28
Al_2O_3	30.87 (0.22)	30.72	19.95 (0.20)	19.70	24.73 (0.25)	24.37
CaO	25.19 (0.13)	25.33	21.61 (0.21)	21.74	24.09 (0.08)	23.64
Sc_2O_3	n.d. –	0.00	0.10 (0.08)	0.00	1.67 (0.05)	1.59
Y_2O_3	3.97 (0.10)	4.08	n.d. (0.00)	0.00	1.16 (0.12)	1.32
La_2O_3	4.34 (0.05)	4.28	12.63 (0.21)	12.38	1.22 (0.04)	1.22
Ce_2O_3	4.09 (0.10)	4.00	12.56 (0.36)	12.03	1.29 (0.08)	1.22
Pr_2O_3	4.45 (0.11)	4.44	n.d. –	0.00	1.20 (0.09)	1.22
Nd_2O_3	n.d. –	0.00	n.d. –	0.00	1.29 (0.08)	1.21
Sm_2O_3	n.d. –	0.00	n.d. –	0.00	1.15 (0.12)	1.21
Eu_2O_3	n.d. –	0.00	n.d. –	0.00	1.30 (0.20)	1.20
Gd_2O_3	n.d. –	0.00	n.d. –	0.00	1.36 (0.15)	1.20
Tb_2O_3	n.d. –	0.00	n.d. –	0.00	1.30 (0.11)	1.19
Dy_2O_3	n.d. –	0.00	n.d. –	0.00	1.15 (0.14)	1.20
Ho_2O_3	n.d. –	0.00	n.d. –	0.00	1.16 (0.11)	1.19
Er_2O_3	n.d. –	0.00	n.d. –	0.00	1.10 (0.11)	1.19
Tm_2O_3	n.d. –	0.00	n.d. –	0.00	1.18 (0.13)	1.19
Yb_2O_3	n.d. –	0.00	n.d. –	0.00	1.21 (0.10)	1.18
Lu_2O_3	n.d. –	0.00	n.d. –	0.00	1.20 (0.08)	1.18

All analyses by SEM-EDS. n = number of analyses; 1σ = one standard deviation; pub = published reference ([1] Drake and Weill, 1972; [2] Roeder, 1985); n.d. = not detected.

in detectable amounts. Of the minor elements present in bastnäsite, only Ca and Sr are ubiquitous with concentrations of up to 1.23 wt% CaO and 1.67 wt% SrO. Chlorine was not detected in bastnäsite and Fe and Th, when present, occur only in trace amounts.

By virtue of the occurrence of calcium layers in its structure, parisite typically has lower ΣRE_2O_3 contents (54–62 wt%) and significantly higher CaO concentrations (9–11 wt%). In addition to the light REE, gadolinium (up to 1.05 wt% Gd_2O_3) is also commonly present in the parisite, as is yttrium (up to 0.68 wt% Y_2O_3). Although Sr is present in a number of parisite analyses, it occurs in much lower concentrations (< 0.42 wt% SrO) than in bastnäsite. Chlorine, with a concentration of up to 0.41 wt%, is a ubiquitous minor element in parisite. The abundance of Fe in parisite is highly variable and can reach quite high concentrations (up to 9.0 wt% FeO), in marked contrast to bastnäsite where it is characteristically below the level of analytical detection. In certain grains of parisite there exist Fe-rich 'domains' which are clearly discernible in BSE photographs and extend over areas as large as $20 \times 20\,\mu m$.

Chondrite-normalized REE patterns for bastnäsite and parisite are shown

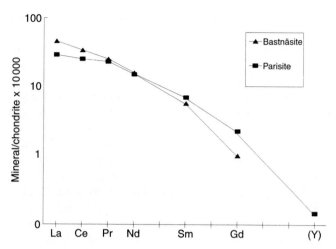

Figure 7.5 Chondrite-normalized REE patterns for bastnäsite and parisite from the zone of highest grade REE mineralization in the Upper Intermediate Zone. (The REE data are normalized to the chondritic values of Taylor and McLennan, 1985.)

in Figure 7.5. Based upon the relative proportions of REE (determined as atomic percentages), bastnäsite and parisite from the Upper Intermediate Zone have, respectively, the following characteristics: La + Ce + Pr values of 76.3–85.9 and 73.6–81.5; La/Nd ratios of 1.05–2.27 and 0.93–1.57; and 100Y(Y + REE) values of 0.0–0.59 and 0.0–1.72. Such ranges of values are similar to those documented for bastnäsite and parisite from a wide range of igneous and hydrothermal settings (Fleischer, 1978). Following the terminology proposed by Bayliss and Levinson (1988), the species names, bastnäsite-(Ce) and parisite-(Ce), can be applied to all of the bastnäsite and parisite analysed thus far. No chemical analyses of synchysite were made because of its rarity and the complex nature and very fine-grained character of its intergrowths.

7.5 $^{40}Ar/^{39}Ar$ geochronology

Standard laboratory techniques were used to purify lithium mica separates taken from three mineralized Upper Intermediate Zone samples (TZ-72, TZ-92, TZ-103). X-ray diffraction and chemical analysis of the mica indicate that it has the composition of polylithionite (1M polytype), and contains up to 6.41 wt% Li_2O. The polylithionite was selected for $^{40}Ar/^{39}Ar$ geochronology because, in terms of the paragenetic sequence of mineral deposition in the Upper Intermediate Zone, it is one of the minerals most closely associated with the REE-fluocarbonates. Incremental heating experiments on the three polylithionite samples were carried out in the

University of Maine argon laboratory. Two irradiation monitors were used. MMhb-1 was the primary standard and additionally a University of Maine interlaboratory standard, SB-51, was also used (Taylor *et al.*, 1987). The samples were irradiated in the Ford Nuclear Reactor at the University of Michigan for 30 h. Samples were heated incrementally in a molybdenum crucible within the ultrahigh vacuum system using radiofrequency induction heating. Sample gas was gettered using Cu–CuO and Zr–Al getters and a molecular sieve desiccant. The isotopic composition of Ar was measured with a Nuclide 6-60-SGA 1.25 mass spectrometer. Individual ages and uncertainties (2σ) were calculated using the decay constants recommended by Steiger and Jäger (1977) and the equations given by Dalrymple *et al.* (1981). Criteria for the determination of plateaus are from Fleck, Sutter and Elliot (1977), and the critical value test (Dalrymple and Lanphere, 1969) was used to test for concordancy between increments. The total gas ages represent a weighted average based on the amount of ^{39}Ar in each increment.

^{40}Ar/^{39}Ar release spectra for the three samples of hydrothermal polylithionite from mineralized Upper Intermediate Zone samples yield plateau ages of 2125 ± 31 Ma (TZ-92), 2118 ± 16 Ma (TZ-72) and 2087 ± 25 Ma (TZ-103). All the polylithionite separates contain trace to minor amounts of very fine-grained intergrown chlorite (<5%). Of the three, the polylithionite from sample TZ-72 has the lowest content of impurities, and its plateau age of 2118 ± 16 Ma (Figure 7.6 and Table 7.4) is considered to reflect most accurately the timing of alteration and REE mineralization in the Upper Intermediate Zone.

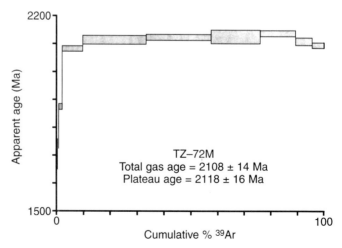

Figure 7.6 ^{40}Ar/^{39}Ar release spectrum for polylithionite mica separate, analyzed by the incremental heating method, from sample TZ-72 from the Upper Intermediate Zone.

Table 7.4 Argon isotope data for hydrothermal polylithionite sample TZ-72 from the Upper Intermediate Zone

Temp (°C)	$^{40}Ar/^{39}Ar$	$^{37}Ar/^{39}Ar$	$^{36}Ar/^{39}Ar$	%^{39}Ar	%$^{40}Ar_R$	K/Ca	Age (Ma)
			$J = 0.006748$				
650	257.25	0.2744	0.1120	0.4	87.1	1.79	1662 ± 14
720	259.90	0.1611	0.0667	0.5	92.4	3.04	1738 ± 16
825	275.75	0.0696	0.0172	1.3	98.2	7.03	1875 ± 9
900	323.53	0.0337	0.0052	7.4	99.5	14.5	2083 ± 11
950	331.32	0.0250	0.0030	23.5	99.7	19.6	2115 ± 14
1000	332.46	0.0221	0.0018	24.1	99.8	22.2	2121 ± 10
1050	331.87	0.0200	0.0016	18.8	99.9	24.52	2119 ± 23
1100	335.84	0.0190	0.0013	13.0	99.9	25.74	2134 ± 10
1150	327.76	0.0266	0.0015	6.5	99.9	18.4	2103 ± 11
Fuse	323.46	0.0307	0.0016	4.5	99.9	16.0	2087 ± 10
Total				100.1			2108 ± 14
Plateau age							2118 ± 16

The $^{40}Ar/^{39}Ar$ plateau age of the hydrothermal polylithionite from TZ-72 is analytically indistinguishable from the U–Pb age of 2094 ± 10 Ma obtained for zircon from carbonate veins in the Lake Zone (Bowring, Van Schmus and Hoffman, 1984), but is significantly younger than the age of emplacement of the Grace Lake Granite determined by Sinclair, Hunt and Birkett (1994). These age data support the contemporaneity of the T-Zone and Lake Zone deposits, and indicate that REE mineralization and alteration clearly postdated the emplacement of the host peralkaline granite and syenite.

7.6 Fluid inclusion analysis

Samples of quartz and fluorite, for fluid inclusion and stable isotope analysis, were hand-picked from large (>1 cm) euhedral–subhedral crystals. Identification of primary and pseudosecondary inclusions in quartz samples from the Upper Intermediate Zone was impeded, in places, by the occurrence of crosscutting planes of secondary inclusions so numerous that the quartz has taken on a milky appearance. The relationship of fluid inclusions to primary growth zones cannot easily be observed in the doubly polished wafers used for analysis, and the inclusions in quartz are classified as primary or pseudosecondary if they occur either in a plane which begins and ends within a single crystal, particularly where this is crosscut by later planes of secondary inclusions, or as isolated individuals and/or groups of inclusions, especially where their composition is different from that of nearby secondary inclusions. Neither of these criteria is entirely definitive and it is possible that some early secondary inclusions have been classified as primary or pseudosecondary.

Heating and freezing experiments were carried out using a USGS-type gas flow heating–freezing stage which was calibrated using natural and synthetic standards. The relative volumes of phases present within the fluid inclusions were estimated visually, and this potentially introduced large errors in calculations of fluid compositions, particularly in CO_2-rich inclusions (Brown and Lamb, 1989). Data reduction for calculating fluid compositions was carried out using the program FLINCOR (Brown and Lamb, 1989). The data summarized here are for those fluid inclusions in samples that are most representative of the REE mineralization in the Upper Intermediate Zone and which have also been the subject of hydrogen and oxygen isotope analysis. They represent only a part of a much larger database.

Primary fluid inclusions in all of the mineralized Upper Intermediate Zone samples used in this study typically consist of three phases at room temperature: H_2O-rich liquid, CO_2 liquid and CO_2 vapour. The volume of the CO_2 phases is highly variable, ranging from 5 to 98% in individual samples. Some inclusions appear to contain 100% CO_2 phases, although it is possible that the inclusion walls are wet by a thin film of H_2O. The melting temperature of CO_2 in primary inclusions in most of the samples is fairly constant (-56.6 to $-56.8°C$), indicating essentially pure CO_2. However, the occurrence of lower CO_2 melting temperatures (-57.7 to $-57.8°C$) in a limited number of samples suggests the local presence of other miscible components such as CH_4 in the CO_2 phase. Salinities for the primary fluid inclusions, calculated from clathrate melting temperturees, range from 2 to 17 wt% equivalent NaCl in the H_2O-rich phase, with values of about 12 wt% being most common.

Daughter minerals are uncommon in the primary fluid inclusions. However, some were observed and the solids contained in opened primary fluid inclusions were studied using SEM-EDS analysis. Halite, and less commonly sylvite, were observed in a limited number (<10%) of opened inclusions. In addition, parisite was also observed in an opened inclusion in one instance. The euhedral nature of the solid phases in the opened inclusions is considered to support their origin as daughter minerals.

Final homogenization to the H_2O or CO_2 phase occurs over a wide range of temperatures between 140 and 420°C, with quite a high proportion of the fluid inclusions (>40%) decrepitating within the same temperature range (Figure 7.7). Of those fluid inclusions which did not decrepitate upon heating, move than 75% homogenized in the much more restricted temperature range from 280–420°C, with a median value close to 340°C. Pressures at homogenization, calculated using the equation of state of Brown and Lamb (1989), range from 1650 to 4500 bars for the H_2O-rich ($X_{H_2O} > 0.75$) primary inclusions. Figure 7.8 is a plot of the mole fraction of CO_2 versus temperature for the primary fluid inclusions whose homogenization temperatures are shown in the histogram in Figure 7.7. Figure 7.8 also shows the solvus, calculated from the data of Bowers and Helgeson

FLUID INCLUSION ANALYSIS

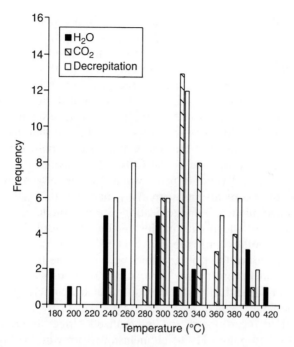

Figure 7.7 Histogram of homogenization temperatures of primary three-phase fluid inclusions in hydrothermal quartz associated with REE-fluocarbonate mineralization from the Upper Intermediate Zone.

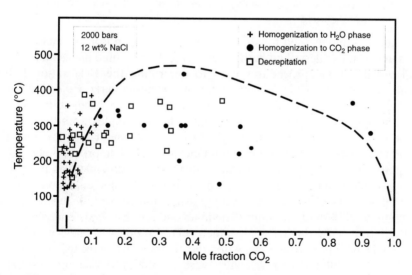

Figure 7.8 A plot of mole fraction CO_2 versus temperature for primary fluid inclusions in hydrothermal quartz from the Upper Intermediate Zone (see text for explanation).

(1983), for a H_2O-CO_2 fluid with 12 wt% NaCl at a pressure of 2000 bars (i.e. median values for salinity and pressure for the primary inclusions).

Those primary inclusions in quartz from the Upper Intermediate Zone that homogenized to the H_2O phase fit closely to the H_2O-rich limb of the 2 kbar solvus (Figure 7.8). In contrast, the distribution of points for those inclusions that homogenized to the CO_2 phase, and those that decrepitated upon heating, shows much more scatter. However, in assessing the fit of the latter two groups of inclusions (i.e. CO_2-rich and those that decrepitated) to the solvus, it should be noted that the temperature measurements for both groups are minima (in the case of the CO_2-rich inclusions this is because of the difficulty of optical estimation of when CO_2 homogenization occurs). Therefore, all of the inclusions with X_{CO_2} between 0.2 and 0.6 could be expected to have higher homogenization temperatures, a feature which would shift many of the points in this region of Figure 7.8 closer to the 2 kbar solvus. Together with the range in CO_2 contents, and the homogenization behaviour of the primary fluid inclusions, the reasonable fit of the data for primary inclusions to the 2 kbar solvus suggests trapping of a boiling $H_2O-NaCl-CO_2$ fluid during the deposition of hydrothermal quartz in the Upper Intermediate Zone.

Secondary fluid inclusions in the mineralized Upper Intermediate Zone samples include two-phase, H_2O-rich inclusions with variable liquid–vapour ratios, one-phase H_2O inclusions and single-phase vapour inclusions of unknown composition.

7.7 Hydrogen and oxygen isotope analysis of fluid inclusion water and host quartz

The quartz samples used for the hydrogen and oxygen isotope analysis were the same as those described in the previous section on fluid inclusion analysis. Full details of the procedures used to determine the D/H ratio of fluid inclusion water, and the $^{18}O/^{16}O$ ratio of the host quartz are contained in Taylor, Fallick and Breaks (1992). Therefore, only a brief outline of the experimental procedures will be presented here. Samples were degassed of labile volatiles by overnight heating (16 h) at 130°C under high vacuum. For the determination of D/H ratio of fluid inclusion H_2O, approximately 500 mg of quartz was first heated *in vacuo* to 800°C. The crucible was previously degassed platinum and heating was by a radiofrequency induction furnace. The released water was cryogenically purified and converted to hydrogen by reaction at 800°C with uranium (Friedman and Smith, 1958). The yield of H_2 was determined in a mercury manometer. D/H ratios were measured using a VG 602D mass spectrometer. After the release of the inclusion fluids for D/H analysis, the quartz residues were stored in a vacuum desiccator. Silicate $\delta^{18}O$ was subsequently determined on a 10–12 mg powdered aliquot of each of these quartz residues, which was reacted with a stoichiometric excess of BrF_5 at 650°C for 12 h. The oxygen released during the fluorination process was converted to CO_2 for analysis, by reaction with a hot platinized graphite rod (Clayton and Mayeda, 1963), and the yield determined manometrically. All yields indicated high-purity quartz. Replicate analyses of pure isotopically homogeneous material suggest analytical precision (1σ) of 0.2‰ for silicate $\delta^{18}O$, 2.0‰ for δD, and 0.02 wt% for H_2O. The hydrogen or oxygen isotope composition of a sample is reported in per mil (‰), relative to the Standard Mean Ocean Water (SMOW) standard, using the δ notation (cf. O'Neil, 1986).

A summary of the hydrogen and oxygen isotope data is presented in Table 7.5. The total water content of the Upper Intermediate Zone quartz samples varies only slightly from 0.13 to 0.16 wt%. These concentration data indicate that the distribution of H_2O in the quartz is homogeneous. In addition, they are consistent with derivation via the release of H_2O from fluid inclusions only (cf. Roedder, 1984), with the fluid inclusions comprising approximately 0.001% by mass of the host quartz.

The δD values of fluid inclusion H_2O released by thermal decrepitation range from -49 to -76‰ (Table 7.5). Such values are very close to the range of δD values (-39 to -79‰) for fluid inclusion H_2O in quartz in miarolitic cavities in layered aplite-pegmatite dykes reported by Taylor, Foord and Friedrichsen (1979), and typical of those of exsolved magmatic H_2O (Taylor, 1986).

The oxygen isotope compositions of the quartz residues range from 8.5 to 11.5‰. Using these data and the fractionation equation of Zhang et al. (1989), the $\delta^{18}O$ value of the coexisting hydrothermal fluid can be calculated for a given temperature. The results of the fluid inclusion analysis summarized in the previous section indicate that the temperature range, 350–400°C, was the most likely for quartz deposition and REE-fluocarbonate

Table 7.5 Hydrogen and oxygen isotope analyses of fluid inclusion water and host quartz from the Upper Intermediate Zone

Sample	H_2O (wt%)	δD (‰)$_{SMOW}$ H_2O	$\delta^{18}O$ (‰)$_{SMOW}$ Quartz
TZ-23	0.14	−64	11.5
TZ-119	0.13	−73	9.2
TZ-122	0.16	−57	11.0
TZ-132/2A	0.13	−76	8.5
TZ-132/2B	0.13	−49	n.d.

All values are reported in per mil (‰) relative to the Standard Mean Ocean Water (SMOW) reference standard; n.d. = not determined.

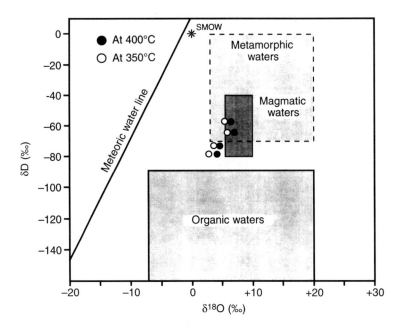

Figure 7.9 A plot of $\delta^{18}O_{H_2O}$ versus δD_{H_2O} depicting the stable isotope geochemistry of the hydrothermal fluid responsible for REE-fluocarbonate mineralization in the Upper Intermediate Zone of the North T-Zone rare-metal deposit. (The data for the Meteoric Water Line and for the fields of Magmatic and Metamorphic Water are from Craig (1961) and Sheppard (1986) respectively.)

mineralization in the Upper Intermediate Zone. The hydrothermal fluid responsible for REE mineralization in the Upper Intermediate Zone thus has calculated values for $\delta^{18}O_{H_2O}$ of 3.9–6.9‰ (at 400°C) and of 2.7–5.7‰ (at 350°C).

The δD and $\delta^{18}O$ values, at temperatures of 350 and 400°C, of the hydrothermal fluid that was responsible for the alteration and REE-mineralization in the Upper Intermediate Zone are plotted in Figure 7.9. The values are very suggestive of a magmatic origin for the mineralizing fluid, although at the lower temperature they fall outside of the field of magmatic water. The trend to lower $\delta^{18}O$ is not unexpected, given that interaction of the magmatic fluid with the wall rocks in the Upper Intermediate Zone, under either closed- or open-system conditions, would result in a significant depletion in ^{18}O in the hydrothermal fluid after isotopic exchange (cf. Pollard, Andrew and Taylor, 1991).

7.8 Carbon and oxygen isotope analysis of carbonates and sulphur isotope analysis of sulphide minerals

The analyses of carbon and oxygen isotope ratios in carbonates and sulphur isotope ratios in sulphide minerals were determined using conventional techniques (Craig, 1957; Faure, 1986). Eight carbonate, and 10 sulphide, specimens were separated from rocks of the Upper Intermediate Zone and concentrated using conventional procedures. The purity of all of the mineral separates was checked by X-ray diffraction analysis. The analytical precision (2σ) for the carbonate $\delta^{13}C$ and sulphide $\delta^{34}S$ values is 0.1‰ and 0.2‰, respectively.

The $\delta^{13}C$ and $\delta^{18}O$ values of the hydrothermal carbonates from the Upper Intermediate Zone are listed in Table 7.6. All of the carbonate specimens are ankerite, with the exception of one siderite which was found to coexist with ankerite in sample TZ-93. The carbonate $\delta^{13}C$ values have a restricted range, from -2.1 to -5.9‰ (relative to the PDB standard). Such values are essentially similar to those of $\delta^{13}C$ in carbonate minerals from carbonatites (i.e. -3.4 to -7.1‰; Deines and Gold, 1973).

The sulphur isotopic data for the ten sulphide separates (pyrite, chalcopyrite, sphalerite) from the Upper Intermediate Zone are presented in Table 7.7. The isotopic data include two sulphide pairs, and examples of core and rim samples isolated from sphalerite, which is commonly zoned. The $\delta^{34}S$ values have a very restricted range, from 0.4 to 1.6‰ (Table 7.7), and thus show very little deviation from the Canyon Diablo troilite (CDT) standard. Such $\delta^{34}S$ values are very close to those of mid-ocean ridge and ocean-island basalts (ranges of 0.5 to -0.5‰ and 0.5 to 1.0‰, respectively; Taylor, 1986), and are entirely consistent with a single 'magmatic' source for the sulphur.

Table 7.6 Carbon and oxygen isotope analyses of carbonates from the Upper Intermediate Zone

Sample	Mineral	$\delta^{13}C$ (‰)$_{PDB}$	$\delta^{18}O$ (‰)$_{SMOW}$
TZ-23	Ankerite	-3.6	14.0
TZ-29	Ankerite	-4.8	11.7
TZ-36	Ankerite	-4.0	12.3
TZ-42	Ankerite	-2.1	15.2
TZ-87	Ankerite	-2.4	13.7
TZ-93	Ankerite	-3.9	12.2
TZ-93	Siderite	-5.9	11.8
TZ-131	Ankerite	-5.4	12.2

Mineral identification by X-ray diffraction analysis. All values are reported in per mil (‰) relative to the Peedee Belemnite (PDB) or Standard Mean Ocean Water (SMOW) reference standards.

Table 7.7 Sulphur isotope analyses of sulphides from the Upper Intermediate Zone

Sample	Mineral	$\delta^{34}S$ (‰)$_{CDT}$
TZ-73	Chalcopyrite	0.6
TZ-73	Pyrite	0.9
TZ-76	Pyrite	0.5
TZ-76	Sphalerite (core)	0.5
TZ-76	Sphalerite (rim)	0.4
TZ-93	Pyrite	0.4
TZ-129	Pyrite	1.0
TZ-132/1	Sphalerite (core)	1.5
TZ-132/1	Sphalerite (rim)	1.6
TZ132/2	Sphalerite	1.2

Mineral identification by X-ray diffraction analysis. All values are reported in per mil (‰) relative to the Canyon Diablo Troilite (CDT) reference standard.

7.9 Conclusions and summary

1. Rare earth element mineralization is most highly concentrated in the Upper Intermediate Zone of the North T-Zone rare-metal deposit, where ΣRE_2O_3 grades of >10% are not uncommon. The REE mineralization is disseminated to locally massive, and is characterized by the extensive development of REE-fluocarbonate mineral pseudomorphs.
2. Complex lithological, textural and paragenetic relationships are responsible for the irregular distribution of REE mineralization within the Upper Intermediate Zone. Locally, areas of REE-fluocarbonate mineralization are intercalated with mica- and feldspar-rich lithologies which are generally poor in REE. In addition, REE-fluocarbonate mineralization is commonly overprinted, and partially replaced, by later ankerite + fluorite + sulphide- or phenakite + fluorite-bearing hydrothermal assemblages.
3. Bastnäsite-(Ce) and parisite-(Ce) are the two most abundant REE-fluocarbonate minerals in the zones of hydrothermal REE mineralization. Typically, they occur as syntactic intergrowths in polycrystals in mineral pseudomorphs. The REE-fluocarbonate polycrystals show a variety of complex textures, which are most easily differentiated by the use of backscattered electron images. Total RE_2O_3 contents of 71–74 wt% for bastnäsite and 54–62 wt% for parisite are typical. The value for La + Ce + Pr (at.%) ranges from 76.3 to 85.9 in bastnäsite, and from 73.6 to 81.5 in parisite. Aside from its high Ca content (9–11 wt% CaO), parisite is also distinguished by the ubiquitous presence of Cl and Fe.
4. $^{40}Ar/^{39}Ar$ geochronology yields plateau ages of 2125 ± 31 Ma, 2118 ±

16 Ma and 2087 ± 25 Ma for hydrothermal lithium mica (viz. polylithionite) in the Upper Intermediate Zone. These ages indicate that hydrothermal alteration and REE mineralization in the Upper Intermediate Zone clearly postdated the emplacement of the peralkaline granite and syenite that are the hosts of the North T-Zone deposit. In addition, they support the interpretation that the North T-Zone and Lake Zone rare-metal deposits are coeval, with both deposits having formed at about 2100 Ma.
5. Primary three-phase fluid inclusions (H_2O-rich liquid, CO_2 liquid and CO_2 vapour) in hydrothermal quartz intergrown with REE-fluocarbonate mineralization have highly variable CO_2 contents (5–98%). Salinities in the primary fluid inclusions range from 2 to 20 wt% equivalent NaCl in the H_2O-rich phase. Final homogenization temperatures vary from 200 to 400°C, with the majority of values in the range 320–360°C. The variation in the CO_2 contents and homogenization behaviour suggest trapping of a boiling H_2O–NaCl–CO_2 fluid at c. 350°C.
6. Total water contents of the hydrothermal quartz samples vary from 0.13 to 0.16 wt%, consistent with the derivation of H_2O only from fluid inclusions. The δD values for fluid inclusion H_2O in hydrothermal quartz range from −49 to −76‰, values that are consistent with an origin as exsolved magmatic H_2O.
7. The $\delta^{18}O$ values of the hydrothermal quartz range from 8.5 to 11.5‰. These data, when incorporated into published fractionation equations for oxygen isotope fractionation in the quartz–water–salt system, yield calculated values for $\delta^{18}O_{H_2O}$ of 2.7 to 5.7‰ (at 350°C) and 3.9 to 6.9‰ (at 400°C) for the mineralizing fluid. The higher values calculated for $\delta^{18}O_{H_2O}$ overlap with the field of magmatic water, whilst the lower $\delta^{18}O_{H_2O}$ values, which fall outside of the field of magmatic water, are interpreted to have resulted from isotopic exchange of the magmatic–hydrothermal fluid with the wall rocks during alteration.
8. The $\delta^{13}C$ values of hydrothermal ankerite, and $\delta^{34}S$ values of hydrothermal sulphide, from the Upper Intermediate Zone vary from −2.1 to −5.9‰ and from 0.4 to 1.6‰, respectively. Such values, together with their restricted ranges, indicate a single magmatic source for both the carbon and sulphur.

The majority of the data presented here are thus consistent with a magmatic origin for the hydrothermal fluids that produced REE mineralization in the North T-Zone deposit. However, the data do not uniquely define the nature of the magmatic source. In particular the $^{40}Ar/^{39}Ar$ ages of hydrothermal lithium mica suggest that alteration and REE mineralization may have occurred well after the emplacement of the peralkaline units that are the host to the North T-Zone deposit. If this is the case, it is conceivable that the REE mineralization in the North T-Zone might be related to some as yet unexposed agpaitic intrusive unit, such as the nepheline syenite recently identified in diamond drill-core from below the Lake Zone (Pinckston, 1989).

Acknowledgements

The authors would like to express their sincere appreciation to the organizers of the 1993 Rare Earth Minerals conference in London, and to the volume editors for their interest and assistance during the preparation of this chapter. Also, our sincere thanks go to Dave Sinclair of the Geological Survey of Canada for his unflagging support during all of the stages of the Thor Lake project, which was initially funded by an Energy, Mines & Resources Mineral Development Agreement. Tony Fallick (Scottish Universities Research and Reactor Centre), Dan Lux (University of Maine) and Peter Jones (Carleton University) provided invaluable assistance in the acquisition of the stable isotope, Ar isotope and SEM-EDS data, respectively. The comments of Wayne Taylor and an anonymous reviewer were most helpful. One final note of appreciation to Louise de St Jorre (formerly of the University of Alberta) who did a remarkable job of synthesis and interpretation in her MSc thesis on the T-Zone deposit, which greatly assisted us in the early part of this study. Financial support for parts of this study were provided by NSERC Operating Grant OGP-0001877 (to R. P. Taylor).

References

Bayliss, P. and Levinson, A. A. (1988) A system of nomenclature for rare-earth mineral species; revision and extension. *Am. Mineral.*, **73**, 422–3.

Bowden, P. (1985) The geochemistry and mineralization of alkaline ring complexes in Africa. *J. Afr. Earth Sci*, **3**, 17–40.

Bowers, T. S. and Helgeson, H. C. (1983) Calculation of thermodynamic and geochemical consequences of nonideal mixing in the system H_2O–CO_2–NaCl on phase relations in geologic systems: equation of state for H_2O–CO_2–NaCl fluids at high pressures and temperatures. *Geochim. Cosmochim. Acta*, **47**, 1247–75.

Bowring, S. A., Van Schmus, W. R. and Hoffman, P. F. (1984) U–Pb zircon ages from Athapuscow Aulacogen, East Arm of Great Slave Lake, N.W.T., Canada. *Can. J. Earth Sci.*, **21**, 1315–24.

Brown, P. E. and Lamb, W. M. (1989) P–V–T properties of fluids in the system $H_2O \pm CO_2 \pm$ NaCl: new graphical presentations and implications for fluid inclusions. *Geochim. Cosmochim. Acta*, **53**, 1209–22.

Clayton, R. N. and Mayeda, T. K. (1963) The use of bromine pentafluoride in the extraction of oxygen from oxides and silicates for isotopic analysis. *Geochim. Cosmochim. Acta*, **27**, 43–52.

Craig, H. (1957) Isotopic standards for carbon and oxygen and correction factors for mass spectrometric analysis of carbon dioxide. *Geochim. Cosmochim. Acta*, **12**, 133–49.

Craig, H. (1961) Isotopic variations in meteoric waters. *Science*, **133**, 1702–3.

Dalrymple, G. B. and Lanphere, M. A. (1969) *Potassium–Argon dating*. W. H. Freeman, San Francisco, 258 pp.

REFERENCES

Dalrymple, G. B., Alexander E. C., Jr, Lanphere, M. A. and Kraker, G. P. (1981) Irradiation of samples for $^{40}Ar/^{39}Ar$ dating using the Geological Survey TRIGA Reactor. *US Geol. Surv. Prof. Pap.*, **1176**, 55 pp.

Davidson, A. (1978) The Blatchford Lake intrusive suite–an Aphebian plutonic complex in the Slave Province, Northwest Territories. *Geol. Surv. Can. Pap.*, **78–1A**, 119–27.

Davidson, A. (1981) Petrochemistry of the Blatchford Lake complex, District of Mackenzie. *Geol. Surv. Can. Open File*, **764**.

Davidson, A. (1982) Petrochemistry of the Blatchford Lake complex near Yellowknife, Northwest Territories. *Geol. Surv. Can. Pap.*, **81–23**, 71–80.

de St Jorre, L. (1986) Economic mineralogy of the North T-Zone deposit, Thor Lake, Northwest Territories. Unpublished MSc thesis, University of Alberta, 247 pp.

de St Jorre, L. and Smith, D. G. W. (1988) Cathodoluminescent gallium-enriched feldspars from the Thor Lake rare-metal deposits, Northwest Territories. *Can. Mineral.*, **26**, 301–8.

Deines, P. and Gold, D. P. (1973) The isotopic composition of carbonatite and kimberlite carbonates and their bearing on the isotopic composition of deep-seated carbon. *Geochim. Cosmochim. Acta*, **37**, 1709–33.

Donnay, G. and Donnay, G. D. H. (1953) The crystallography of bastnäsite, parisite, roentgenite, and synchysite. *Am. Mineral.*, **38**, 932–63.

Drake, M. J. and Weill, D. F. (1972) New rare earth element standards for electron microprobe analysis. *Chem. Geol.*, **10**, 179–81.

Fleck, R. D., Sutter, J. F. and Elliot, D. (1977) Interpretation of discordant $^{40}Ar/^{39}Ar$ agespectra of Mesozoic tholeiites from Antarctica. *Geochim. Cosmochim. Acta*, **41**, 15–32.

Faure, G. (1986) *Principles of Isotope Geology*, John Wiley, New York, 589 pp.

Fleischer, M. (1978) Relative proportions of the lanthanides in minerals of the bastnäsite group. *Can. Mineral.*, **16**, 361–3.

Friedman, I. and Smith, R. L. (1958) The deuterium content of water in some volcanic glasses. *Geochim. Cosmochim. Acta*, **15**, 218–28.

Hoffman, P. F., Bell, I. R., Hildebrand, R. S. and Thorstad, L. (1977) Geology of the Athapuscow Aulacogen, east arm of Great Slave Lake, District of Mackenzie. *Geol. Surv. Can. Pap.*, **77–1A**, 117–29.

Jackson, N. J. (1986) Mineralization associated with felsic plutonic rocks of the Arabian Shield. *J. Afr. Earth Sci.*, **4**, 213–28.

Kinnaird, J. (1985) Hydrothermal alteration and mineralization of the alkaline anorogenic ring complexes of Nigeria. *J. Afr. Earth Sci.*, **3**, 229–52.

Neumann, E. R. (1978) Petrology of plutonic rocks. *Norges Geologiske Undersokelse Bulletin*, **45**, 25–34.

O'Neil, J. R. (1986) Terminology and standards. *Reviews in Mineralogy*, **16**, Mineralogical Society of America, 561–70.

Pinckston, D. R. (1989) Mineralogy of the Lake Zone Deposit, Thor Lake, Northwest Territories. Unpublished MSc Thesis, University of Alberta, 155 pp.

Pollard, P. J., Andrew, A. S. and Taylor, R. G. (1991) Fluid inclusion and stable isotope evidence for interaction between granites and magmatic hydrothermal fluids during formation of disseminated and pipe-style mineralization at the Zaaiplaats Tin Mine. *Econ. Geol.*, **86**, 121–41.

Roedder, E. (1984) Fluid inclusions. *Reviews in Mineralogy*, **12**, Mineralogical Society of America, 644 pp.

Roeder, P. L. (1985) Electron-microprobe analysis of minerals for rare earth elements: use of calculated peak-overlap corrections. *Can. Mineral.*, **23**, 263–71.

Sheppard, S. M. F. (1986) Characterization and isotopic variations in natural waters. *Reviews in Mineralogy*, **16**, Mineralogical Society of America 165–83.

Sinclair, W. D., Jambor, J. L. and Birkett, T. C. (1992) Rare earths and the potential for rare earth deposits in Canada. *Explor. Mining Geol.*, **1**, 265–81.

Sinclair, W. D., Hunt, P. A. and Birkett, T. C. (1994) U–Pb zircon and monazite ages of the Grace Lake Granite, Blatchford Lake Intrusive Suite, Slave Province, Northwest Territories. *Geol. Surv. Can.*, Current Research 1994-F, 15–20.

Steiger, R. H. and Jäger, E. (1977) Subcommission on geochronology: convention on the use of decay constants in geo- and cosmochronology. *Earth Planet. Sci. Lett.*, **36**, 359–62.

Stoeser, D. B. (1986) Distribution and tectonic setting of plutonic rocks of the Arabian Shield. *Jour. Afr. Earth Sci.*, **4**, 21–46.

Taylor, B. E. (1986) Magmatic volatiles: isotopic variation of C, H and S. *Reviews in Mineralogy*, Mineralogical Society of America, **16**, 185–225.

Taylor, B. E., Foord, E. E. and Friedrichsen, H. (1979) Stable isotope and fluid inclusion studies of GEM-bearing granitic pegmatite-aplite dykes, San Diego Co., California. *Contrib. Mineral. Petrol.*, **68**, 187–205.

Taylor, R. P., Lux, D. R., MacLellan, H. E. and Hubacher, F. (1987) Age and genesis of granite-related W–Sn–Mo mineral deposits, Burnthill, New Brunswick, Canada. *Econ. Geol.*, **82**, 2187–98.

Taylor, R. P., Fallick, A. E. and Breaks, F. (1992) Volatile evolution in Archean rare-element granitic pegmatites: evidence from the hydrogen isotopic composition of channel H_2O in beryl. *Can. Mineral.*, **30**, 877–93.

Taylor, S. R. and McLennan, S. M. (1985) *The Continental Crust: Its Composition and Evolution*. Blackwell Scientific Publications, Oxford, 312 pp.

Trueman, D. L., Pedersen, J. C. and de St Jorre, L. (1984) Geology of the Thor Lake beryllium deposits; an update. *Contrib. Geol. Northwest Territories*, **1**, 115–120.

Trueman, D. L, Pedersen, J. C., de St Jorre, L. and Smith, D. G. W. (1988) The Thor Lake rare metal deposits, Northwest Territories, in *Recent Advances in the Geology of Granite-related Mineral Deposits* (eds Taylor, R. P. and Strong, D. F.), CIM Special Volume, **39**, 280–90.

Wanless, R. K., Stevens, R. D., Lachance, G. R. and DeLabio, R. N. (1979) Age determinations and geological studies, K–Ar isotopic ages, report 14. *Geol. Surv. Can. Pap.*, **79-2**.

Zhang, L., Liu, J., Zhou, H. and Chen, Z. (1989) Oxygen isotope fractionation in the quartz–water–salt system. *Econ. Geol.*, **84**, 1643–1650.

CHAPTER EIGHT
Rare earth minerals in carbonatites: a discussion centred on the Kangankunde Carbonatite, Malawi

F. Wall and A. N. Mariano

8.1 Introduction

It is a characteristic of carbonatites that they contain higher levels of REE than almost all other rock types. Concentrations of REE are often sufficiently high to produce rare earth minerals and can occasionally attain tenors of commercial importance. Two of the world's major REE sources are the carbonatite-related deposits at Mountain Pass, California and Bayan Obo, Inner Mongolia, China. An understanding of the mineralogy and petrogenesis of rare earth minerals in carbonatites is therefore of economic as well as academic importance.

This chapter gives an introduction to rare earth minerals in carbonatites and details an example at Kangankunde, Malawi to show some typical – and not so typical – features of rare earth-rich carbonatites. We then consider some replacement reactions which can produce RE minerals in carbonatites.

8.2 Rare earth elements and their mineral hosts in carbonatites

Carbonatites are igneous rocks comprising more that 50% modal carbonate (Woolley and Kempe, 1989; Le Maitre *et al.*, 1989). Ideally, they are named according to their carbonate mineralogy (e.g. calcite carbonatite, dolomite carbonatite, ankerite carbonatite) and chemically they may be divided into the three main varieties of calcio-, magnesio- and ferrocarbonatite. Average REE contents for calcio-, magnesio- and ferrocarbonatites were calculated by Woolley and Kempe (1989) as 0.37 wt%, 0.42 wt% and 1 wt% respectively, showing, as is commonly found, a progression in levels of REE. The term 'rare earth-rich carbonatite' has no precise definition but rocks suitable for such description would contain levels of REE in excess of a couple of weight per cent and contain RE minerals.

Rare Earth Minerals: Chemistry, origin and ore deposits. Edited by Adrian P. Jones, Frances Wall and C. Terry Williams. Published in 1996 by Chapman & Hall. ISBN 0 412 61030 2

Enrichment in the light REE with chondrite-normalized values for La at least two orders of magnitude higher than Lu is typical of all carbonatites (e.g. Loubet et al., 1972; Möller, Morteani and Schley, 1980; Woolley and Kempe, 1989) and this is reflected in the predominance of minerals of the light rare earths. Hogarth (1989) listed 27 RE minerals from carbonatites, (excluding the Bayan Obo rare earth deposit) and although this list would expand if the Bayliss and Levinson (1988) nomenclature rules were applied, it would still fall far short of the over 200 RE minerals now known (see Appendix). It is probable that the low silica activity also restricts the range of RE minerals in carbonatites. The most abundant minerals, as shown in Table 8.1, are the three fluocarbonate minerals, bastnäsite-(Ce), parisite-(Ce) and synchysite-(Ce); the hydrated carbonate, ancylite; and the phosphate, monazite-(Ce). Other carbonatite minerals include Ba-RE-fluocarbonate minerals such as cordylite, huanghoite and zhongacerite, crandallite-group aluminium phosphates such as florencite, Na,Ca,Sr,Ba-bearing carbonates such as burbankite and carbocernaite, and the silicate, allanite. Heavy REE minerals are rare but two examples are mckelvyite, reported in late zeolite veins from the Khibina carbonatite, Kola Peninsula (Voloshin et al., 1990), and microcrystalline aggregates of xenotime in fluorite from the Hicks Dome carbonatite complex, USA, and Mt Weld, Western Australia (Mariano, 1984, 1987).

8.3 'Late carbonatites'

The concept of carbonatite 'stages' is useful when considering the mineral hosts for REE in carbonatites because, when a variety of carbonatites is present, it is almost always rocks emplaced late in the sequence that contain the highest levels of REE and therefore the greatest abundance of RE minerals. These 'late' carbonatites are often the ferrocarbonatites; dark-coloured, iron-rich dolomite or ankerite carbonatites which, according to Le Bas (1989), may have the field and textural characteristics of either igneous or hydrothermal rocks and often contain RE minerals such as bastnäsite-(Ce) and monazite-(Ce). Baryte, fluorite, U–Th mineralization and silica in the form of quartz or chalcedony are also common components. However, as shown in Table 8.1, calcite carbonatites can also host RE minerals. These calcite carbonatites also tend to be 'late' and do not contain the 'lozenge-shaped' apatite and magnetite assemblage usual in early carbonatites. A compilation of the mineralogical constituents of 'early' and 'late' varieties of carbonatites, mainly from observations of Russian carbonatites, is given in Kapustin (1980).

In the 'early' predominantly calcio- and magnesiocarbonatites, the REE are hosted by the rock-forming minerals apatite and calcite and by accessory phases such as perovskite and pyrochlore, and sometimes monazite which often forms at the end of apatite crystallization. Levels of REE in the rock-

Table 8.1 Examples of carbonatite occurrences that contain rare earth minerals

Carbonatite	Rare earth mineral assemblage	References
Adiounedj, northeast Mali	Synchysite and quartz in a groundmass of fluorite, calcite and ferric oxides; hydrothermal	Mariano (1983), Sauvage and Savard (1985)
Araxá, Minas Gerais, Brazil	Monazite and crandallite-group minerals in lateritic weathered carbonatite.	see Morteani and Preinfalk (Chapter 9)
Bayan Obo, Inner Mongolia, China	Bastnäsite, monazite, other RE fluocarbonates including Ba fluocarbonates, aeshynite and many other RE minerals in dolomite host rock; origin is controversial	see Wu, Yuan and Bai (Chapter 11)
Bear Lodge, Wyoming, USA	Ancylite intergrown with calcite in calcite carbonatite (hydrothermal)	Mariano (1978)
Fen, Norway	Bastnäsite-(Ce), parisite-(Ce), monazite-(Ce), monazite-(Nd) in ankerite carbonatite and hematite-rich carbonatite (rødberg); suggested that rødberg formed by alteration of ankerite carbonatite.	Andersen (1986), Andersen (1987a, 1987b)
Gatineau, Quebec, Canada	Parisite-(Ce), synchysite-(Ce), some monazite-(Ce) in calcite carbonatite; formed by reworking of apatite by low-temperature solutions	Hogarth et al. (1985)
Gem Park, Colorado, USA	Monazite intergrown with apatite in dolomite carbonatite, monazite plus quartz; hydrothermal	Mariano (1985)
Kangankunde, Malawi	Monazite-(Ce), bastnäsite-(Ce), florencite–goyazite in ferroan dolomite carbonatite and quartz rocks.	This chapter
Khibina, Kola Peninsula, Russia	Burbankite, ancylite-(Ce), synchysite-(Ce), carbocernaite, cordylite and other Ba, RE fluocarbonates in calcite, ankerite carbonatites; mckelvyite in late zeolite veins	Zaitsev et al. (1990), Minakov, Dudkin and Kamenev (1981), Voloshin et al. (1990)
Kizilçaören, Turkey	Bastnäsite with baryte and fluorite in calcite carbonatite; alteration of earlier calcite carbonatite at high HF fugacity	Hatzl (1992)
Mountain Pass, California, USA	Bastnäsite, parisite with Sr-baryte in calcite and dolomite carbonatite (probably magmatic); synchysite-(Ce), sahamalite-(Ce) and monazite-(Ce) with quartz, fluorite, low-Sr baryte and strontianite; hydrothermal	Olson et al. (1954), Wyllie, Jones and Jinfu Deng, (Chapter 4) Interpretation of assemblages, this chapter

Table 8.1 *Continued*

Carbonatite	Rare earth mineral assemblage	References
Mt Weld, Western Australia	Lateritic weathered carbonatite; monazite, rhabdophane, churchite, crandallite group including florencite, and minor cerianite; replacing apatite, monazite, synchysite from the fresh carbonatites	Lottermoser (1990)
Nkombwa, Zambia	Monazite-(Ce) with baryte, isokite, occasional daqingshanite, in silicified carbonatite and dolomite carbonatite	Deans and McConnell (1955) Appleton et al. (1992)
Sarnu Rajasthan, India	Carbocernaite as exsolved lamellae in calcite and (magmatic) cotectic texture with calcite and strontianite in 10 cm-wide carbonatite dykes; up to 5.5 wt% REE	Wall, Le Bas and Srivastava (1993)
Cerro Manomó, Bolivia	Cerianite, monazite, bastnäsite, La,Nd phosphates and silicates in silicified carbonatite	Fletcher et al. (1981)
St Honoré, Quebec, Canada	Bastnäsite, parisite and lesser monazite in the most evolved ferrocarbonatite, precipitated from $F-CO_3$-bearing aqueous phase which brecciated earlier carbonatite	Fournier, Williams-Jones and Wood (1993)
Tundulu, Malawi	Synchysite-(Ce), parisite-(Ce) and bastnäsite-(Ce) in hydrothermal quartz–baryte veins in calcite and ankerite carbonatite	Ngwenya (1994) Garson (1966)
Uyaynah, United Arab Emirates	Allanite in extrusive silicocarbonatite lapilli; formed, at least in part, during metamorphism of the original carbonatite	Woolley et al. (1991)
Wigu Hill, Tanzania	Bastnäsite with quartz, strontianite and baryte in dolomite carbonatite; hydrothermal	Mariano (1973)

forming minerals are usually low (e.g. <0.5 wt%) but because they are major carbonatite components they can host significant amounts of REE (e.g. Mariano, 1979; Hornig, 1988). The mineralogy of perovskites is considered by Mitchell (Chapter 2) and apatite and pyrochlore in carbonatites were reviewed by Hogarth (1989).

Therefore, a 'typical' intrusive carbonatite sequence would be early calcite carbonatite with apatite hosting much of the REE at low levels, accessory pyrochlore; then, sometimes, a dolomite carbonatite with slightly higher levels of REE and finally subordinate veins of an RE-rich carbonatite, often ferrocarbonatite, containing RE minerals such as the fluocarbonates and monazite. Examples of this kind of evolution include the carbonatites at Chilwa Island, Malawi (Garson and Campbell-Smith, 1958; Garson, 1966) and Wasaki, Kenya (Chapter 10 in Le Bas, 1977). An isotope study of carbonatite differentiation at Qaqarrsuk is reported by Knudsen and Buchardt (1991). Although these sequences are common in carbonatites, it is interesting to note that many of the most RE-rich carbonatites, e.g. Mountain Pass (Olson et al., 1954) and Kangankunde (below), do not show this kind of evolution.

8.4 Extrusive carbonatites

A rare example of a rare earth-rich extrusive carbonatite occurs at Uyaynah in the United Arab Emirates, where silicate-rich calcite carbonatite contains abundant allanite-(Ce) (Table 8.1; Woolley et al., 1991). Unfortunately, this carbonatite has undergone low-grade metamorphism and at least some of the allanite appears to be a product of this alteration so that the identity of the original primary extrusive RE mineral is unclear. The allanite is a pleochroic, low-Th variety typical of carbonatites (Kapustin, 1980; authors' unpublished data on the Mountain Pass and Sarun carbonatites).

8.5 Formation of rare earth minerals in carbonatites

The tendency for RE minerals to form in the final stages of carbonatite emplacement leads to particular problems in the interpretation of their mode of formation. It is often unclear whether they have precipitated from a carbonatite magma or from hydrothermal or carbothermal solutions. The presence of particular rare earth minerals is rarely a reliable indicator of the way in which a deposit has developed because most RE minerals can precipitate under a variety of conditions. For example, monazite crystallizes in magmatic, hydrothermal and supergene environments (Overstreet, 1967); experimental work indicates that it is possible for the RE fluocarbonate minerals to crystallize hydrothermally or from a melt (Wyllie, Jones and Deng, Chapter 4) and the RE fluocarbonates are common constituents of hydrothermal deposits (Gieré, Chapter 5).

Rare earth mineral deposits produced by primary crystallization from a carbonatite magma are rare. Mariano (1989a, 1989b) gave the bastnäsite and parisite at Mountain Pass as the main example (Wyllie, Jones and Deng, Chapter 4) and suggested the monazite–strontianite carbonatite at Kangankunde, Malawi as another possible candidate.

Hydrothermal RE mineralization is much more common, producing minerals such as bastnäsite, parisite, synchysite, monazite and ancylite. Often these minerals occur as fracture or void fillings where they are usually accompanied by quartz, fluorite, baryte, hematite or sulphides. Alternatively, they may occur as disseminated, fine-grained, polycrystalline aggregates overprinting earlier minerals. Concentrations of rare earths in these rocks are often about 1–3 wt%. From an economic point of view it is noteworthy that although such deposits are common, even if the quantities are of commercial interest, the fine-grained nature of the minerals without compensating very high REE tenors usually precludes their economic beneficiation by conventional mineral-winning techniques. Low-temperature hydrothermal solutions may cause the breakdown of primary minerals, thus releasing REE to form secondary RE minerals in a manner similar to that caused by weathering (below) and this mechanism is believed to be responsible for many of the RE minerals in carbonatites. Not all hydrothermal mineralization associated with carbonatites contains RE minerals.

During the weathering of carbonatites primary minerals, such as apatite, calcite and dolomite, may break down with the consequent release of REE which, together with any RE minerals from the fresh rock, may become concentrated in the weathering products. RE minerals formed in this way include monazite, rhabdophane, churchite, and florencite and other RE-bearing members of the crandallite group (goyazite, gorceixite and crandallite). Morteani and Preinfalk describe a study of two weathered Brazilian carbonatites in Chapter 9.

8.6 Rare earth minerals in Kangankunde Carbonatite Complex

The Kangankunde Carbonatite Complex is one of the largest carbonatites in the Chilwa Alkaline Province of southern Malawi (Garson, 1965; Woolley, 1991) and is notable as the richest in rare earths. It consists of ferroan dolomite and ankerite carbonatites, including some manganese-rich varieties, surrounded by a fenite aureole of inner potassic and outer sodi-potassic fenites (Figure 8.1). There are no major associated silicate rocks and no calcite carbonatite. The major proportion of the Kangankunde carbonatites are magnesio- and ferrocarbonatites that would be thought of as 'late stage'. 'Early' carbonatites are present only as two small intrusions of apatite-rich dolomite carbonatite (beforsite) (Figure 8.1). Kangankunde was the subject of a detailed petrographic study by Garson and Campbell-Smith (1965) and the specimens described in their memoir, now part of the collection at the Natural History Museum London (specimens BM 1962,73 (1)–(146)), form the basis of this study. Rare earth minerals form 5–10% of some Kangankunde rocks and a number of economic investigations have been carried out, including those of (Holt 1965) and, most recently, BRGM, France. Isotopic investigations of Kangankunde material have been made by Snelling (1965) who dated the complex at 129 Ma, Nelson *et al.* (1988), Ziegler (1992) and Wall, Barreiro and Spiro (1994). Magnesite–siderite series carbonates from Kangankunde were described by Buckley and Woolley (1990) and secondary strontianite by Garson and Morgan (1978).

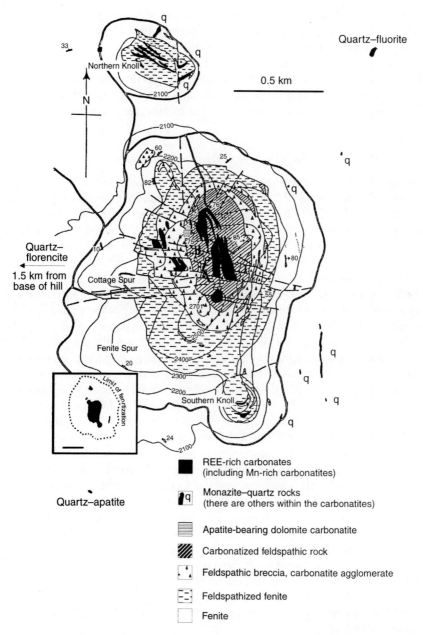

Figure 8.1 Geological map of the Kangankunde carbonatite complex. (Modified from Garson, 1965 and Garson and Campbell-Smith, 1965.)

Kangankunde is remarkable for the amount of bright green monazite-(Ce) which occurs throughout much of the carbonatite. Crystallographic studies of this monazite have been made by Ni, Hughes and Mariano (1995) and Cressey *et al.* (in preparation). In addition to the carbonatites there are a number of quartz rocks which also host rare earth mineralization. Some of these occur within the main area of carbonatite whilst others have been found within, and beyond, the outer fenite aureole (Garson and Campbell-Smith, 1965).

8.6.1 Rare earth mineral assemblage in carbonatites

Practically all the ankerite and ferroan dolomite carbonatites at Kangankunde contain some RE minerals and, although Garson and Campbell-Smith (1965) were able to distinguish a number of carbonatite types by colour, grain size and the amount of Fe and Mn, the assemblage of RE and directly associated minerals remains similar throughout. The most abundant components are monazite-(Ce), bastnäsite-(Ce) and, to a lesser extent, florencite–goyazite, which are part of a distinctive assemblage associated with one or more of strontianite, baryte and occasional quartz. Limonitic iron oxides are also sometimes abundant and minor phases include secondary calcite and apatite and rare pyrochlore and daqingshanite-(Ce).

The distribution of the RE mineral assemblage within the carbonatites is heterogeneous. Sometimes it occurs as radiating, prismatic pseudomorphs with a hexagonal cross-section (Figure 8.2a) which are sometimes revealed by the contrast between dark surrounding carbonate and the light-coloured RE assemblage in hand specimen, or if the surrounding carbonate is light they are shown by backscattered electron imagery or by cathodoluminescence. Segregations of the RE mineral assemblage, sometimes containing drusy cavities, are also common within the carbonatites. Veins of monazite–baryte–ferroan dolomite and monazite–baryte–strontianite, usually several centimetres wide, are also observed cross-cutting Fe-rich carbonatites. Later 1–6 mm-wide veins of Fe-rich carbonatite are barren of RE minerals.

Monazite-(Ce) is green in hand specimen and pale green to colourless and slightly pleochroic in thin section. It occurs in clusters of euhedral crystals, 30–400 μm in diameter, which often appear to have co-crystallized with strontianite and are sometimes partially enclosed by baryte. Bastnäsite-(Ce) always forms narrow laths in thin section (these are probably cross-sections through small platelets) and is often in clusters or sheaves. Its modal proportion does not exceed 1%. Other members of the fluocarbonate group are present but rare.

8.6.2 Rare earth mineral assemblage in quartz rocks

Quartz rocks occur as patches, ranging from one to tens of metres in diameter, within the main area of carbonatite. They contain an assemblage

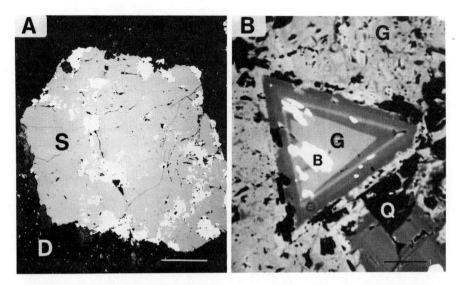

Figure 8.2 Backscattered electron images. (a) Hexagonal pseudomorph of strontianite (S) with monazite-(Ce) with a little baryte (both white, very similar backscatter coefficient) in ferroan dolomite (D) carbonatite. Scale bar = 1 mm. BM 1962, 73 (76). (b) Zoned florencite–goyazite (G) with inclusions of bastnäsite-(Ce) and later massive goyazite (G) in quartz (Q) rock. Scale bar = 10 μm; BM 1962, 73 (114).

of RE and associated minerals which is similar to that in the carbonatites except for the absence of strontianite. The most abundant RE mineral is monazite-(Ce), which is similar to the carbonatite monazite-(Ce), although at 200 μm or more diameter, the crystals tend to be larger and of a less intense green. Florencite-goyazite is more abundant than in the carbonatites. It occurs as 10–200 μm rhombs grouped in aggregates up to 1 cm diameter and comprising 5–15 modal % of the rock. Many crystals are concentrically zoned in Sr versus REE and such crystals sometimes contain laths of bastnäsite-(Ce) (Figure 8.2b) which may cut across the concentric zones. There are occasional 5 μm monazite-(Ce) crystals amongst the florencite–goyazite rhombs and sometimes two varieties, zoned and massive, florencite–goyazite are intergrown (Figure 8.2b). Baryte is ubiquitous. Apatite is rare but has been observed intergrown with, and possibly being replaced by, goyazite.

8.6.3 *Comparison of the RE mineral assemblage and mineral chemistry in carbonatites and quartz rocks*

The similarity of the mineral assemblages in the carbonatites and quartz rocks suggests that the two are closely related and there is a continuum

Table 8.2 Selected electron microprobe analyses of monazite-(Ce) from Kangankunde

Rocks:	From REE-rich carbonatites					From quartz rocks					From apatite-bearing dolomite carbonatite		
BM 1962,73()	1	2	3	4	5	6	7	8	9	10	11	12	13
	115	115	78	87	122	114	114	114	109	109	59	59	65
SiO_2	–	–	–	0.11	–	–	0.18	0.17	0.14	0.23	–	–	0.15
P_2O_5	30.47	31.68	30.65	29.01	30.25	30.30	29.52	29.63	30.25	30.17	31.12	30.79	28.92
CaO	0.41	0.44	0.46	0.21	–	–	0.15	0.19	0.18	0.15	0.44	4.08	0.17
SrO	2.14	2.97	1.89	1.14	1.52	1.44	1.49	1.73	1.48	1.75	1.98	1.30	1.31
Y_2O_3	–	–	–	–	–	–	–	0.18	0.08	0.09	–	–	0.08
La_2O_3	22.42	18.50	20.63	23.11	26.70	18.58	21.21	15.88	19.99	15.34	17.06	24.76	16.16
Ce_2O_3	34.60	33.97	35.45	36.13	34.29	35.61	33.76	34.63	35.91	36.39	36.88	31.94	40.71
Pr_2O_3	2.16	2.95	2.12	2.64	2.05	3.24	2.70	3.55	3.33	3.61	2.49	1.16	3.24
Nd_2O_3	7.62	9.15	7.79	7.78	5.49	9.12	7.80	11.50	9.69	12.23	9.72	4.92	9.03
Sm_2O_3	–	–	–	0.48	0.27	0.50	0.76	1.30	0.66	1.16	–	–	0.79
Gd_2O_3	–	–	–	–	–	–	0.28	0.46	–	–	–	–	–
ThO_2	–	–	–	–	–	–	0.05	0.29	–	0.21	–	–	–
Total	99.82	99.66	98.99	100.61	100.57	98.79	97.90	99.51	101.71	101.33	99.69	98.95	100.56

Formula calculated to 16(O)

	1	2	3	4	5	6	7	8	9	10	11	12	13
Ca	0.068	0.072	0.077	0.036	–	–	0.025	0.032	0.030	0.025	0.073	0.668	0.029
Sr	0.193	0.263	0.171	0.105	0.137	0.131	0.138	0.158	0.133	0.157	0.177	0.115	0.121
Y	–	–	–	–	–	–	–	0.015	0.007	0.007	–	–	0.007
La	1.287	1.043	1.186	1.350	1.534	1.075	1.247	0.925	1.139	0.876	0.971	1.394	0.946
Ce	1.970	1.901	2.022	2.094	1.955	2.043	1.970	2.001	2.030	2.062	2.081	1.784	2.363
Pr	0.122	0.164	0.120	0.152	0.116	0.185	0.157	0.204	0.188	0.204	0.140	0.065	0.187
Nd	0.423	0.500	0.434	0.440	0.305	0.511	0.444	0.649	0.535	0.676	0.535	0.268	0.512
Sm	–	–	–	0.026	0.014	0.027	0.042	0.071	0.035	0.062	–	–	0.043
Gd	–	–	–	–	–	0.031	0.015	0.024	–	–	–	–	–
Th	–	–	–	–	–	–	0.002	0.010	–	0.007	–	–	–
Total	4.064	3.944	4.009	4.203	4.063	4.006	4.040	4.089	4.096	4.077	3.977	4.294	4.207
Si	–	–	–	0.017	–	–	–	–	0.022	0.036	–	–	0.024
P	4.014	4.101	4.044	3.890	3.990	4.022	3.980	3.994	3.957	3.956	4.064	3.980	3.884
Total	4.014	4.101	4.044	3.908	3.990	4.022	3.980	3.994	3.979	3.991	4.064	3.980	3.908

Analyses 1,2,3,5,11,12 by energy-dispersive electron microprobe analysis using a Hitachi SEM with Link An10/55 S EDS system operated at 15 kV and 1 nA with ZAF correction. Analyses 4,6,7,8,9,10,13 by wavelength-dispersive electron microprobe analysis using either a Cambridge Instruments Microscan 9 at 20 kV and 25 nA with ZAF correction or a Cameca SX 50 at 20 kV and 20 or 25 nA with a øpz correction programme. Synthetic REE-bearing glasses were used as standards for the REE (Williams, Chapter 13).
All analyses were made at the Natural History Museum, London: – = below detection limits. Analyses 7 and 8 are from the same crystal.

between them, from monazite–baryte–strontianite to florencite–goyazite and quartz.

The close relationship between the carbonatites and quartz rocks is further supported by an electron microprobe study of the monazite-(Ce). Its composition is the same in both rock types (Table 8.2). Kangankunde monazite is strongly enriched in the light REE with a La/Y$_{cn}$ ratio of about 1500 (Figure 8.3), has a low ThO$_2$ content, usually below 0.5 wt%, and consistently contains a small and variable amount of SrO which averages 1.5 wt% (Table 8.2). These features are typical of carbonatite monazites. For example, Fleischer and Altschuler (1969) showed that monazite from carbonatites is usually more enriched in the light REE than monazites from granitic rocks or granitic pegmatites; Overstreet (1967) noted that monazite from 14 carbonatites contained an average of only 1.8 wt% ThO$_2$, which is lower than monazites formed in other geological environments, and Sr is reported in monazite from Nkombwa, Zambia (authors' unpublished data), Uyaynah, UAE (Woolley et al., 1991) and Kizilçaören, Turkey (Hatzl, 1992).

Although the abundance of monazite at Kangankunde is unusual, green monazite is reported from other carbonatites. The Sr-bearing monazite at Nkombwa Hill (Deans and McConnell, 1955) is green and some also forms pseudomorphs, although it is different from Kangankunde monazite in that some is spherulitic. The monazite-bearing rocks at Nkombwa are siliceous but it is difficult to determine whether the silica is a product of late hydrothermal activity of silicification due to weathering. Associated phases are

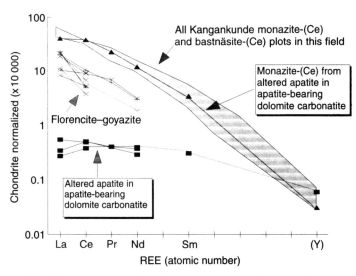

Figure 8.3 Chondrite-normalized plots of RE and RE-bearing minerals at Kangankunde. (Chondrite values from Wakita, Rey and Schmitt, 1971.)

baryte and isokite. A second example is at Katete, Zimbabwe where Lee (1974) describes 'lime green' monazite associated with baryte and quartz in circular patches up to 15 mm in diameter in dolomite carbonatite.

A second unusual feature of the Kangankunde monazite crystals is that they are sector zoned. The degree of light REE-enrichment varies between sectors, as does Sr. The greatest variation is in the amount of La_2O_3, which ranges between 11 and 25 wt%, whereas Ce_2O_3 is nearly constant and this produces a 'pivoting' effect on a chondrite-normalized graph (Figure 8.3). An X-ray map of one of these zoned crystals is given in Williams (Chapter 13, Figure 13.4) and the crystallography is being described by Cressey, Wall and Cressey (in preparation). It is believed to be this sector zoning, rather than precipitation from an evolving carbonatite magma or fluid, which has caused most of the variation in point analyses of the monazite crystals. A plot of the main variants in Kangankunde monazite, Sr and La/Ce, emphasizes the similarity between monazite from quartz rocks and carbonatites and demonstrates the variation caused by this sector zoning (Figure 8.4).

The Kangankunde composition and lack of compositional evolution is in marked contrast to monazite reported by Paterson and Cooper (1993) from carbonatite dykes at Haast River, New Zealand, where compositions become successively more depleted in the light REE from calcite through ankerite to siderite and dolomite–hematite-rich carbonatites. These authors modelled this change as a product of fractional crystallization.

The florencite–goyazite series minerals are all midway between the Sr and

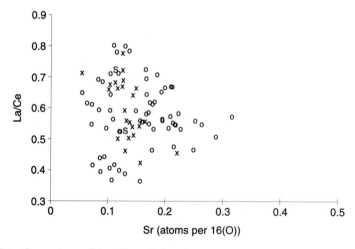

Figure 8.4 Comparison of La, Ce and Sr, the greatest variables in monazite-(Ce) from Kangankunde carbonatities (circles) and quartz rocks (crosses; S = example of two analyses on one crystal).

Table 8.3 Selected electron microprobe analyses of florencite–goyazite and bastnäsite-(Ce) from Kangankunde

	Florencite–goyazite									Bastnäsite-(Ce)			
	From quartz rocks					From REE-rich carbonatite				From quartz rock	From REE-rich carbonatite		From apatite-bearing dolomite carbonatite
Rock: BM 1962.73()	1 109	2 109	3 109	4 128	5 128	6 100	7 100	8 123	9 123	10 117	11 103	12 103	13 59
Al_2O_3	31.30	31.60	31.60	30.40	31.60	31.50	31.70	30.10	30.40	–	–	–	–
SiO_2	–	–	–	0.51	–	–	–	–	–	0.38	0.41	0.33	0.30
P_2O_5	29.10	29.60	30.00	27.80	29.30	29.30	29.30	28.60	28.90	–	–	–	–
CaO	0.13	0.21	0.12	0.81	0.24	–	–	0.14	–	0.29	0.48	0.27	0.60
FeO	0.40	0.40	0.50	1.00	–	–	–	–	–	–	–	–	–
SrO	6.60	11.20	15.40	7.80	14.10	11.50	10.20	9.00	8.50	0.50	0.37	0.61	0.80
BaO	–	–	–	0.50	2.30	0.70	–	–	–	–	–	–	–
La_2O_3	8.10	8.80	4.60	3.60	3.40	9.20	8.70	4.20	7.50	23.10	27.70	25.30	22.59
Ce_2O_3	11.20	5.70	4.10	8.30	5.70	5.40	7.40	9.80	10.00	33.40	29.40	31.30	32.77
Pr_2O_3	0.90	–	–	0.90	–	–	–	1.00	–	3.20	2.10	2.60	1.49
Nd_2O_3	2.30	–	–	3.50	1.40	–	1.10	2.40	1.10	7.80	5.60	7.10	7.81
Sm_2O_3	–	–	–	0.40	–	–	–	–	–	–	–	–	–
ThO_2	–	–	–	1.40	–	–	–	–	–	–	–	–	–
SO_3	0.30	0.27	0.88	0.55	0.23	0.26	1.50	0.39	0.44	–	–	–	–
Total	90.33	87.78	87.20	87.47	88.27	87.86	89.90	85.63	86.84	68.67	66.06	67.51	66.36

	Formula calculated to 22 (O)									Formula calculated to 24(O)			
Al	6.045	6.107	6.042	6.017	6.138	6.128	6.008	6.026	6.015	–	–	–	–
Si	–	–	–	0.086	–	–	–	–	–	–	0.260	0.210	0.001
P	4.037	4.109	4.120	3.952	4.088	4.094	3.989	4.113	4.107	–	–	–	–
Ca	0.023	0.037	0.021	0.146	–	–	–	0.025	–	0.240	0.330	0.180	0.252
Fe	0.055	0.055	0.068	0.140	0.042	–	–	–	–	0.190	–	–	–
Sr	0.627	1.064	1.448	0.759	1.347	1.100	0.951	0.886	0.827	0.180	0.140	0.230	0.153
Ba	–	–	–	0.033	0.149	0.045	–	–	–	–	–	–	–
La	0.490	0.532	0.275	0.223	0.207	0.560	0.516	0.263	0.464	5.320	6.600	5.940	5.464
Ce	0.672	0.342	0.243	0.510	0.344	0.326	0.435	0.609	0.614	7.640	6.950	7.290	7.853
Pr	0.054	–	–	0.055	–	–	–	0.062	–	0.730	0.490	0.600	0.355
Nd	0.135	–	–	0.210	0.082	–	0.063	0.146	0.066	1.740	1.290	1.610	1.891
Sm	–	–	–	0.023	–	–	–	–	–	–	–	–	–
Th	–	–	–	0.054	–	–	–	–	–	–	–	–	–
S	0.037	0.033	0.107	0.069	0.028	0.032	0.181	0.050	0.055	–	–	–	–
Total	12.173	12.280	12.325	12.276	12.425	12.287	12.143	12.179	12.149	16.050	16.070	16.070	15.896

Analyses by EDS (see Table 8.2); total Fe as FeO; F, CO_2 not analysed; – = below detection limit; All rocks are from BM 1962.73(1)–(146), e.g. 109 is BM 1962.73(109)

Analyses 1,4,8, and 9 are florencite-(Ce); analysis 7 is florencite-(La); analyses 2,3,5 and 6 are goyazite; rock BM 1962.73(128) occurs 1.5 km from the carbonatite (see Figure 8.1).

REE end members of the florencite–goyazite series, $SrAl_3(PO_4)_2(OH)_5 \cdot H_2O$–$(La,Ce)Al_3(PO_4)_2(OH)_6$ (Table 8.3). Their SrO content ranges from 6.6 to 18.4 wt% in a reciprocal relationship with REE so that the proportion of the florencite end member ranges from 0.32 to 0.85. Florencite and goyazite are members of the crandallite group of minerals of which there are two other common end members, gorceixite (Ba) and crandallite (Ca). However, the Ba and Ca contents as Kangankunde are both low (Table 8.3).

Like monazite, florencite is a light REE-enriched mineral (synthesis of pure REE florencites with any REE heavier than Gd was found impossible by Schwab *et al.*, 1990) but the Kangankunde florencite–goyazite REE chondrite-normalized patterns are more variable than those of the monazite (Figure 8.3). A comparison of the La/Ce ratio demonstrates the difference (Figure 8.5). The greater degree of variation coupled with the concentric zoning and variety of textures (Figure 8.2) suggests that the compositional variation in florencite–goyazite represents changes in fluid chemistry, in contrast to the crystal chemical control on the monazite-(Ce). Again there are no systematic differences between florencite–goyazite in the carbonatites and quartz rocks (Figure 8.5).

Florencite–goyazite series minerals have never been described as magmatic products. They are best known as carbonatite weathering products (e.g. Mariano, 1979; Laval, Johan and Tourlière, 1988; Lottermoser, 1990), have been synthesized hydrothermally (Schwab *et al.*, 1990) and have been described from retrograde metamorphic deposits (Ek and Nysten, 1990).

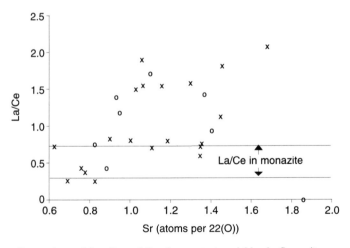

Figure 8.5 Comparison of La, Ce and Sr, the greatest variables in florencite–goyazite from Kangankunde carbonatites (circles) and quartz rocks (crosses). The range of La/Ce in monazite is indicated for comparison.

Differential thermal analyses reported by McKie (1962) suggest that florencite from Kangankunde is stable to 535 ± 10°C at 200 bars P_{H_2O} and 565 ± 10°C at 2500 bars P_{H_2O}. The Kangankunde florencite–goyazite contrasts with reported carbonatite weathering products in that the weathered florencite–goyazite usually contains larger proportions of the Ba and Ca end members (e.g. Laval, Johan and Tourlière, 1988; Lottermoser, 1990). The Kangankunde material is believed to be of hydrothermal origin.

It is difficult to obtain reliable electron microprobe analyses of the bastnäsite-(Ce) at Kangankunde because the crystals are only 1–2 μm wide and best results were obtained by energy-dispersive spectrometry (Table 8.3). There is little variation in the composition of bastnäsite-(Ce), the REE content of which is similar to that of the monazite. Kangankunde fluocarbonates almost always occur as acicular crystals, often in sheaves. Similar habits are common in other carbonatites such as Tundulu (Ngwenya, 1994) and Chilwa Island (Garson and Campbell-Smith, 1958). Mariano (1989b) has suggested that this form is indicative of a hydrothermal origin. The predominance of bastnäsite-(Ce) rather than the Ca-bearing RE fluocarbonates is probably a reflection of the low Ca content throughout the RE mineral assemblage.

The composition of baryte is also the same in quartz and carbonate rocks. It is close to the Ba end member with 0.6–1.2 wt% Na_2O, <0.5 wt% CaO and 0.4–1 wt% SrO. Baryte in carbonatites has been reported to contain up to 5 wt% SrO (e.g. Kapustin, 1980) but such baryte is reported as an early or 'first' generation phase and there is usually also a later, low-Sr baryte in the same complex. Examples of this contrast in the composition of early and late baryte occur at Vuoriyarvi, Sallanlatva and Kovdor (Kapustin, 1980), at Mountain Pass (see below) and in the carbonatites peripheral to the Paraná Basin. The Kangankunde baryte is consistent with the late baryte reported from elsewhere. Strontianite in the carbonatites is also close to the end-member composition, containing only 1–3 wt% CaO.

It is concluded that the RE minerals in both the carbonatites and quartz rocks at Kangankunde belong to the same phase of major mineralization, which was predominantly a subsolidus replacement of earlier carbonatite.

8.6.4 Incipient mineralization in apatite dolomite carbonatites

The earliest carbonatite at Kangankunde consists of small intrusions of medium-grained, apatite-rich dolomite carbonatite (apatite beforsite of Garson and Campbell-Smith, 1965; Figure 8.1) which is petrologically distinct from the RE-rich carbonatite. There are relict silicate phases such as large olivine crystals, now replaced by serpentine, and possibly nepheline. Garson and Campbell-Smith (1965) speculated that these rocks may represent carbonated nephelinite. The main carbonate mineral is ferroan dolomite: apatite occurs as the lozenge-shaped crystals typical of many early carbonatites. Occasional perovskite crystals are a Th-bearing variety, with an outer REE-bearing overgrowth.

If the main phase of RE mineralization is metasomatic, it might be expected to have invaded these earlier carbonatites and, indeed, there is evidence of incipient RE mineralization. The best example is in a rock which has also been phlogopitized (BM 1962, 73 (59)). Monazite overgrowths emanate from the margins of the larger apatite crystals and cathodoluminescence imagery shows alteration in some apatite to produce zones with higher levels of REE. In one turbid area, cathodoluminescence and backscattered electron imagery revealed altered apatite with development of a neighbouring assemblage of monazite-(Ce), bastnäsite-(Ce), fluorite, baryte and a more REE-rich apatite and monazite at the corroded crystal margin (Figure 8.6). Veinlets of calcite cut the apatite. The mineral assemblage and composition of the minerals formed (Tables 8.2 and 8.3) are very similar to those of the main RE mineral assemblage (Figure 8.3). Again there is some variation in monazite, pivoting around Ce as in the sector-zoned crystals. Altered apatite contains up to $1\,wt\%$ RE_2O_3, which is more than in the unaltered material (Table 8.3) but much less than can be accommodated in the apatite structure. Monazite appears therefore to be formed in preference to a REE-rich apatite. Chondrite-normalized REE plots for the apatite are less light REE-enriched than those of the monazite-(Ce) and bastnäsite-(Ce) (Figure 8.3).

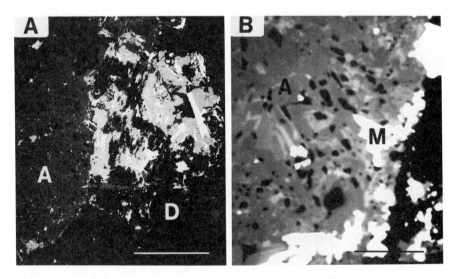

Figure 8.6 Backscattered electron images. (a) Bastnäsite-(Ce) needles (white) in strontianite (fluorite not visible) next to altered apatite (A) in dolomite (D) in apatite-bearing dolomite carbonatite BM 1962, 73 (59). Scale bar = $250\,\mu m$. (b) Monazite (M) at the edge of altered apatite (A) featured in Figure 6a. Variations in the backscatter coefficient of apatite (grey colour) relate to Sr and REE content. Scale bar = $10\,\mu m$.

8.6.5 Quartz rocks outside the carbonatite

There are also REE-bearing quartz rocks outside of the main area of carbonatite at Kangankunde (Figure 8.1). Quartz–monazite rocks close to the carbonatite are similar to those described above but further away Garson and Campbell-Smith (1965) described and mapped quartz–apatite, quartz–fluorite and quartz–florencite rocks.

The quartz–apatite rock occurs 1 km from the carbonatite. It comprises 40% quartz, 40% apatite and 20% iron oxides. In hand specimen the apatite has a pinkish tint, in thin section it has turbid cores and clear, euhedral overgrowths where it is in contact with the quartz. The iron oxides, now limonitic material, have rhombohedral outlines, again where in contact with quartz, and were termed rhombohedral carbonate by Garson and Campbell-Smith (1965), who believed them to be remnants of an earlier carbonate. Quartz is the last phase. Fluorapatite in the turbid cores contains 1–2 wt% SrO (Table 8.4) and is not zoned. In contrast, Sr-bearing fluorapatite overgrowths exhibit fine oscillatory zoning and have occasional zones which are particularly rich in REE (Figure 8.7 and Table 8.4). Individual zones are too small to be analysed by electron microprobe but, as indicated by analysis of the bright bars, the zoning is believed to be due to variation in the REE. Towards the edge of many overgrowths there is a zone consisting of small REE-rich phosphate crystals plus quartz (Figure 8.7). Analyses from three such crystals are reasonably consistent, with about 25 wt% P_2O_5 and 50 wt% RE_2O_3, the main variation being in ThO_2, which ranges from 6.8 to 15.7 wt%; however, the identity of this phase is not known. An overgrowth of xenotime-(Y) (Table 8.5; Figure 8.7) was found on niobian rutile. Other minor phases in the rock include rare 1 μm crystals of boulangerite ($Pb_5Sb_4S_{11}$) in the quartz and rare columbite. Similar-looking apatite overgrowths from the Tundulu carbonatite complex, Malawi, were described by Ngwenya (1994), also as a product of carbonatitic hydrothermal activity. However, these overgrowths and the associated RE mineral, synchysite-(Ce), are all light REE enriched.

The Kangankunde minerals described above do not have the strongly light REE-enriched chondrite-normalized patterns characteristic of the rest of the carbonatite complex. Instead, they show varying degrees of enrichment in the mid-REE (Figure 8.8). Xenotime-(Y) is more enriched in the heavy REE than the other phases but its chondrite-normalized pattern peaks at Tb rather than the more usual Y. The xenotime and apatite results plotted on Figure 8.8 were determined on neighbouring overgrowths of xenotime and apatite and, assuming that they co-crystallized, which seems reasonable, the xenotime–apatite partition coefficients range from 0.26 for Ce, which is favoured by the apatite, to 486 for Y, which is strongly partitioned into the xenotime-(Y). Thus it is likely that the peak at Tb on the xenotime pattern is not due to the signature of the fluid from which it formed but is consistent

Table 8.4 Selected electron microprobe analyses of apatite from Kangankunde

Rock Description	From apatite-bearing dolomite carbonatite BM 1962.37(59)						From quartz-apatite rock BM 1962.73(131)					
	Unaltered			Altered (see Figure 8.6)			Core		Overgrowth		See Figure 8.7 Brightest zones in overgrowth	
	1	2	3	4	5	6	7	8	9	10	11	12
Na_2O	0.17	–	0.29	–	–	0.09	–	0.45	0.36	0.26	1.27	0.75
MgO	0.17	0.17	–	–	–	–	–	–	–	–	–	–
Al_2O_3	0.06	–	0.50	0.12	–	–	0.23	0.04	–	–	–	–
SiO_2	0.20	0.15	0.07	–	0.07	–	0.07	–	0.07	0.15	–	–
P_2O_5	40.95	41.81	39.41	39.72	40.78	40.46	39.13	39.99	39.71	38.61	38.86	39.71
CaO	54.42	54.66	53.47	54.31	51.52	51.72	54.62	52.70	50.75	50.81	46.75	48.33
MnO	–	0.09	–	–	0.05	0.05	–	–	–	–	–	–
FeO	0.18	0.16	0.29	0.33	–	–	0.56	0.40	0.40	0.42	0.61	0.41
SrO	0.32	0.21	1.69	1.21	3.37	4.02	1.20	1.47	2.74	2.72	2.78	4.02
Y_2O_3	–	–	–	–	–	–	–	–	0.07	0.07	–	–
La_2O_3	–	–	0.14	0.12	0.11	0.22	–	–	0.10	0.11	–	–
Ce_2O_3	0.10	–	0.23	0.19	0.41	0.53	–	0.06	0.33	0.31	1.51	1.06
Pr_2O_3	–	–	–	–	–	–	–	–	–	–	0.60	–
Nd_2O_3	0.10	–	–	0.12	0.30	0.22	–	–	0.29	0.22	2.14	1.04
Sm_2O_3	–	–	–	–	–	–	–	–	0.14	0.16	0.62	–
Gd_2O_3	–	–	–	–	–	–	–	–	0.13	0.17	–	–
ThO_2	–	–	–	–	–	–	–	0.12	–	–	–	0.55
F	2.63	2.90	3.49	3.11	3.24	3.15	3.76	4.35	3.95	4.36	–	–
Total	99.30	100.15	99.58	99.23	99.85	100.46	99.57	99.58	99.04	98.37	95.14	95.87
O = F	1.11	1.22	1.47	1.31	1.36	1.33	1.58	1.83	1.66	1.84	–	–
Total	98.19	98.93	98.11	97.92	98.49	99.13	97.99	97.75	97.38	96.53	95.14	95.87

Formula calculated to 25(O)

	1	2	3	4	5	6	7	8	9	10	11	12
Na	0.056	–	–	–	–	0.03	–	0.153	0.124	0.091	0.447	0.260
Mg	0.043	0.043	0.098	–	–	–	–	–	–	–	–	–
Al	0.012	–	0.103	0.025	0.014	–	–	–	–	–	–	–
Ca	9.966	9.908	10.010	10.143	9.625	9.674	10.241	9.891	9.649	9.808	9.091	9.261
Mn	–	0.013	–	–	–	0.007	0.012	0.024	–	0.021	–	–
Fe	0.026	0.002	0.042	0.048	0.007	–	0.082	0.059	0.059	0.063	0.093	0.061
Sr	0.021	0.013	0.111	0.079	0.221	0.264	0.079	0.097	0.183	0.184	0.189	0.270
Y	–	–	–	–	–	–	–	–	0.007	0.007	–	–
La	–	–	0.009	0.008	0.007	0.014	–	–	0.007	0.007	–	–
Ce	0.006	–	0.015	0.012	0.026	0.034	–	0.004	0.021	0.020	0.100	0.069
Pr	–	–	–	–	–	–	–	–	–	–	0.040	–
Nd	0.006	–	–	0.008	0.019	0.014	–	–	0.018	0.014	0.139	0.066
Sm	–	–	–	–	–	–	–	–	0.009	0.010	0.039	–
Gd	–	–	–	–	–	–	–	–	0.008	0.010	–	–
Th	–	–	–	–	–	–	–	0.005	–	–	–	0.022
Total	10.136	9.980	10.388	10.322	9.919	10.037	10.414	10.079	9.964	10.145	10.138	10.010
P	5.925	5.988	5.829	5.861	6.019	5.979	5.796	5.930	5.965	5.888	5.971	6.012
Si	0.034	0.025	0.012	–	–	–	–	–	–	–	–	–
Total	5.959	6.013	5.841	5.861	6.019	5.979	5.797	5.930	5.965	5.889	5.971	6.012
F	1.422	1.552	1.927	1.715	1.787	1.739	2.081	2.410	2.217	2.484	–	–

Analyses 7, 8 by EDS, all others by WDS using Cameca SX 50 (see Table 8.2); – = below detection limits; FeO = total iron.

Table 8.5 Selected electron microprobe analyses of xenotime-(Y) from Kangankunde

	1 K	2 J	3 D	4 C	5 Average of 4	SD
SiO_2	3.52	1.47	2.64	1.39	2.26	1.02
P_2O_5	30.19	31.07	30.45	30.93	30.66	0.41
CaO	0.48	0.46	0.54	0.48	0.49	0.03
Sc_2O_3	0.66	0.74	0.75	0.72	0.72	0.04
TiO_2	0.96	0.53	0.43	0.61	0.63	0.23
FeO(T)	0.66	1.32	1.58	1.40	1.24	0.40
Y_2O_3	33.61	36.76	38.34	36.14	36.21	1.97
La_2O_3	–	–	0.30	–	–	
Ce_2O_3	0.10	0.11	–	0.13	0.09	0.06
Pr_2O_3	–	0.13	–	–	0.03	
Nd_2O_3	0.40	0.39	0.34	0.69	0.46	0.16
Sm_2O_3	2.43	1.99	1.34	2.25	2.00	0.48
Eu_2O_3	2.00	1.41	1.32	1.67	1.60	0.31
Gd_2O_3	9.41	6.96	6.11	8.45	7.73	1.48
Tb_2O_3	2.81	2.08	1.93	2.61	2.36	0.42
Dy_2O_3	9.07	8.57	8.58	8.83	8.76	0.24
Ho_2O_3	0.59	1.00	1.06	0.49	0.79	0.29
Er_2O_3	1.61	2.04	2.03	1.65	1.83	0.23
Yb_2O_3	0.20	0.12	0.65	0.16	0.28	0.25
ThO_2	0.82	1.21	1.27	1.13	1.11	0.20
Total	99.52	98.36	99.66	99.73	99.24	
	\multicolumn{6}{c}{Formula calculated to 24(O)}
Ca	0.108	0.105	0.121	0.109	0.111	0.007
Sc	0.060	0.068	0.068	0.066	0.066	0.004
Ti	0.151	0.084	0.067	0.097	0.100	0.036
Fe	0.116	0.234	0.275	0.248	0.218	0.070
Y	3.755	4.151	4.256	4.072	4.058	0.216
La	–	–	0.023	–	–	
Ce	0.008	0.009	–	0.010	0.007	0.004
Pr	–	0.010	–	–	0.002	–
Nd	0.030	0.030	0.025	0.052	0.034	0.012
Sm	0.176	0.146	0.096	0.164	0.145	0.035
Eu	0.143	0.102	0.094	0.121	0.115	0.022
Gd	0.655	0.490	0.422	0.593	0.540	0.104
Tb	0.194	0.145	0.132	0.182	0.163	0.029
Dy	0.613	0.586	0.577	0.602	0.594	0.016
Ho	0.039	0.067	0.070	0.033	0.053	0.019
Er	0.106	0.136	0.133	0.110	0.121	0.015
Yb	0.013	0.008	0.041	0.010	0.018	0.016
Th	0.039	0.058	0.060	0.054	0.053	0.010
Total	6.206	6.428	6.462	6.524	6.399	0.138
P	5.364	5.580	5.376	5.543	5.465	0.112
Si	0.739	0.312	0.551	0.294	0.475	0.212
Total	6.103	5.892	5.927	5.838	5.940	0.115

Analyses by WDS on Cambridge Instruments Microscan 9 at 20 kV and 25 nA; – = below detection; FeO(T) = total Fe as FeO; SD = standard deviation; all analyses are from rock BM 1962,73(131), (see Figure 8.7).

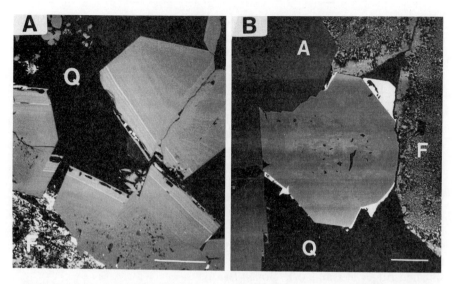

Figure 8.7 Backscattered electron images of overgrowths in quartz–apatite rock. BM 1962, 73 (131) (a) Apatite overgrowths, zoned in Sr and REE in quartz (Q). Scale bar = 50 μm. (b) Overgrowth of xenotime-(Y) (white) on niobian rutile in quartz (Q) and neighbouring iron oxide (F). A = apatite. Scale bar = 10 μm.

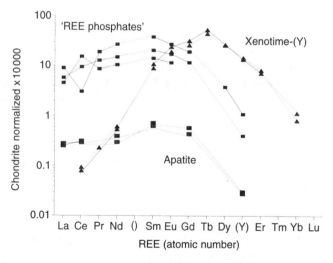

Figure 8.8 Chondrite-normalized plot of apatite in the overgrowths which contain REE detectable by electron microprobe (Figure 8.7a), RE phosphates near the edge of the apatite overgrowth (Figure 8.7a) and xenotime-(Y) (Figure 8.7b).

with the crystal chemically controlled preferential uptake of heavy REE from the mid-REE-enriched environment.

Florencite in a quartz–florencite rock collected 1.5 km from the carbonatite also has a 'flatter' chondrite-normalized distribution of REE than florencite within the main carbonatite (analyses 4 and 5, Table 8.3; Figure 8.3).

Quartz–fluorite occurs about the same distance as the quartz–apatite from the carbonatite in the opposite direction (Figure 8.1). However, although fluorite mineralization is commonly associated with carbonatites, Sm–Nd isotope data (Wall, Barreiro and Spiro, 1994) are not consistent with the rest of the carbonatite complex and therefore it is thought that this must belong to a separate vent or phase of activity. There are many small vents in the Chilwa Province (Garson, 1965) and so this is not an unreasonable possibility.

RE mineralization occurs outside of the carbonatite at a number of other carbonatite localities. For example, it invades the silicified country rock ridges at Araxá, Brazil, and Chiriguelo, Paraguay, and occurs outside the silicified ridges at Catalão I, Goiàs, Brazil.

8.7 Comparison with the Mountain Pass carbonatite

The primary magmatic origin of the Mountain Pass carbonatite has been questioned by a number of authors, for example Andersen (1987a), Bailey (1993), Vlasov (1966) and Semenov (1974). Having decided that Kangankunde is not a candidate for primary RE mineralization, it is worth a further brief consideration of Mountain Pass.

There are a variety of forms of RE mineralization in the Mountain Pass carbonatites. Bastnäsite-(Ce) and parisite-(Ce) in many of the calcitic and dolomitic units show textural evidence of having crystallized from a magmatic stage. For example, bastnäsite-(Ce) grains range from anhedral to well-defined hexagonal prisms with an aspect ratio of 2–3. Parisite-(Ce) forms short, stubby prisms with prominent basal partings. Rarely, these two minerals occur as a syntaxial intergrowth. Baryte in these carbonatites occurs as crystals which are zoned in Sr and have evidence of exsolution of Sr to form a graphic intergrowth of baryte and celestine. The pervasive occurrence of quartz druse and films of quartz on the primary minerals has led others to assume that the bastnäsite mineralization is of hydrothermal origin. However, textural evidence established that quartz has a late paragenesis relative to the RE minerals, calcite and baryte. There is, however, an association of RE minerals including synchysite-(Ce), sahamalite-(Ce) and monazite-(Ce) with quartz, fluorite, late low-Sr baryte and strontianite which certainly appears to have been deposited from a hydrothermal solution. There are some key differences between Mountain Pass and Kangankunde, notably the mineral textures, including the lack of pseudomorphous structures, the presence of Sr in baryte, the abundance of Ca-bearing RE fluocarbonates such as parisite and the difference in crystal form of the fluocarbonates at Mountain Pass. The interpretation of the Mountain Pass carbonatite as primary remains, in our view, reasonable. A carbonatite with comparable mineralogy may be that of Maoniuping, Sichuan, China (Wu, Yuan and Bai, Chapter 11).

8.8 Other carbonatites with RE mineralization in polycrystalline pseudomorphs

Large, hexagonal, prismatic, polycrystalline pseudomorphs with no obvious precursor, such as those at Kangankunde, have been described from a number of other RE-rich carbonatites. Some examples are illustrated in Figure 8.9 and the similarity to Kangankunde is easily seen. Ferroan dolomite at Gem Park, Colorado, contains pseudomorphs of polycrystalline monazite, apatite, quartz and minor calcite (Mariano, 1985). At Wigu Hill, Tanzania, the pseudomorphs are filled with polycrystalline bastnäsite, dolomite, quartz, strontianite and baryte in a groundmass of creamy-white dolomite (Mariano, 1973). In the Adiounedj carbonatite complex of northeast Mali (Mariano, 1983; Sauvage and Savard, 1985) the pseudomorphs comprise polycrystalline, microscopic platelets of synchysite and quartz in a groundmass of fluorite, calcite and ferric iron oxides. Finally at Bear Lodge, Wyoming, USA, hexagonal pseudomorphs are again present in calcite carbonatite; they consist of ancylite and strontianite in a graphic intergrowth (Mariano, 1978).

There are no remnants of the original material at any of these localities, including Kangankunde, and therefore it is impossible to prove the identity of the original mineral. However, it is likely that a similar process would have been responsible for the pseudomorph production in each case and two possible processes, replacement of apatite and replacement of the 'A-B' type REE-bearing carbonate, burbankite, are considered below.

8.8.1 Replacement of apatite

Reaction of apatite to form a RE mineral assemblage including monazite is observed at Kangankunde and necessitates reaction of apatite with RE- and Sr-bearing fluids, possibly with introduction of further F, and would release Ca, some of which may have been deposited as the calcite veinlets observed cross-cutting the apatite.

In a study of the solubility and dissolution kinetics of apatite in felsic melts (at rather higher temperatures than expected in carbonatites, 850–1500°C and 8 kbar), Harrison and Watson (1984) showed that Ca has a transport rate generally two orders of magnitude faster than P, data which may help to explain the selective removal of Ca from apatite in the presence of REE-rich hydrothermal solutions and the relative immobility of P.

Complete replacement of apatite by monazite has been observed in

Figure 8.9 Pseudomorphs of a hexagonal prismatic mineral which has been replaced by polycrystalline RE mineral assemblages. (a) Gem Park, USA; (b) Wigu Hill, Tanzania;

RE MINERALIZATION IN POLYCRYSTALLINE PSEUDOMORPHS

Figure 8.9 *Continued* (c) Adiounedj, Mali; (d) Kangankunde, Malawi, See text for details of the mineral assemblages.

dolomite carbonatite at Araxá, Brazil, where pseudomorphs of monazite after the original apatite have been produced. This occurs below the zone of weathering and is believed to be a hydrothermal replacement (Mariano, 1979). Pseudomorphs of monazite after apatite are also reported from Mount Weld and the association of small amounts of monazite at the margins of apatite has been observed in other carbonatites, e.g. at Bayan Obo (Le Bas *et al.*, 1992). Hogarth *et al.* (1985) also suggests the alteration of apatite during a hydrothermal event, possibly due to waning metamorphism sometime after emplacement of the carbonatite, as the explanation for RE fluocarbonates and possibly monazite in carbonatites near Gatineau, Quebec. However, the main problem with an apatite precursor for the Kangankunde, and other, pseudomorphs is the amount of apatite that the pseudomorphs now contain. Although it is not expected that such a reaction would be a closed system with respect to any of the components, at Araxá, for example, the whole pseudomorph comprises monazite whereas at Kangankunde monazite is found within the altered apatite but the rest of the assemblage occurs outside the apatite boundary and therefore would not produce a pseudomorph structure.

8.8.2 *Replacement of burbankite*

A second possibility is that the pseudomorphs at Kangankunde represent earlier crystals of burbankite $((Na,Ca)_3(Sr,Ba,Ce)_3(CO_3)_5)$, because both burbankite and its subsequent replacement to form pseudomorphous RE mineral assemblages have been observed in two Russian carbonatites. At Khibina, Kola Peninsula (Zaitsev *et al.*, 1990) hexagonal prisms of burbankite with cross-sections of 1–3 mm are present in calcite carbonatite. These crystal forms are found in other Khibina carbonatites replaced by ancylite and other RE minerals. Another example occurs at East Sayan, East Siberia, Russia (Zdoric, 1960), where large, hexagonal prisms of burbankite in dolomite and ankerite carbonatites are replaced first by carbocernaite, then by ancylite and then by strontianite, calcite and bastnäsite, preserving the hexagonal forms as pseudomorphs. Monazite, allanite and calcite are produced in a post-carbonatitic stage. The composition of burbankite can be rather variable (Kapustin, 1980) and this could partly account for the diversity of mineral assemblages observed in the pseudomorphs. However, in order to account for the known examples and the other pseudomorphs, the removal and introduction of various phases is required. Sodium must always be removed and hydration and the addition of one or more of phosphate, sulphur and silica is necessary. An open-system hydrothermal process would be required. Burbankite was first described by Pecora and Kerr (1953) in hydrothermal carbonate veins in shonkinite from the Bearpaw Mountains, where it had reacted to form the hydrous carbonate, ancylite.

Carbonates such as burbankite and the closely related mineral, carbocernaite ((Ca,Na)(Sr,Ce,Ba)(CO_3)$_2$) (called A-B carbonates by Burt, 1989), are rare but have been observed as magmatic minerals in carbonatites at a few other localities besides the Russian carbonatites described above. For example, Platt and Woolley (1990) noted blebs of burbankite in dolomite carbonatite at Chipman Lake, Ontario, and Wall, Le Bas and Srivastava (1993) report a carbocernaite–calcite cotectic texture in carbonatite at Sarnu, Rajasthan. Shi Li and Tong Wang (1993) report burbankite in biotite–aegirine carbonatite in the Shaxiongdong carbonatite complex, Hubei Province, China. Estimates of the temperature of formation of these rocks are above 500°C (e.g. 550–600°C from carbonates at Chipman Lake (Platt and Woolley, 1990), 550°C for carbonates at Sarnu (Wall, Le Bas and Srivastava, 1994) and 535–570°C for inclusions in calcite and biotite at Shaxiongdong (Shi Li and Tong Wang, 1993), higher than many estimates for RE fluocarbonate mineralization. Kapustin (1980) places burbankite earlier in the carbonatite sequence than the RE fluocarbonates and notes that it is an unstable mineral.

It is suggested that the 'A-B' type RE carbonates such as burbankite and carbocernaite may have a more important role in the evolution of RE minerals in carbonatites than has previously been realized.

8.8.3 Comparison with other proposed replacement processes

Other replacement processes have been proposed as mechanisms for the production of rare earth-rich carbonatites. For example at Fen (Andersen 1986, 1987a, 1987b) an RE mineral assemblage is believed to have formed as a result of alteration of ankerite carbonatite in which the REE remained immobile. In order to produce a sufficient enrichment in REE, it was necessary to propose a 70% volume reduction. At Kizilçaören, Turkey (Hatzl 1992), it is thought that an earlier calcite carbonatite was subject to fluorination and decalcification to produce a bastnäsite-(Ce)–baryte assemblage. Both of these models suggest the involvement of external fluids in the alteration process. At Kangankunde strontianite in the RE mineral assemblage is very close to (C and O) isotopic equilibrium with ferroan dolomite and there is no isotopic evidence of the introduction of any external fluids to form the quartz rocks in the carbonatite. It is therefore likely that, in contrast to the above models, the formation of pseudomorphs and related mineralization takes place very close to the magmatic boundary, possibly a form of deuteric alteration.

8.9 Conclusions

1. Mineralogical and mineral chemical evidence demonstrates that subsolidus, hydrothermal processes were responsible for the RE mineralization at Kangankunde and confirms that such processes are predominant in the formation of RE minerals in carbonatites.

2. Large, hexagonal, prismatic polycrystalline pseudomorphs are reported from seven RE-rich carbonatites with hydrothermal mineral assemblages and in two of these the precursor mineral is still present and seen to be burbankite. Although replacement of apatite to form some of these pseudomorphs is also a possibility and the reaction of apatite is important in carbonatites, both burbankite and its close chemical relation, carbocernaite, are reported as products of magmatic crystallization and may play a more important role than was previously supposed in the crystallization of RE from magmatic carbonatites.
3. RE mineralization can occur outside of the main area of carbonatite and the greatest changes in the relative concentrations of the REE are seen in these rocks at Kangankunde. This is an interesting area for further study of the transport of REE in hydrothermal fluids and may be a useful guide for exploration.

Acknowledgements

Anatoly Zaitsev is thanked for useful discussions and information; Alan Woolley, Gordon Cressey and Terry Williams helped to improve the manuscript and the Geological Survey of Malawi provided kind assistance with fieldwork.

References

Andersen, T. (1986) Compositional variation of some rare earth minerals from the Fen complex (Telmark, SE Norway): implications for the mobility of rare earths in a carbonatite system. *Mineral. Mag.*, **50**, 503–9.

Andersen, T. (1987a) A model for the evolution of hematite carbonatite, based on whole-rock major and trace element data from the Fen complex, southeast Norway. *Appl. Geochem.* **2**, 163–80.

Anderson, T. (1987b) Mantle and crustal components in a carbonatite complex, and the evolution of carbonatite magma: REE and isotopic evidence from the Fen complex, southeast Noway. *Chem. Geol. (Isotope Geosci. Sect.)*, **65**, 147–66.

Appleton, J. D., Bland, D. J., Nancarrow, P. H. *et al.* (1992) The occurrence of daqingshanite-(Ce) in the Nkombwa Hill carbonatite, Zambia. *Mineral. Mag.*, **56**, 419–22.

Bailey, D. K. (1993) Carbonatite magmas. *J. Geol. Soc., Lond.*, **150**, 637–51.

Bayliss, P. and Levinson, A. A. (1988) A system of nomenclature for rare-earth mineral species: revision and extension. *Am. Mineral.*, **73**, 422–3.

Buckley, H. A. and Woolley, A. R. (1990) Carbonates of the magnesite–siderite series from four carbonatite complexes. *Mineral. Mag.*, **54**, 413–8.

Burt, D. M. (1989) Compositional and phase relations among rare earth element minerals, in *Geochemistry and Mineralogy of Rare Earth Elements* (eds B. R. Lipin and G. A. McKay), *Reviews in Mineralogy*, **21**, Mineralogical Society of America, 259–307.

Deans, T. and McConnell, J. D. C. (1955) Isokite, $CaMgPO_4F$, a new mineral from Northern Rhodesia. *Mineral. Mag.*, **30**, 681–90.

REFERENCES

Ek, R. and Nysten, P. (1990) Phosphate mineralogy of the Hålsjöberg and Hökensås kyanite deposits. *Geologiska Föeningens i Stockholm Förhandlingar*, **112**, 9–18.

Fleischer, M. and Altschuler, Z. S. (1969) The relationship of rare-earth composition of minerals to geological environment. *Geochim. Cosmochim. Acta*, **33**, 725–32.
Fletcher, C. J. N., Appleton, J. D., Webb, B. C. and Basham, I. R. (1981) Mineralization in the Cerro Manomó carbonatite complex, eastern Boliuia. *Trans. Instn Min. Metall. (Sect. B: Appl. Earth Sci.)*, **90**, B37–B50.
Fournier, A., Williams-Jones, A. E. and Wood, S. A. (1993) Magmatic and hydrothermal controls of LREE mineralization of the St. Honoré Carbonatite, Quebec. *TERRA Abstracts. Abstract Supplement No. 3 to TERRA nova*, **5**, 15.

Garson, M. S. (1965) Carbonatites in Southern Malawi. *Bull. Geol. Surv. Malawi*, **15**, Government Press, Zomba, 128 pp.
Garson, M. S. (1966) Carbonatites in Malawi. In *Carbonatites* (eds O. F. Tuttle and J. Gittins), John Wiley & Sons, New York, pp. 33–71.
Garson, M. S. and Campbell-Smith, W. (1958) Chilwa Island. *Mem. Geol. Surv. Malawi*, **1**, Government Press, Zomba, 127 pp.
Garson, M. S. and Campbell-Smith, W. (1965) Carbonatite and agglomeratic vents in the Western Shire Valley. *Mem. Geol. Surv. Malawi*, **3**, Government Press, Zomba, 167 pp.
Garson, M. S. and Morgan, D. J. (1978) Secondary strontianite at Kangankunde carbonatite complex, Malawi. *Inst. Mining Metall. Trans Sect. B*, **87**, B70–B73.

Harrison, M. T. and Watson, E. B. (1984) The behaviour of apatite during crustal anatexis: equilibrium and kinetic consideration. *Geochim. Cosmochim. Acta*, **48**, 1467–77.
Hatzl, T. (1992) Die Genese Karbonatite- und Alkalivulkanit-assoziierten Fluorit–Baryt–Bastnäsit-Vererzung bei Kizilçaören (Türkei). *Münchner Geol. Hefte*, **8**, 271 pp.
Hogarth, D. D. (1989) Pyrochlore, apatite and amphibole: distinctive minerals in carbonatite, in *Carbonatites: Genesis and Evolution* (ed. K. Bell), Unwin Hyman, London, 105–48.
Hogarth, D. D., Hartree, R., Loop, J. and Solberg, T. N. (1985) Rare-earth element minerals in four carbonatites near Gatineau, Quebec. *Am. Mineral.*, **70**, 1135–42.
Holt, D. N. (1965) The Kangankunde Hill rare earth prospect. *Bull. Geol. Surv. Malawi*, **20**, Government Press, Zomba, 130 pp.
Hornig, I. (1988) Spurenelementunterssuchungen an Karbonatiten mit hilfe der ICP-Atomemissionsspektroskopie. Doctoral Dissertation, Albert-Ludwigs-Universität, Freiburg i. Br., 238 pp.

Kapustin, Yu. L. (1980) *Mineralogy of Carbonatites*, Amerind Publishing, New Dehli, 259 pp.
Knudsen, C. and Buchardt, B. (1991) Carbon and oxygen isotope composition of carbonates from Qaqarssuk Carbonatite Complex, southern West Greenland. *Chem. Geol. (Isotope Geosci. Sect.)*, **86**, 263–74.

Laval, M., Johan, V. and Tourlière, B. (1988) La carbonatite de Mabounié: exemple de formation d'un gîte résiduel à pyrochlore. *Chron. Rech. Min.*, **491**, 125–36.
Le Bas, M. J. (1977) *Carbonatite–Nephelinite volcanism*. John Wiley & Sons, London, 347 pp.
Le Bas, M. J. (1989) Diversification of carbonatite, in *Carbonatites: Genesis and Evolution* (ed. K. Bell), Unwin Hyman, London, pp. 1–14.
Le Bas, M. J., Keller, J., Tao Kejie *et al.* (1992) Carbonatite dykes at Bayan Obo, Inner Mongolia, China. *Mineral. Petrol.*, **46**, 195–228.
Le Maitre, R. W., Bateman, P., Dudek, A. *et al.* (1989) *A Classification of Igneous Rocks and Glossary of Terms*. Blackwell Scientific Publications, Oxford, 193 pp.
Lee, C. A. (1974) The geology of the Katete carbonatite, Rhodesia. *Geol. Mag.*, **111**, 133–42.

Lottermoser, B. G. (1990) Rare-earth element mineralisation within the Mt Weld carbonatite laterite, Western Australia. *Lithos*, **24**, 151–67.

Loubet, M., Bernat, M., Javoy, M. and Allegre, C. J. (1972) Rare earth contents in carbonatites. *Earth. Planet. Sci. Lett.*, **14**, 226–32.

Mariano, A. N. (1973) Carbonatite investigations in Tanzania – Wigu Hill, Ofime and Kerimasi. Report to the Geological Survey of Tanzania (Idara Ya Madini), 105 pp.

Mariano, A. N. (1978) A petrographic examination of selected drill core from the Bear Lodge Carbonatite Project, Crook County, Wyoming. Confidential Report to Molycorp Inc., 34 pp.

Mariano, A. N. (1979) Rare earth mineralization at Araxá Minas Gerais, Brasil. Confidential Report to Molycorp Inc. and CBMM, 88 pp.

Mariano, A. N. (1983) A petrographic description of carbonatite and related rocks from the Adrardes iforas, Mali. Oued Anesrouf, Adiounedj, In Imanal. Confidential Report to UNRFNRE.

Mariano, A. N. (1984) On the mineralogy of Na, Ta and REE in the Mount Weld laterite. Confidential Report to Molycorp Inc., 86 pp.

Mariano, A. N. (1985) Selected mineralogic, petrographic and paragenetic studies on some rocks from the Gem Park Complex Colorado. Confidential Report to Molycorp Inc. 54 pp.

Mariano, A. N. (1987) Analytical report on 4 regolith samples and 4 pieces of drill core from Hicks Dome, Hardin County, Illinois. Confidential Report to John Lee Carroll. 24 pp.

Mariano, A. N. (1989a) Nature of economic mineralization in carbonatites and related rocks. In *Carbonatites: Genesis and Evolution* (ed. K. Bell). Unwin Hyman, London, pp. 149–176.

Mariano, A. N. (1989b) Economic geology of rare earth minerals, in *Geochemistry and Mineralogy of Rare Earth Elements* (eds B. R. Lipin and G. A. McKay), *Reviews in Mineralogy*, Mineralogical Society of America, **21**, 307–37.

McKie, D. (1962) Goyazite and florencite from two African carbonatites. *Mineral. Mag.*, **33**, 281–97.

Minakov, F. V., Dudkin, O. B. and Kamenev, Ye. A. (1981) The Khibiny carbonatite complex. *Trans. (Doklady) USSR Acad. Sci.: Earth Sci. Sect.*, **259**, 59–60.

Möller, P., Morteani, G. and Schley, F. (1980) Discussion of REE distribution patterns of carbonatites and alkalic rocks. *Lithos*, **13**, 171–9.

Nelson, D. R., Chivas, A. R., Chappell, B. W., and McCulloch, M. T. (1988) Geochemical and isotopic systematics in carbonatites and implications for the evolution of ocean-island sources. *Geochim. Cosmochim. Acta*, **52**, 1–17.

Ngwenya, B. T. (1994) Hydrothermal rare earth mineralisation in carbonatites of the Tundulu complex, Malawi: Processes at the fluid/rock interface. *Geochim. Cosmochim. Acta*, **58**, 2061–72.

Ni, Y., Hughes, J. M. and Mariano, A. N. (1995) The crystal chemistry of monazite and xenotime structures. *Am. Mineral.*, in press.

Olson, J.C., Shawe, D. R., Pray, L. C. and Sharp, W. N. (1954) Rare-earth mineral deposits of the Mountain Pass District San Bernardino County California. *U.S. Geol Surv. Prof. Pap.*, **261**, 75 pp.

Overstreet, W. C. (1967) The geologic occurrence of monazite. *U.S. Geol Surv. Prof. Pap.*, **530**, 327 pp.

Paterson, L. A. and Cooper, A. F. (1993) Variation of the REE content of monazites from carbonatites at Haast River, New Zealand: implications for the behaviour of REE in carbonatite melts. *Rare Earth Minerals: Chemistry, Origin and Ore Deposits*. Mineralogical Society Spring Meeting, Natural History Museum, London, UK, Conference Extended Abstracts, pp. 104–6.

REFERENCES

Pecora, W. T. and Kerr, J. H. (1953) Burbankite and calkinsite, two new carbonate minerals from Montana. *Am. Mineral.*, **38**, 1169–83.

Platt, R. G. and Woolley, A. R. (1990) The carbonatites and fenites of Chipman Lake, Ontario, *Can. Mineral.*, **28**, 241–50.

Sauvage, J. F. and Savard, R. (1985) Les complexes alcalins sous-saturés à carbonatites de la région d'In Imanal (Sahara malien): une présentation. *J. Afr. Earth Sci.*, **3**, 143–9.

Schwab, R. G., Hardy, H. Götz, C. and Pinto de Oliveira, N. (1990) Compounds of the crandallite type: synthesis and properties of pure rare earth element-phosphates. *N. Jb. Miner. Mh.*, **1990**, 241–54.

Semenov, E. I. (1974) Economic mineralogy of alkaline rocks, in *The Alkaline Rocks* (ed. H. Sorensen). John Wiley and Sons, New York, pp. 543–52.

Shi Li and Tong Wang (1993) Origin of burbankite in biotite–aegirine carbonatite from Hubei, China. *Rare Earth Minerals: Chemistry, Origin and Ore Deposits*. Mineralogical Society Spring Meeting, Natural History Museum, London, UK, Conference Extended Abstracts, p. 72.

Snelling, N. J. (1965) Age determinations on three African carbonatites. *Nature*, **205**, 491.

Vlasov, K. A. (ed.) (1966) *Geochemistry and mineralogy of rare elements and genetic types of their deposits. Vol. II, Mineralogy of Rare Elements*. Israel Program for Scientific Translations, Jerusalem, 945 pp.

Voloshin, A. V., Subbotin, V. V., Yakoventchuk, V. N. *et al.* (1990) Mckelveyite from carbonatites and hydrothermalites of alkaline rocks, the Kola Peninsula (the first findings in the USSR). *Zap. Vses. Mineral. Obsh.*, **119**, 76–86 (in Russian).

Wakita, H., Rey, P. and Schmitt, R. A. (1971) Abundances of the 14 rare earth elements and 12 other elements in Apollo 12 samples: five igneous and one breccia rocks and four soils. *Proc. 2nd Lunar Sci. Conf., Geochim. Cosmochim. Acta, Suppl.*, **2**, 1319–29.

Wall, F., Barreiro, B. A. and Spiro, B. (1994) Isotopic evidence for late-stage processes in carbonatites: rare earth mineralization in carbonatites and quartz rocks at Kangankunde, Malawi. V. M. Goldschmidt Conference Extended Abstracts, *Mineral. Mag.*, **58A**, 951–2.

Wall, F., Le Bas, M. J. and Srivastava, R. K. (1993) Calcite and carbocernaite exsolution and cotectic textures in a Sr,REE-rich carbonatite dyke from Rajasthan, India. *Mineral. Mag.*, **57**, 495–513.

Woolley, A. R. (1991) The Chilwa Alkaline Igneous Province of Malawi: a review, in *Magmatism in Extensional Structural Settings. The Phanerozoic Plate.* (eds A. B. Kampunzu and R. T. Lubala), Springer-Verlag, Berlin Heidelberg, 377–409.

Woolley, A. R. and Kempe, D. R. C. (1989) Carbonatites: nomenclature, average chemical compositions, and element distribution, in *Carbonatites: Genesis and Evolution* (ed. K. Bell), Unwin Hyman, London, 1–14.

Woolley, A. R., Barr, M. W. C., Din, V. K. *et al.* (1991) Extrusive carbonatites from the Uyaynah area, United Arab Emirates. *J. Petrol.*, **32**, 1143–67.

Zaitsev, A. N., Menshikov, Yu. P., Polezhaeva, L. I. and Latysheva, L. G. (1990) Ba, Sr and REE-bearing minerals from the latest carbonatites of the Khibina Alkaline Massif, in *New in mineralogy of Karelian–Kola region* (ed. Yu. N. Yakovlev) Karelian Science Centre of the USSR Academy of Sciences, Petrozavodsk, 76–89 (in Russian).

Zdorik, T. B. (1966) Evolution of rare earth mineralization in carbonatites using the example of one of the deposits from East Siberia, in *Geology of Ore Deposits of Rare Elements* (ed. A. I. Ginzburg). Nedra, Moscow, **30**, 121–31 (in Russian).

Ziegler, U. R. F. (1992) Preliminary results of geochemistry, Sm–Nd and Rb-Sr studies of post-Karoo carbonatite complexes in Southern Africa. *Schweiz. Mineral. Petrogr. Mitt.*, **72**, 135–42.

CHAPTER NINE
REE distribution and REE carriers in laterites formed on the alkaline complexes of Araxá and Catalão (Brazil)

G. Morteani and C. Preinfalk

9.1 Introduction

The overwhelming majority of REE are still produced from primary bastnäsite deposits or from monazite-bearing marine placers, the main exception to this being the REE produced in undefined quantities from the so-called ionic ores in the People's Republic of China. The most important bastnäsite mines are those of Mountain Pass, California (USA) and Bayan Obo, Inner Mongolia (People's Republic of China) and because of these two sources, China and the USA are the world leaders in REE production, with 62% of the market in 1994 (Preinfalk and Morteani, in preparation). However, in recent years there has been an increasing economic interest in the REE found in the laterites formed on REE-rich alkaline complexes. REE contents of up to 42 wt% have been described in laterites on the Mt Weld alkaline complex in Western Australia (Lottermoser, 1991) and the Australian Nuclear Science and Technology Organization (ANSTO) has set up a pilot plant at Lucas Heights, which could enable Australia to capitalize on the rare earth market. The plant started operating in 1993 (Anon., 1992, 1993). The so-called ionic ores produced at Longnan and Xunwu in the Jiangxi province in south China can also be considered to be laterites (Clark and Zheng, 1991; Wu, Yuan and Bai, chapter 11). The ores from Longnan show REE distribution patterns very similar to those of xenotime, i.e. an enrichment of the heavy REE with respect to the light ones (Morteani, 1991; Hedrick, 1993).

The most comprehensive information on the REE contents in laterites and bauxites is given by Bronevoi, Zhilberminc and Teniakov (1985). They report the average REE contents of about 6700 samples collected from 48 main bauxite districts. Unfortunately they did not distinguish between different types of bauxites, e.g. lateritic or karst-bauxites and did not refer to

Rare Earth Minerals: Chemistry, origin and ore deposits. Edited by Adrian P. Jones, Frances Wall and C. Terry Williams. Published in 1996 by Chapman & Hall. ISBN 0 412 61030 2

the economic importance of the bauxite occurrences. According to Bronevoi, Zhilberminc and Teniakov (1985), the total REE contents in bauxites vary between 20 and 3600 ppm and they point out that the enrichment of the REE in the laterites, compared to the parent rocks, is weaker than for Al, the main component of most of the laterites.

When searching for high REE contents, the most promising laterites are likely to be those developed on top of alkaline complexes; the RE_2O_3 contents of laterites found on top of alkaline rocks, such as nepheline syenites, melteigites, ijolites and carbonatites, often amount to 10–25 wt%, the crustal average being 145.3 ppm REE (Deans, 1978; Greenwood and Earnshaw, 1984; Preinfalk, Morteani and Fuganti, 1992).

According to the literature the REE in laterites are predominantly bound to residual magmatic pyrochlore and apatite or to secondary phosphate minerals of the crandallite group, monazite, secondary apatite, plumbogummite, rhabdophane and/or iron hydroxides (Laval, Johan and Tourlière, 1988; Lottermoser, 1989a, 1990; Duncan and Willett, 1990; Morteani and Preinfalk, 1993).

In this chapter the REE contents, the REE distribution and the REE-carrying minerals in the laterites formed on top of the alkaline complexes of Araxá and Catalão, Brazil, are described and discussed in detail using data, produced by the authors, on the primary rocks and laterites. In addition, selected data from the literature on the laterites and alkaline rocks at Mt Weld (Australia), Mabounié (Gabon) and Lueshe (Zaire) are included and discussed for comparison.

9.2 Laterites and the laterite profile

Laterites or lateritic bauxites are the most important residual formations in tropical areas. According to the definition given by Schellmann (1982, 1983) laterites are

> Products of intense subaerial rock weathering. They consist predominantly of mineral assemblages of goethite, hematite, Al-hydroxides, kaolinite minerals and quartz. The $SiO_2:(Al_2O_3 + Fe_2O_3)$ ratio of a laterite must be lower than that of the lateritized parent rock. This definition includes all highly weathered materials, strongly depleted in silica and enriched in Fe and Al, regardless of their morphological and physical properties (fabric, colour, consistency).

The term laterite was coined by Buchanan (1807) for weathered material he observed in Southern India although he did not specify the composition of the laterites. Details on laterite formation, chemistry and mineralogy are given in Maignien (1966), Valeton (1967, 1972), McFarlane (1976), Schellmann (1981, 1982), Bocquier, Müller and Boulangé (1984) and Bardossy and Aleva (1990). A nomenclature of the different laterites as well

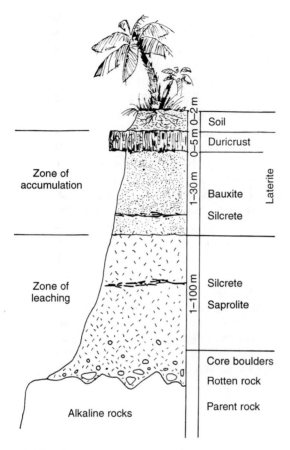

Figure 9.1 The typical laterite–saprolite weathering profile at Catalão and Araxá. This profile corresponds closely to the classical profile given by Bardossy and Aleva (1990).

as that of kaolinitic, bauxitic and ferritic rock types in terms of the mineralogical modal composition is given by Bardossy and Aleva (1990).

Lateritization processes are the consequence of an environment in which the metals Al, Mg, Fe, Cu, Ni and the REE are totally or partially freed from their parent rock minerals and then accumulate in the laterite–saprolite weathering profile. Figure 9.1 gives a typical laterite–saprolite weathering profile as observed at Catalão and Araxá. It corresponds closely to the classical profile given by Bardossy and Aleva (1990). From top to bottom it consists of a soil, a duricrust, a bauxite, a saprolite horizon and the fresh parent rock. Sometimes one or more silcrete layers are also present.

The residual soil horizon is composed of the mechanical and chemical weathering products of the underlying horizon mixed with plant remains, e.g. roots, and humic matter. Its colour is related to that of the underlying horizon but is often also black due to humic matter.

The residual soil horizon is followed to depth by the duricrust horizon, which is generally the hardest one in the profile due to newly formed or recrystallized Fe-minerals. In most profiles this horizon shows the strongest colour ranging from intense red to red-brown.

The bauxite horizon represents the lowest section of the lateritic part of the total weathering profile and may be distinguished from the upper duricrust horizon by the lower iron mineral content. The colour of the bauxite horizon may change from almost white to yellowish, orange-red to brown. Sometimes it is also violet, yellow-green or greenish. The duricrust and the bauxite horizon form the laterite *sensu stricto*.

The saprolite horizon is composed mostly of weathered silicate products of the parent rock. The saprolite is usually grey to brown and violet in colour. Minerals such as rutile and zircon are highly resistant in the saprolite environment. Sometimes, less evolved materials as well as relict structures and textures of the parent rocks are found in the lower parts of the saprolite horizon.

The lateritic weathering profile ends with the parent rocks. Obviously the parent rocks influence the nature of the weathering profile but when laterites are 100 m or more in thickness it is not always possible to establish their nature.

Silcretes are found in the open cast mines of Catalão, usually in the saprolite horizon. The miners call them 'silexites', but this is a misuse of the

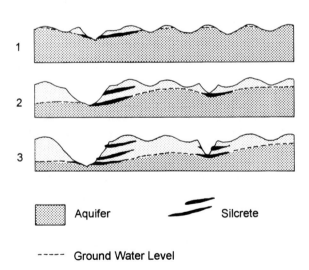

Figure 9.2 Schematic model of silcrete formation during successive stages (1, 2, 3) of erosion and consequent changes in ground water level; according to Milnes and Thircy (1992).

term silexite, which describes a magmatic rock consisting predominantly of quartz (Bates and Jackson, 1980). Silcretes should also not be confused with microcrystalline, brownish, postmagmatic, sharply delimited quartz veins preserved as weathering relics in the saprolite. These quartz veins are very low in REE, in contrast to the very REE-rich silcretes (see below). Silcretes are rocks produced by cementation, usually in supergene conditions, of unconsolidated regolith materials, e.g. saprolite, bauxite and soils by various forms of secondary silica including microcrystalline or well-crystallized quartz (Milnes and Thiry, 1992). Sometimes, silcrete is erroneously classed as duricrust (Woolnough, 1927). Silcretes may be associated with palaeosurfaces but are typically formed at different levels of a weathering or sedimentary profile due to successive changes of the water table, leaching and deposition of silica. In Figure 9.2 the formation of different groundwater silcrete layers is shown schematically according to Milnes and Thiry (1992).

The relative thickness of the soil, the laterite and the saprolite horizons may vary laterally due to mechanical erosion. The phenomenon of total removal of the soil horizon is described as truncation (Bardossy and Aleva, 1990).

9.3 The behaviour of REE during lateritization

The lateritization process dramatically changes the mineralogical and chemical composition of the primary rocks. In magmatic rocks the REE usually either substitute as REE^{3+} into Ca^{2+}-containing minerals or form accessory RE minerals (Möller, Morteani and Schley, 1980; Möller and Morteani, 1983; Preinfalk and Morteani, in preparation). In alkaline rocks the REE are mostly bound by substitution to apatite, pyrochlore, zircon and calcite or form monazite.

Investigations of weathering profiles indicate that REE mobility and deposition in surface environments are distinctly influenced by changes in the ground water pH conditions (Ronov, Valashov and Migdisov, 1967; Burkov and Podporina, 1967; Nesbitt, 1979; Duddy, 1980; Middleburg, van der Weijden and Woittiez, 1988; Lottermoser, 1989a; McLennan, 1989). REE are thought to be found in aqueous solutions as complexes including fluoro-, carbonato-, and phosphato-complexes (Möller, 1963; Möller et al., 1965; Cantrell and Byrne, 1987).

In the deeper part of the laterite and to a lesser extent also in the saprolite, secondary phosphates are formed and adsorb most of the REE. Alteration of magmatic apatite during lateritization and the formation of secondary Al-phosphates is generally characterized by decreasing alkali contents and increasing Al–Fe/P ratios in the weathering products (Flicoteaux and Lucas, 1984). This trend is illustrated, for example, in the case of the Mt Weld laterite by a decrease in P_2O_5, CaO, Na_2O and K_2O

values and an increase in the Fe_2O_3 and Al_2O_3 contents from the apatite-rich to the Al-phosphate-rich laterite horizons (Lottermoser, 1989a). The dissolution of apatite and the formation of Al-phosphate minerals are essentially independent of Eh and start at a pH of about 7.5 (Nriagu, 1976; Lottermoser, 1989b). The Al-phosphates are stable phases in the acidic to neutral environment. The Al needed for the formation of the secondary Al-phosphates is derived from the decomposition of the igneous aluminosilicates. The silica liberated by the decomposition of the silicates may form kaolinite or, if there are changes in the water table, silcretes may be deposited.

There is no debate about the fact that the REE contents found in the laterites on top of alkaline complexes are derived from the underlying parent rocks, but whether the REE are leached and fractionated during lateritization still seems to be an open question which is discussed in detail, for example, by Lottermoser (1989a, 1989b) and Walter (1991).

9.4 Geology

A short geological description of the alkaline rock complexes of Catalão and Araxá is given in the following section. For details see Barbosa et al. (1970) and Woolley (1987).

The outcrops which are presently accessible and the available drill cores show large changes in composition on a scale of 1 to tens of metres so that the units given by Barbosa et al. (1970) and Woolley (1987) are very much a simplification of the magmatic lithologies. The lack of fresh outcrops together with the extreme inhomogeneity make it impossible to refer the composition of the laterites directly to that of the fresh rocks (by using normalization, for example).

The town of Catalão (Goiás State) gives its name to the alkaline complexes of Catalão I and Catalão II. The two complexes are situated about 15 km from each other and Catalão I, which is the subject of this study, is approximately 10 km northeast of the town of Catalão. Catalão I is a plug of about 6 km in diameter comprising alkaline, ultramafic and carbonatitic rocks. It is emplaced within a fold belt of Brasiliano age in mica schists and quartzites of the Precambrian Araxá Group, which it fenitizes (Figure 9.3) (Barbosa et al., 1970). Extensive drilling has indicated that a carbonatitic core is surrounded by pyroxenites, serpentinized peridotites and glimmerites (Woolley, 1987). According to Woolley (1987) the lateritic cover overlying the intrusive rocks of the Catalão I complex reaches depths of 15–250 m and contains variable amounts of phosphates, pyrochlore, Ti- and RE-minerals.

The Araxá or Barreiro alkaline complex is situated 2 km south of the town of Araxá (Minas Gerais State) in the same Brasiliano-age fold belt as Catalão (Figure 9.3). It is intruded into quartzites and mica schists of the Precambrian metasedimentary sequences of the Araxá Group (Barbosa et al., 1970) and is a circular complex of about 4.5 km diameter.

Figure 9.3 Geological sketch map of Brazil showing Araxá and Catalão and other important alkaline complexes of Brazil.

Drilling has proved the central area of the complex to be principally alkaline rocks with an outer collar of glimmerites. The alkaline rocks are coarse- to medium-grained sövites (calcite carbonatites) and rauhaugites (dolomitic carbonatites) (Woolley, 1987). The glimmerites are the products of an alteration of original pyroxenites and peridotites by an interaction with metasomatic alkali-rich fluids released from the carbonatitic magma (Woolley, 1987). The aureole of fenitization in the country rocks is extensive, up to 2.5 km wide, and quartz–feldspathic fenites have been formed. According to Woolley (1987), the lateritic mantle concealing the alkaline complex is up to 230 m thick and consists of an upper lateritic soil (2–30 m) and an intermediate residuum of goethite, magnetite, baryte, monazite and gorceixite. Bariopyrochlore and cerian pyrochlore are worked as a source of niobium.

The lateritic profiles at both Araxá and Catalão correspond closely to the general description given by Bardossy and Aleva (1990). The uppermost part of the laterite profile is a deep reddish soil horizon up to 10 m thick. At Araxá the lower part of this horizon contains thick, hard lumps of iron oxide which arise as a result of duricrust formation.

The laterite *sensu stricto* is 30–60 m thick and is characterized by yellowish to red-brown colours and lower iron mineral contents. The typical bauxite composition is missing. There has been localized development of lenses and layers of whitish kaolinite.

The lowermost horizon in the weathering profile is a dark red-brown to dark grey saprolite with partially relict structures of the original rock and local strong enrichment in vermiculite. The occurrence of baryte is typical of the saprolite horizon at both localities. Sometimes it forms massive vein-like bodies of coarse-grained sparry baryte. In the open cast mines the saprolite crops out for at least for 30–50 m but the real thickness cannot be determined due to the lack of a continuous outcrop of the alkaline rocks.

Silcretes are found only at Catalão, where they occur as very irregular layers up to 1 m thick in the laterite and the saprolite horizon. They consist of fine-grained brownish quartz.

The mineralogical composition of the different horizons of the laterite–saprolite weathering profile as determined by XRD, optical and scanning electron microscopy is given in Table 9.1.

Table 9.1 Mineralogical composition of the laterite horizons of Catalão and Araxá

Horizon	Minerals
Soil	Gorceixite, goethite, hematite, magnetite, gibbsite, kaolinite, quartz
Laterite	Gorceixite, goethite, hematite, magnetite, kaolinite, quartz
Saprolite	Apatite, gorceixite, goethite, magnetite, quartz, vermiculite, baryte, boehmite
Alkaline rocks	Apatite, monazite, magnetite, hematite, pyrite, silicates, carbonates

9.5 Geochemistry

In order to investigate the main and rare earth element (REE) contents through the laterite profile, samples have been taken at a vertical interval of 5–10 m along the benches of the open cast mines of Araxá and Catalão. The profiles start from the soil horizon and end at the contact with the fresh rock, or in the lowermost part of the saprolite profile if the fresh rock was not exposed in the mines. In the diagrams the duricrust horizon is treated separately from the soft laterite.

9.5.1 Analytical methods

The main and trace elements of the laterites, including La, Ce, Nd and Y, were analysed in Li-tetraborate glass discs with a fully automated Siemens SRS 303 XRF. The XRD work was performed on a computerized Phillips diffractometer using a copper tube and graphite monochromator.

The REE were determined partly by routine INAA, partly by ICP-OES. Irradiation for INAA was performed in the nuclear experimental reactor of the Technical University of München. Only La, Ce, Nd, Sm, Eu, Tb, and Yb can be easily determined by routine INAA method and therefore in the REE distribution diagrams based exclusively on INAA analyses only these REE are given on the x-axis; in all figures promethium is not shown. The natural abundance of this element, an unstable fission product of ^{235}U, is exceedingly low (4.5×10^{-20} ppm) and it is undetectable by the analytical methods used here.

Microprobe analyses were obtained at the Mineralogical Institute of the University of Kiel on a Camebax Microbeam electron microprobe with four wavelength-dispersive spectrometers using fluorapatite as standard. Operating conditions were 15 kV accelerating voltage and 15 nA beam current. For the SEM images and chemical phase identification, a JEOL 35 C machine with an ORTEC EDS was used.

9.5.2 Major element geochemistry

The major element chemistry of the laterites and alkaline rocks of Catalão and Araxá is presented and discussed using ternary ($Al_2O_3 + Fe_2O_3$)–SiO_2–($CaO + K_2O$) and SiO_2–Al_2O_3–Fe_2O_3 diagrams as well as a CaO versus P_2O_5 plot.

In the ($Al_2O_3 + Fe_2O_3$)–SiO_2–($CaO + K_2O$) diagram the alkaline rocks plot in a well-defined field with ($Al_2O_3 + Fe_2O_3$) contents between 10 and 70% and SiO_2 contents of up to 50%. The saprolite formation as well as the lateritization starts from the field of the alkaline rocks (shaded) and trends towards the ($Al_2O_3 + Fe_2O_3$) edge (Figure 9.4). A second trend is indicated by the silcrete composition plotting near the SiO_2 edge.

In the SiO_2–Al_2O_3–Fe_2O_3 plot the alkaline rocks and the saprolite compositions are found in a narrow field (shaded) stretching along the SiO_2–Fe_2O_3 side between 10 and 75% SiO_2 (Figure 9.5). The soil shows higher Al_2O_3 contents and a corresponding typical depletion of SiO_2 and Fe_2O_3 with respect to the alkaline rocks. The laterite shows two main trends. One is characterized, as in the soils, by an enrichment of Al_2O_3 and a depletion in SiO_2 and Fe_2O_3, the other trend is a strong enrichment in Fe_2O_3. Many of the laterites plot near the Fe_2O_3 corner. The saprolites plot mostly in the field close to the primary rocks but show a very slight increase in Al_2O_3 content. The silcretes plot very near to the upper SiO_2 corner of the diagram

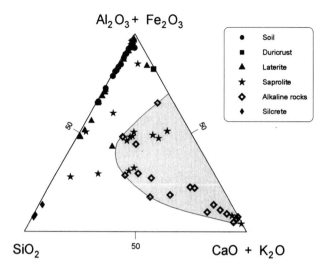

Figure 9.4 $(Al_2O_3 + Fe_2O_3) - SiO_2 - (CaO + K_2O)$ diagram for the laterite–saprolite profiles at Araxá and Catalão. The compositional field of the alkaline rocks is shaded.

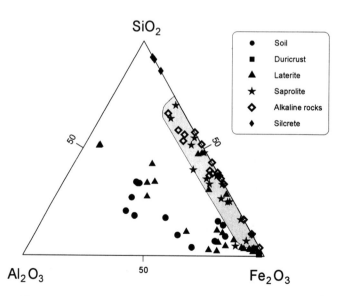

Figure 9.5 $SiO_2 - Al_2O_3 - Fe_2O_3$ diagram for the laterite–saprolite weathering profiles at Araxá and Catalão. The compositional field of the alkaline rocks is shaded.

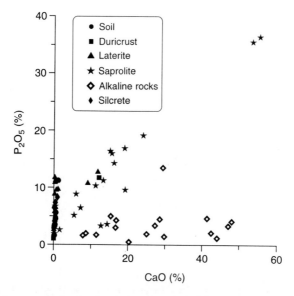

Figure 9.6 CaO versus P_2O_5 plot for the laterite–saprolite weathering profiles of Araxá and Catalão. Notice the ratio of about 1:1 for CaO and P_2O_5 in the saprolite.

with very low Al_2O_3 and rather low Fe_2O_3 contents. The carbonatitic rocks are typically low in Al_2O_3. The analyses of the duricrust horizon obviously plot in the Fe_2O_3 corner of the diagram.

In the CaO versus P_2O_5 diagram it can be seen that the alkaline rocks of both Araxá and Catalão have CaO contents ranging between 10 and 50% and, with one exception, P_2O_5 maximum contents of 5%, but there is no correlation between these two oxides (Figure 9.6). This indicates that Ca is present mainly as a carbonate. In contrast, a strong correlation between the P_2O_5 and CaO values can be seen for most of the saprolite samples, corresponding to the P_2O_5:CaO ratio of about 1:1 in apatite. This indicates that with three exceptions Ca is predominantly bound to apatite in the saprolite, and carbonates are missing (compare Table 9.1). The P_2O_5 contents in the saprolite may reach 35%. The CaO contents of the silcretes, laterites and soils are, with three exceptions, near to zero but the maximum P_2O_5 contents are about 13%. The low CaO and high P_2O_5 contents indicate that the high phosphorus contents are bound to the Ca-free Al-phosphates and not to apatite, as in the saprolite.

9.5.3 REE contents of the alkaline rocks and the laterites of Catalão

In Catalão the total REE contents in the soil horizon range between 0.4 and 3.4% with an average of 1.9%; in the laterite they range between 0.7 and

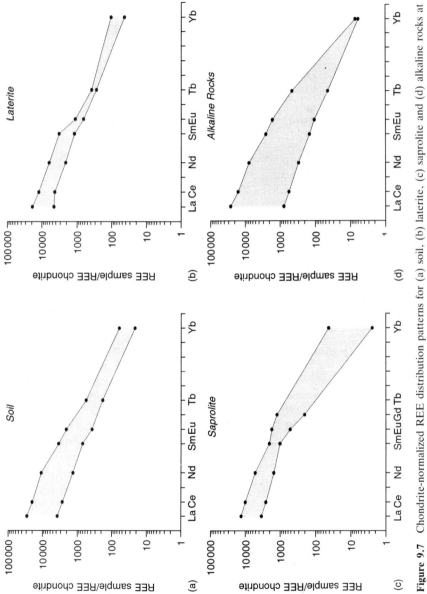

Figure 9.7 Chondrite-normalized REE distribution patterns for (a) soil, (b) laterite, (c) saprolite and (d) alkaline rocks at Catalão.

3.7% with an average of 1.3%, and in the saprolite they are between 0.4 and 1.5% with an average of 1.0%. The average total REE content of the alkaline rocks is 0.9%. The silcretes have REE contents of between 4.7 and 12.7% total REE with an average of about 10%, i.e. the highest REE contents determined.

Chondrite-normalized REE distribution patterns (Haskin *et al.*, 1968) for representative samples of the soil, laterite and saprolite, as well as those of the unweathered alkaline rocks from Catalão are given in Figure 9.7. The REE patterns of the soil horizon start with chondrite-normalized La values of between 4000 and 30 000, decreasing to Yb values of between 20 and 60; the La:Yb ratio is about 750. The REE patterns of the laterite horizon start with a chondrite-normalized La value of between 4500 and 20 000 and show decreasing values to Yb (45–110); the La:Yb ratio is about 190.

Compared to the laterite, the REE patterns of the saprolite horizon start with a slightly lower chondrite-normalized La value of between 3500 and 13 000, decreasing to Yb values of between 2 and 40 with a La:Yb ratio of about 380.

The REE pattern of the alkaline rocks starts with a chondrite-normalized La value of between 800 and 26 000 and ends with Yb values of about 7; the La:Yb ratio is about 2100.

Figure 9.8 shows the REE distribution pattern for two selected samples of the silcretes. The content of light REE is particularly high and the patterns are steeper than those of the soils, laterites, saprolites and alkaline rocks, though showing the same trend. The La:Yb ratio is about 2800.

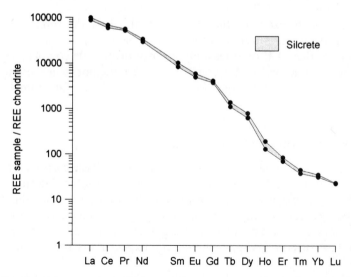

Figure 9.8 Chondrite-normalized REE distribution patterns for two selected silcrete samples from Catalão.

The different laterite horizons, including the very REE-rich silcretes of Catalão, also show the same REE distribution patterns, enriched in light REE and very low in heavy REE, like the underlying alkaline rocks. It is remarkable that during lateritization in a highly oxidizing environment no Ce anomaly developed.

9.5.4 REE contents of the alkaline rocks and the laterites of Araxá

At Araxá the total REE contents in the soil horizon range between 0.3 and 1.3% with an average of 0.9%, in the laterite horizon between 0.5 and 4.4% with an average of 1.8%, and in the saprolite horizon between 0.4 and 5.8% with an average of 1.7%. The duricrust has REE contents between 0.7 and 1.0%, with an average of 0.9%. The average total REE content of the alkaline rocks is 0.5%.

The chondrite-normalized REE distribution patterns of the different laterite horizons as well as those of the unweathered alkaline rocks are shown in Figure 9.9. The REE patterns of the soil horizon start with high chondrite-normalized La values of between 6000 and 10000, decreasing to Lu values of between 35 and 65. The La:Yb ratio is about 170. The laterite horizon shows La values between 4000 and 30000, the pattern decreasing to Lu values of between 50 and 400; the La:Yb ratio is about 80. The chondrite-normalized La values of the saprolite are between 3000 and 80000, the patterns ending with Lu values of between 50 and 70; the La:Yb ratio is about 670. The chondrite-normalized La values of the alkaline rocks scatter between 1500 and 9000, decreasing to Lu values of between 1 and 30; the La:Yb ratio is about 320.

The different horizons of the laterite profile show the same light REE-enriched distribution patterns as the underlying alkaline rocks, and also an enrichment of the REE of at least an order of magnitude compared with the underlying alkaline rocks. At Araxá the REE enrichment in the laterites is less evident than at Catalão but, as at Catalão, no Ce or Eu anomalies were found.

9.6 REE carriers at Catalão and Araxá

In most of the unweathered alkaline rocks the main REE carriers are apatite and calcite. From the XRD determinations on the different laterite horizons of Catalão and Araxá and the above considerations, it is thought that the main REE carriers in the saprolite are primary and possibly secondary apatites as well as Al-phosphates like gorceixite. In the overlying laterite and soil horizons the Al-phosphates are expected to be the main REE carriers.

Nevertheless iron hydroxides are also candidates. It is well known that iron hydroxides can adsorb considerable amounts of REE. In fact, copre-

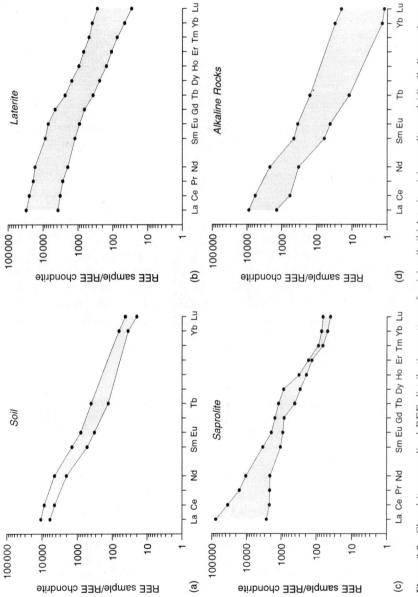

Figure 9.9 Chondrite-normalized REE distribution patterns for (a) soil, (b) laterite, (c) saprolite and (d) alkaline rocks at Araxá.

cipitation of the REE with $Fe(OH)_3$ is used as a preconcentration technique in analytical chemistry (Hogdahl, Bowen and Melson, 1968; Elderfield and Greaves, 1982). From a technological point of view (i.e. in the design of a beneficiation and extraction process) it is important to determine whether the REE are bound to iron hydroxides or to phosphates. Therefore, the REE distribution between Al-phosphates and the iron hydroxides has been investigated by selective leaching experiments which selectively dissolve the iron hydroxides and consequently free the adsorbed REE.

9.6.1 Apatite and calcite in the alkaline rocks

The REE contents of apatite and calcite, which are the main REE-bearing minerals in the alkaline rocks studied, have been determined on selected samples from both Catalão and Araxá. Mineral separation was made by magnetic separation, hand picking or by heavy liquids.

Figures 9.10 and 9.11 show the chondrite-normalized REE distribution patterns of calcite and coexisting apatite from carbonatitic rocks at Catalão and Araxá. The purity of the concentrates was checked carefully by XRD and microscopy. It can be seen that the apatites are enriched in REE by up to an order of magnitude with respect to the calcites. Both minerals display the typical light REE-enriched patterns of primary calcites and apatites from alkaline rocks. Some of the Catalão calcites and apatites have markedly higher REE contents than those of Araxá. The high REE contents of these primary REE carriers are reflected in the higher REE contents found in the alkaline rocks at Catalão compared with those at Araxá. The average REE contents of the calcite of Catalão and Araxá are 936 ppm and 889 ppm respectively, those of apatite are 5721 ppm and 6973 ppm respectively.

9.6.2 Phosphates in the laterites

During lateritization, of the alkaline rocks first calcite and then apatite are completely dissolved, due to the acidic pH of the weathering solution (Lottermoser, 1989b). During advanced lateritization the REE freed by the dissolution of the calcite and apatite are incorporated in Al-phosphates such as gorceixite.

Table 9.1 shows the mineralogical composition of the soil, the laterite, the saprolite and the alkaline rocks as determined by XRD, optical and scanning electron microscopy. Optical and scanning electron microscopy showed that a consistent amount of primary residual as well as secondary apatite occurs in the saprolite of Araxá and Catalão. The older residual apatite is always strongly corroded. As shown in Figure 9.12, the corrosion starts along definite planes giving a peculiar appearance to the most corroded residual apatites (Figure 9.13). The younger, secondary, euhedral, tabular generation is often growing into open spaces (Figure 9.14). Scanning electron

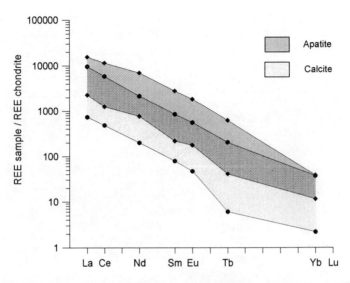

Figure 9.10 Chondrite-normalized REE distribution patterns for calcite and coexisting apatite from Catalão. On average, apatite is much richer in REE than calcite.

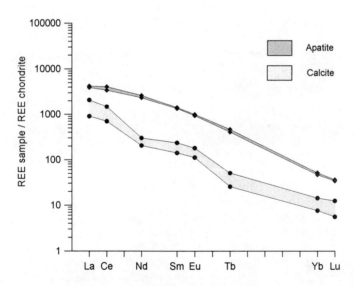

Figure 9.11 Chondrite-normalized REE distribution patterns for calcite and coexisting apatite from Araxá. On average, apatite is much richer in REE than calcite.

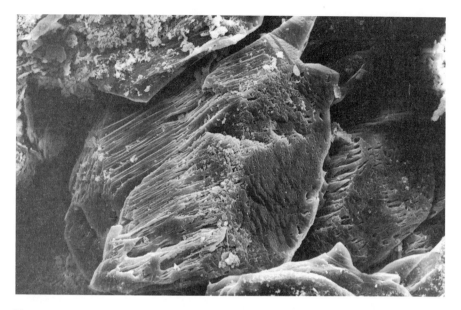

Figure 9.12 Scanning electron micrograph showing incipient corrosion of a magmatic apatite from the saprolite from Araxá. Width of photograph: 0.2 mm.

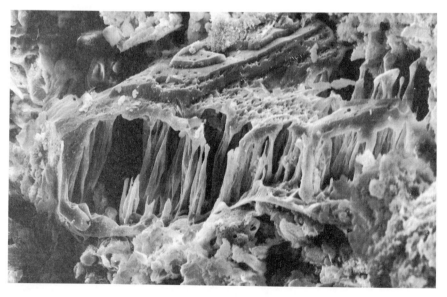

Figure 9.13 Scanning electron micrograph showing intensely corroded magmatic apatite from the saprolite from Catalão. Width of photograph: 0.2 mm.

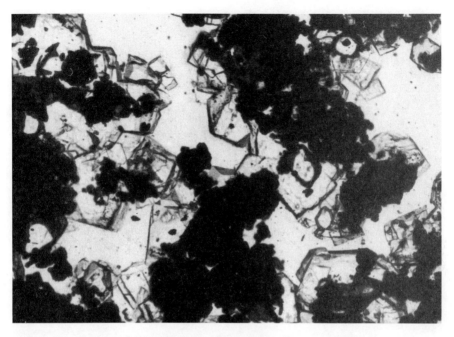

Figure 9.14 Microphotograph showing euhedral apatite crystals lining an open space in the saprolite from Araxá. Plane-polarized light (PPL); width of photograph: 0.7 mm.

microscopy reveals that the young tabular apatite is associated with large euhedral baryte crystals (Figure 9.15). Careful preparation of thin sections of the very friable earthy saprolite showed that in some cases the residual primary apatite is not only associated with a secondary euhedral apatite but is also rimmed by a younger anhedral apatite generation (Figure 9.16).

Microprobe analysis of the different types of apatite revealed that all three textural varieties of apatite (primary, secondary rimming primary and growing into cavities) are fluorapatites with fluorine contents of between 2.6 and 3.2 wt%. However, there is a remarkable difference in REE content between the corroded magmatic apatites and the younger apatite rims (Figure 9.17). The magmatic apatites have lower total REE than the younger rims. The higher REE content of the younger apatite probably results from the incorporation of the REE freed by the dissolution of magmatic calcites. An increase in the La, Ce and Nd oxides in the rims of the younger apatite (Figure 9.16) results in a decrease in CaO, confirming the coupled substitution of these elements (Figure 9.17). The charge balance in the apatite structure is probably due to a substitution of PO_4^{3-} by CO_3^{2-}. The typical substitution of $2Ca^{2+} \rightarrow Na^+ + REE^{3+}$ is not likely, due to the very low Na contents shown by microprobe analyses of the apatites.

Figure 9.15 Scanning electron micrograph showing secondary euhedral tabular apatite crystals surrounding a large euhedral baryte from the saprolite from Catalão. Width of photograph: 0.5 mm.

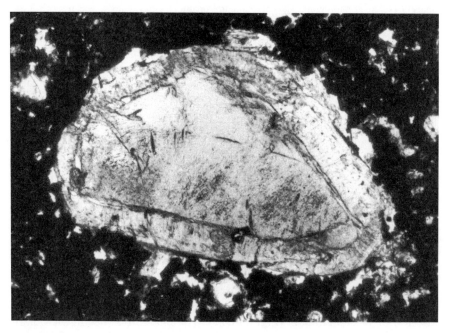

Figure 9.16 Microphotograph of primary magmatic apatite rounded by dissolution and overgrown by secondary apatite from the saprolite from Catalão. PPL; width of photograph: 0.7 mm.

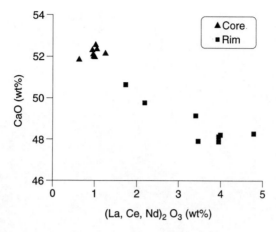

Figure 9.17 Plot of $(La,Ce,Nd)_2O_3$ versus CaO in apatite (wt%) from Catalão. The decrease of CaO with increasing RE_2O_3 content indicates a substitution of Ca by REE in the apatite structure.

In both the laterite and soil horizons at Araxá and Catalão gorceixite was detected by XRD but no apatite was found. Figure 9.18 shows typical oviform euhedral gorceixite crystals found in association with larger euhedral baryte and groups of needles of manganese oxides. The association of gorceixite and baryte is typical and the high concentration and mobility of Ba is thought to be the reason why, in the cases studied, no crandallite or goyazite (the Ca and Sr analogues of gorceixite) were found.

Silcretes were found only at Catalão. They have the highest REE contents of all analysed lateritic rocks. Maximum REE contents are approximately 13 wt%. In contrast to the laterite, the REE carrier in the silcretes is monazite, which occurs as concentrically structured aggregates (Figure 9.19). In some cases SEM-EDS analysis suggests that the rim of the monazite aggregates consists of cerianite. The formation of monazite instead of an Al-phosphate in the silcrete shows that during the formation of the silcretes the Al content in the weathering solution was probably very low. The high REE contents indicate trapping of the REE by phosphates during silica precipitation caused by changing pH conditions.

9.6.3 Iron hydroxides in the laterites

The iron hydroxides are very fine grained and intimately intergrown with all other minerals in the laterites, making a mechanical separation for REE determination impossible. In order to test the REE distribution between iron hydroxides and phosphates, a selective leaching technique was used

Figure 9.18 Scanning electron micrograph of oviform, euhedral gorceixite (upper left), baryte (lower left) and manganese oxide needles (upper right) from the saprolite from Araxá. Width of photograph: 0.1 mm.

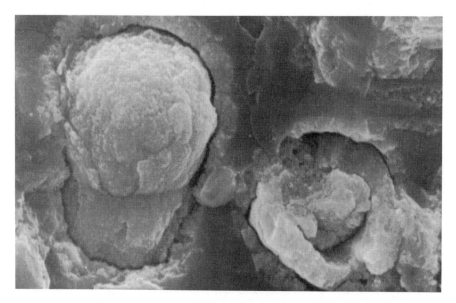

Figure 9.19 Scanning electron micrograph of monazite from the silcretes of Catalão showing a concentric structure and a cerianite $[(Ce^{4+}, Th)O_2]$ rim. Width of photograph: 0.02 mm.

Table 9.2 La, Ce, Nd and Y contents (ppm) in the insoluble residue and in the leachate of a soil, a laterite and a saprolite sample from Araxá

Sample	Element	Insoluble residue	Leachate	Total REE
Soil	La	3 119	93	3 212
	Ce	4 189	384	4 573
	Nd	1 701	36	1 737
	Y	250	12	262
Laterite	La	5 819	728	6 547
	Ce	12 425	3 920	16 345
	Nd	4 754	408	5 162
	Y	791	168	959
Saprolite	La	21 407	980	22 387
	Ce	20 945	2 310	23 255
	Nd	4 186	260	4 446
	Y	65	25	90

to dissolve selectively all the iron hydroxides and consequently free the adsorbed REE whilst avoiding the dissolution of the phosphates, silicates and most of the hematite and magnetite. The method of Mehra and Jackson (1960) was found to be very suitable. Following this procedure the sample is leached in a sodium dithionite – sodium citrate solution buffered with $NaHCO_3$. The addition of $NaHCO_3$ as a source of hydroxyl stabilizes the oxidation potential and pH. This gave an effective removal of iron hydroxides and, with suitable choice of reaction time, most of the iron oxides (e.g. magnetite and hematite), phosphates and silicates were not attacked. The insoluble residues were checked by XRD.

Table 9.2 shows the total La, Ce, Nd and Y values in the insoluble residue and in the leachate of a soil, a laterite and a saprolite sample from Araxá. It can be seen that only a small amount, between at maximum 11% (728 ppm) La and 8% (408 ppm) Nd, but rather consistent amounts of 23% (3920 ppm) Ce and 28% (25 ppm) Y can be bound to the iron hydroxides. The elevated amounts of Ce and Y in the leachate indicate a preferential adsorption of these elements by the iron hydroxides. As an example, Figure 9.20 shows the REE (La, Ce, Nd, Y) contents of the residue and the leachate of one soil sample from Araxá. The predominant amount of REE was bound to insoluble phosphates in the residue and only 3% of La, 8% of Ce, 2% of Nd and 5% of Y were liberated by the dissolution of the iron hydroxides.

9.7 Discussion and conclusions

1. The total REE contents at Araxá and Catalão range between 0.3 and 3.4 wt% in the soil horizons, between 0.5 and 4.4 wt% in the laterite

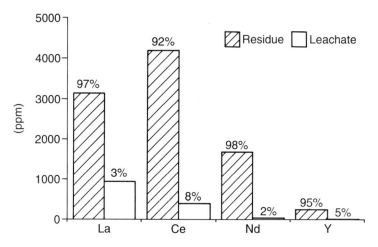

Figure 9.20 La, Ce, Nd and Y contents (ppm) in the leachate and in the residue of a soil sample from Araxá.

horizon and between 0.4 and 5.8 wt% in the saprolite horizon. The underlying alkaline rocks have average REE contents between 0.7 and 1 wt%. The maximum enrichment of total REE in the laterite profile compared to the alkaline rocks is one order of magnitude.

A detailed discussion of the enrichment of the REE, by normalization for example, versus the alkaline rocks would not be valid because the original rocks from which the laterites are formed are no longer available and may have been very different from the rocks which now underlie the laterites.

The highest total REE contents in the laterite profiles, up to 12.7 wt%, are found in the silcrete layers of Catalão.

2. At least in the case of the laterites of Araxá and Catalão, the laterization process produces no consistent fractionation of individual REE and the REE behave as rather immobile elements. The REE distribution patterns in the laterites mirror those of the underlying alkaline rocks.

Unfortunately, no REE patterns for the alkaline rocks of the Mt Weld complex are available so that it is not possible to make direct comparison between the REE fractionation patterns of the alkaline rocks and those of the overlying laterites. However, the REE patterns given in Figure 10 (RC 207) of Lottermoser (1989b) have been redrawn as Figure 9.21 and show that the REE patterns from the soils, laterites and saprolites are mostly sub-parallel and that the laterites are enriched in REE compared with the soil and the saprolite. The REE patterns of the laterite profiles do not suggest a strong fractionation with depth. The slight flattening of

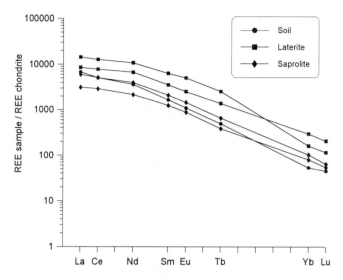

Figure 9.21 REE distribution patterns for the laterite profile (RC 207) of the Mt Weld carbonatite complex redrawn from Lottermoser (1989b).

the slope of the distribution patterns in the range of the light REE may have been inherited from the alkaline rocks. Some carbonatitic rocks do show a very flat slope in the range of the light REE (Figure 60 in Maravic, 1983).

3. Leaching experiments on the soil, the laterite *sensu stricto* and the saprolite from Araxá and Catalão show that if the main REE carrier is apatite and/or gorceixite, only about 2–30% of the individual REE are bound to iron hydroxides.

4. Magmatic apatite is still present as corrosion relics in the saprolite at Araxá and Catalão, but it is often associated and partly overgrown by a younger apatite formed by reprecipitation in the context of the lateritization processes. Both apatites are fluorapatites but the secondary apatite is richer in REE. The REE substitute for Ca in both cases.

5. The main REE carrier in the soil and in the laterite above the saprolite at Catalão and Araxá is gorceixite. Leaching of Ca and the high concentration of Ba favours the crystallization of gorceixite rather than apatite or crandallite. The presence of baryte is further confirmation of the high Ba concentration. Rhabdophane, which has been proposed previously as an REE carrier at these localities, was not found.

Different phosphate phases are present in the laterites at other localities. For example, at Mabounié (Gabon) the phosphate phase is crandallite (Laval, Johan and Tourlière, 1988) and at Mt Weld, Lottermoser (1989a) reports the formation of plumbogummite (crandallite-group minerals), monazite and churchite at the expense of the primary apatite. Von Maravic, Morteani and Roethe (1989) describe the formation of crandallite, goyazite and florencite in the lateritic cover above the alkaline complex of Lueshe.

In contrast, in the saprolite at all the above localities the main phosphate is apatite.

6. In the silcretes of Catalão the main REE carrier is a concentrically structured monazite. The presence of monazite instead of gorceixite in the silcretes is due to the fact that in the silcretes Al was not available for the formation of minerals of the crandallite group.

Acknowledgements

The investigations at Araxá and Catalão were financed through contract CLI*-CT9L-0882(MNLA) by the European Community. The aim of the EC project was to assess the economic REE potential of the laterites found on top of these alkaline complexes. Project partners were the University of Trento in Italy (coordinator) and the Centro de Tecnologia Mineral (CETEM) in Rio de Janeiro, Brazil, and the Technical University of München (TUM), Germany. Italian and Brazilian institutions were in charge of the preparation and extraction tests of REE from the lateritic ores. The mineralogical and chemical characterization of ores was carried out by the TUM. For help during the field work in Brazil we thank F. Lapido Loureiro and R. Neumann (CETEM). The ICP-OES analyses of the leachate were performed by A. Fuganti (University of Trento). We wish to thank the Arafertil and CBMM companies at Araxá and Goiasfertil and Minerçao Catalão companies at Catalão for their cooperation during sampling and for helpful discussions in the field. We acknowledge the kind help of D. Ackermand from the University of Kiel during the microprobe analysis.

References

Anonymous (1992) *Rare Earth Information Center News*, **XXVII**, **2**, 8 pp.
Anonymous (1993) *Rare Earth Information Center News*, **XXVIII**, **4**, 8 pp.

Barbosa, O., Braun, O. P. G., Dyer, R. C. and da Cunha, C. A. B. R. (1970) Geologia da regiao do triangulo Mineiro. *Boletim, Departamento Nacional de Producao Mineral, Divisao de Fomento da Producao Mineral. Rio de Janeiro*, **136**, 140 pp.
Bardossy, G. and Aleva, G. J. J. (1990) *Lateritic Bauxites*. Developments in Economic Geology, **27**, Elserier, Amsterdam, 639 pp.
Bates, R. L. and Jackson, J. A. (1980) *Glossary of Geology*, 2nd edn, American Institute of Geology, 751 pp.
Bocquier, G., Müller, J. P. and Boulangé, B. (1984) Les latérites. Connaissances et perspectives actuelles sur les mécanismes de leur differenciation. In *Livre jubilair du cinquantenaire 1934–1984*, Assoc. Franc. pour l'Etude du Sol, Paris, 123–38.
Bronevoi, V. A., Zhilberminc, A. V. and Teniakov, V. A. (1985) Average chemical com-

REFERENCES

positions of bauxites and their evolution in time. *Geokhimiya*, **4**, 435–46 (in Russian).

Buchanan, F. (1807) A journey from Madras through the countries of Mysore, Canara and Malabar. *East India Co., London*, **2**, 436–61.

Burkov, V. V. and Podporina, Y. K. (1967) Rare earths in granitoid residuum. *Dokl. Akad. Nauk USSR, Earth Sci. Sect.*, **177**, 214–6.

Cantrell, K. J. and Byrne, R. H. (1987) Rare earth element complexation by carbonate and oxalate ions. *Geochim. Cosmochim. Acta*, **51**, 597–605.

Clark, A. L. and Zheng, C. (1991) China's rare earth potential, industry and policy. International Conference on Rare Earth Mineral and Minerals for Electronic Uses, Hat Yai, Thailand, January 1991.

Deans, T. (1978) Mineral production from carbonatite complexes: a world review. Proceedings of the 1st International Symposium on Carbonatites, Poços de Caldas, 1976, Brasil, pp. 123–33.

Duddy, I. R. (1980) Redistribution and fractionation of rare-earth and other elements in a weathering profile. *Chem. Geol.*, **30**, 363–81.

Duncan, R. K. and Willett, G. C. (1990) Mount Weld carbonatite, in *Geology of Mineral Deposits of Australia and Papua New Guinea* (ed. F. E. Hughes), The Australasian Institution of Mining and Metallurgy, Melbourne, pp. 591–7.

Elderfield, H. and Greaves, M. J. (1982) The rare earth elements in seawater. *Nature*, **296**, 214–9.

Flicoteaux, R. and Lucas, J. (1984) Weathering of phosphate minerals, in *Phosphate Minerals* (eds. J. O. Nriagu and P. B. Moore) Springer, Berlin, pp. 292–317.

Greenwood, N. N. and Earnshaw, A. (1984) *Chemistry of the elements*, Pergamon Press, Oxford, 1542 pp.

Haskin, L. A., Haskin, M. A., Frey F. A. and Wildemann, T. R. (1968) Relative and absolute terrestrial abundances of the rare earths, in *Origin and Distribution of the Elements* (ed. L. H. Ahrens), Pergamon Press, Oxford, 889–91.

Hedrick, B. (1993) *Rare earth minerals and metals*, United States Bureau of Mines, 20 pp.

Hogdahl, O. T., Bowen, B. T. and Melson, S. (1968) Neutron activation analysis of lanthanide elements in seawater. *Adv. Chem. Ser.*, **73**, 308–25.

Laval, M., Johan, V. and Tourlière, B. (1988) La carbonatite de Mabounié: exemple de formation d'un gite résidual à pyrochlore. *Chron. Recherche Min.*, **491**, 125–36.

Lottermoser, B. (1989a) Rare-earth element behaviour associated with stratabound scheelite mineralisation (Broken Hill, Australia). *Chem. Geol.*, **78**, 119–34.

Lottermoser, B. (1989b) Rare earth elements and ore formation process. Unpublished Thesis, University of Newcastle, New South Wales, Australia, 308 pp.

Lottermoser, B. (1990) Rare-earth element mineralisation within the Mt Weld carbonatite laterite, Western Australia. *Lithos*, **24**, 151–67.

Lottermoser, B. (1991) Rare element resources and exploration in Australia. *Austr. Inst. Mining Metall. Proc.*, **2**, 49–56.

Maignien, R. (1966) *Compte rendu de recherches sur les latérites*. UNESCO Recherches sur les resources naturelles IV, Paris, 155 pp.

McFarlane, M. J. (1976) *Laterite and landscape*, Academic Press, London, 151 pp.

McLennan, S. M. (1989) Rare earth elements in sedimentary rocks: influence of provenance and sedimentary processes, in *Geochemistry and Mineralogy of Rare Earth Elements* (eds B. R. Lipin and G. A. McKay) *Reviews in Mineralogy*, Mineralogical Society of America, **21**, pp. 169–200.

Mehra, O. P. and Jackson, M. L. (1960) Iron removal from soils and clays by a dithionite-citrate system buffered with sodium bicarbonate. *Clays & Clay Minerals*, **7**, 317–27.

Middleburg, J. J., van der Weijden, C. H. and Woittiez, J. R. W. (1988) Chemical processes affecting the mobility of major, minor, and trace elements during weathering of granitic rocks. *Chem. Geol.*, **68**, 253–73.

Milnes, A. R. and Thiry, M. (1992) Silcretes, in *Developments in Earth Surface Processes 2* (eds I. P. Martini and W. Chesworth), Elsevier, New York, 349–77.

Möller, T. (1963) *The Chemistry of the Lanthanides*, Reinhold, New York, 117 pp.

Möller, P. and Morteani, G. (1983) On the geochemical fractionation of rare earth elements during the formation of Ca-minerals and its application to problems of the genesis of ore deposits, in *The Significance of Trace Elements in Solving Petrogenetic Problems and Controversies* (ed. S. S. Augusthitis), Theophrastus Publications, Athens, pp. 747–91.

Möller, P., Morteani, G. and Schley, F. (1980) Discussion of REE distribution patterns of carbonatites and alkalic rocks. *Lithos*, **13**, 171–9.

Möller, T., Martin, D. F., Thompson, L. C. *et al.* (1965) The coordination chemistry of yttrium and the rare earth metal ions. *Chem. Rev.*, **65**, 1–50.

Morteani, G. (1991) The rare earths: their minerals, production and technical uses. *Eur. J. Mineral.*, **3**, 641–50.

Morteani, G. and Preinfalk, C. (1993) The laterites of Araxá and Catalão, Brazil: an example of REE enrichment during lateritization of alkaline rocks. *Abst. Suppl. 3 to Terra Nova*, **5**, 35.

Nesbitt, H. W. (1979) Mobility and fractionation of rare earth elements during weathering of granodiorite. *Nature*, **279**, 206–10.

Nriagu, J. O. (1976) Phosphate–clay mineral relations in soils and sediments. *Can. J. Earth Sci.*, **13**, 717–36.

Preinfalk, C. and Morteani, G. Rare Earths: from Deposits to Uses in Science and Technology, Springer, Heidelberg (in preparation).

Preinfalk, C., Morteani, G. and Fuganti, A. (1992) Die Seltenerdgehalte der laterisierten Alkaligesteine von Poços de Caldas, Catalão und Araxá (Brasilien); eine Vorstudie zur wirtschaftlichen Bewertung. *13. Geowiss. Lateinamerika-Koll.*, Münster, Zusammenf. Geol. Paläont. Inst. u. Museum d. Westf. Wilhelms-Univ., Münster.

Ronov, A. B., Valashov, Y. A. and Migdisov, A. A. (1967) Geochemistry of the rare earths in the sedimentary cycle. *Geochem. Int.*, **4**, 1–17.

Schellmann, W. (1981) Considerations on the definition and classification of laterites, in *Lateritization Processes*, Proceedings of the International Seminar on Lateritization Processes, December 1979, Trivandrum India, Oxford and IBH Publ. Comp., New Dehli, pp. 1–10.

Schellmann, W. (1982) Eine neue Laterit definition. *Geol Jb., Hannover*, **D58**, 31–47.

Schellmann, W. (1983) A new definition of laterite. *Natural Resources and Development, Hannover/Tübingen*, **18**, 7–21.

Valeton, I. (1967) Laterite und ihre Lagerstätten. *Fortschr. Miner.*, **44**, 67–130.

Valeton, I. (1972) Bauxites. *Developments in soil science*, Elsevier, Amsterdam, 226 pp.

von Maravic, H. (1983) Geochemische und petrographische Untersuchungen zur Genese des niobführenden Karbonatit/Cancrinit-Syenitkomplexes von Lueshe, Kivu/NE-Zaire. Thesis. Techn. Univ. Berlin, 330 pp.

REFERENCES

von Maravic, H., Morteani, G. and Roethe, G. (1989) The cancrinite-syenite/carbonatite complex of Lueshe, Kivu/NE-Zaire: petrographic and geochemical studies and its economic significance. *J. Afr. Earth Sci.*, **9**, 341–55.

Walter, A.-V. (1991) Caracterisation geochimique et mineralogique de l'alteration de la carbonatite du complexe alkalin de Juquia (Bresil) – comportement des terres rares dans les mineraux phosphates. Unpublished Thesis, Universite d'Aix-Marseille, 247 pp.

Woolley, A. R. (1987) *Alkaline Rocks and Carbonatites of the World; Part 1: North and South America*, British Museum (Natural History), London 216 pp.

Woolnough, W. G. (1927) Presidential address, Part I. The chemical criteria of peneplanation. Part II. The duricrust of Australia. *J. Proc. R. Soc. NSW*, **61**, 1–53.

CHAPTER TEN
Authigenic rare earth minerals in karst-bauxites and karstic nickel deposits

Z. J. Maksimović and Gy. Pantó

10.1 Introduction

The presence of secondary minerals of the bastnäsite group in karst-bauxites was first discovered by electron microprobe in the San Giovani Rotondo deposit in Italy (Bárdossy and Pantó, 1973). Similar minerals were also reported in the Nagyharsány bauxite deposit in Hungary (Bárdossy, Pantó and Várhegyi, 1976). Geochemical studies of the REE, together with Ni, Co, Mn and Zn in karst-bauxites have shown that these elements were also 'mobile' in the bauxitization process and were highly enriched in the lowermost part of the deposit (Maksimović, 1976, Maksimović and Roaldset, 1976). These studies enabled the rapid discovery of various authigenic rare earth (RE) minerals in many karst-bauxite deposits in Greece and Yugoslavia, and later, in karstic nickel deposits in Greece (Maksimović and Pantó, 1978, 1980, 1981, 1983, 1985a, 1985b, 1987; Rosenberg 1984; Maksimović 1987, 1993; Maksimović, Skarpelis and Pantó 1993; Pantó and Maksimović, in preparation). With relatively high REE concentrations in the initial clayey material which had accumulated in a karstic depression, coupled with an extensive bauxitization process, authigenic RE minerals were readily formed in the contact zone between bauxite and the footwall limestone. Karstic nickel deposits have been formed under similar conditions (Maksimović, 1987). Here, the contribution of weathered ultramafic material greatly increased the average nickel content of the initial clayey material, and after strong leaching in a karstic environment the authigenic nickel clay minerals were formed above the footwall geochemical barrier. In addition, deposits transitional between karst-bauxites and karstic nickel deposits also have an unusual association of authigenic RE minerals and nickel clay minerals (takovite, brindleyite and nepouite). This association of the REE and nickel in these deposits resulted from materials of different origin and

Rare Earth Minerals: Chemistry, origin and ore deposits. Edited by Adrian P. Jones, Frances Wall and C. Terry Williams. Published in 1996 by Chapman & Hall. ISBN 0 412 61030 2

composition being washed down into a karstic depression from the surrounding weathering profiles and forming on various silicate rocks. The well-known deposits of this transitional type are Marmara, Nisi and Marmeiko in Greece, and Grebnik in Serbia.

10.2 Occurrence

RE minerals occur only in deposits where intense leaching has taken place *in situ*. Erosion, transportation and redeposition of karstic deposits may result in the degradation and ultimately in the total destruction of the primary trace element distribution pattern. Such processes occurred in the major karst-bauxite deposits of the Transdanubian Central Range in Hungary (Maksimović, Mindszenty and Pantó, 1991). A typical REE distribution pattern within a bauxite deposit at Montenegro is shown in Figure 10.1, where the RE minerals occur above the contact with karstified footwall limestone.

Most of the authigenic RE minerals found in the Mediterranean region in bauxite deposits and some karstic nickel deposits are of Cretaceous age. These minerals are visible only at high magnification and were initially discovered using a microprobe by a combination of backscattered electron imaging and searching for characteristic wavelengths of the REE, particularly those of neodymium. The RE minerals occur as irregular segregations,

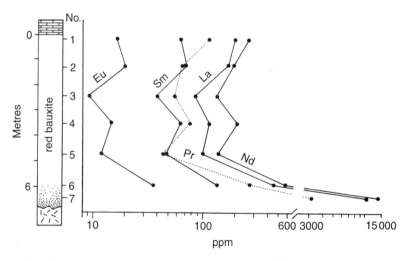

Figure 10.1 Variation of some light REE with depth in the Štitovo Upper Jurassic bauxite deposit, Montenegro; nos 1–7, red boehmitic bauxite; b, zone of occurrence of hydroxylbastnäsite-(Nd). (Analytical data from Maksimović and Roaldset (1976); no. 7, additional data for La, Nd and Pr, Maksimović and Jovic, unpublished results.)

micropore and space fillings, microveins and sometimes as nebular impregnations. Micropore fillings are generally less than 1 μm in diameter and show no sharp boundaries. Larger grains, 30–70 μm in diameter are distributed heterogenously and consist of numerous small grains. The highest concentration of these minerals in the karstic deposits found so far in the world was recorded in the Upper Jurassic bauxites of Montenegro, in the form of hydroxylbastnäsite-(Nd) and hydroxylbastnäsite-(La) (Maksimović and Pantó, 1985a, 1987). These minerals occur as typical authigenic minerals in the red boehmitic bauxite, following the contact zone of the bauxite – footwall limestone, which is up to 2 m in thickness (Figure 10.1). They form whitish irregular aggregates of crystals, usually 0.1–0.2 mm across, but sometimes as clusters 1–5 mm in diameter, in which the hydroxylbastnäsite is intergrown with red bauxite. In some cases the matrix is impregnated by an extremely dispersed form of these minerals. Scanning electron micrographs of the crystal aggregates reveal the hydroxylbastnäsite-(Nd) to have a typically platy habit (Figure 10.2). The plates have crystallized or intergrown to produce a variety of morphologies (Maksimović and Pantó, 1987).

10.3 Chemical properties

The authigenic RE minerals usually occur as very small grains and could only be studied using an electron microprobe. The first analyses were carried out by means of a JEOL JXA-5 electron probe (Maksimović and Pantó, 1980, 1981). Later, all quantitative analyses were obtained using the JEOL Superprobe-733 in the Laboratory for Geochemical Research of the Hungarian Academy of Sciences using the method described by Nagy (1993). Initially, quantitative analysis was performed with pure metal rare earth elements as standards, supplied by Pierce Inorganics, Rotterdam. Subsequently, the standards used were natural apatite for Ca, P and F, and synthetic glass standards of Drake and Weill (1972) for the REE. In all bastnäsites abundant carbon was detected by microprobe analysis. In the hydroxylbastnäsites the carbon content was assessed semiquantitatively to be between 5 and 6 wt%. Only in the bastnäsite-(Nd) from Montenegro, where a sufficient quantity of the mineral was available, could H_2O and CO_2 be analysed quantitatively by thermogravimetric and evolved gas analysis (Maksimović and Pantó, 1985a).

Analyses of authigenic synchysite, bastnäsites, hydroxylbastnäsites and monazites (Tables 10.1–10.3) show wide variations in the contents of the lighter, more abundant REE, i.e. La, Ce and Nd, with Nd often being the most abundant REE in these minerals. Cerium dominates in the F-rich bastnäsites (Table 10.1), but is much lower in the synchysite-(Nd) analysed (Table 10.1) and in all the hydroxylbastnäsites (Table 10.2) and monazites (Table 10.3). Of the other REE, Pr, Sm, Gd and Y usually show little

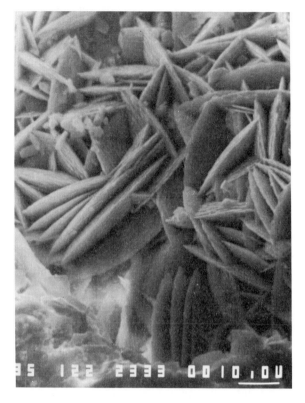

Figure 10.2 SEM photograph of hydroxylbastnäsite-(Nd) from Zagrad deposit, Montenegro.

variation; Eu is detected in eight of the 14 minerals, and Dy, Er and Yb were detected in some minerals, but in all cases at very low concentrations.

Hydroxylbastnäsite is the most frequently observed and the most abundant of the authigenic RE minerals identified in karstic deposits. Most of the bastnäsites analysed show a deficiency of fluorine in the structural formula. This element can be substituted by OH groups giving rise to hydroxylbastnäsites, as in the sample of hydroxylbastnäsite-(Nd), a new mineral from Montenegro (Maksimović and Pantó, 1985a). This mineral has also been found in the karstic deposits of the Lokris area in Greece, where the fluorine content is very variable, ranging from 0.50 atoms per unit cell in the Marmeiko deposit to less than 0.02 atoms in the Nisi deposit (Table 10.2, nos 5 and 6). The composition of this mineral at Nisi corresponds to a hydroxylcarbonate-(Nd), and is possibly a new mineral, being the end member of the isomorphous series of fluocarbonates to hydroxylcarbonates of the REE.

CHEMICAL PROPERTIES

Table 10.1 Microprobe analyses (wt%) of authigenic synchysite and bastnäsites

	1	2	3	4	5	6
La_2O_3	15.5	19.9	0.5	16.8	22.9	17.2
Ce_2O_3	2.4	21.5	72.3	22.2	17.9	38.9
Pr_2O_3	4.4	4.5	0.1	5.0	5.3	2.5
Nd_2O_3	19.0	17.9	1.0	17.0	21.5	7.3
Sm_2O_3	3.6	3.7	–	3.1	2.9	2.8
Eu_2O_3	–	–	–	–	–	0.5
Gd_2O_3	2.5	2.9	–	4.6	2.4	2.1
Dy_2O_3	1.4	0.3	–	–	0.7	0.5
Yb_2O_3	–	0.1	–	–	–	–
Y_2O_3	3.8	3.2	0.2	6.0	2.3	3.1
CaO	18.2	2.7	1.8	1.6	1.6	0.3
F	6.1	7.0	7.0	6.4	6.2	8.1
CO_2*	25.6	19.2	20.0	20.0	18.9	20.1
	102.5	102.9	102.9	102.7	102.6	103.4
$-O = F$	2.5	2.9	2.9	2.7	2.6	3.4
Total	100.0	100.0	100.0	100.0	100.0	100.0

Atomic ratios calculated to

	O = 6		(O,F) = 4			
La	0.29	0.27	0.01	0.23	0.32	0.23
Ce	0.04	0.29	0.98	0.30	0.25	0.52
Pr	0.09	0.06	–	0.07	0.07	0.03
Nd	0.36	0.24	0.01	0.22	0.29	0.10
Sm	0.06	0.04	–	0.04	0.03	0.03
Eu	–	–	–	–	–	<0.01
Gd	0.04 }1.00	0.03 }1.09	– }1.07	0.05 }1.09	0.03 }1.12	0.02 }1.00
Dy	0.02	<0.01	–	–	0.01	<0.01
Yb	–	<0.01	–	–	–	–
Y	0.10	0.06	<0.01	0.12	0.04	0.06
Ca	1.05	0.10	0.07	0.06	0.06	0.01
F	1.04	0.82	0.82	0.75	0.75	0.94
C	1.90	0.98	1.01	1.01	0.98	1.01

*Calculated to sum of 100%.
1, synchysite-(Nd), Grebnik bauxite deposit, Serbia (Maksimović and Pantó, 1981); 2, bastnäsite-(Ce), Nagyharsány bauxite deposit, Hungary (Maksimović and Pantó, 1983); 3, bastnäsite-(Ce), Nagyharsány bauxite deposit, Hungary (Maksimović and Pantó, 1983); 4, bastnäsite-(Ce), Nazda bauxite deposit, Bosnia (Maksimović and Pantó, 1983); 5, bastnäsite-(La), Marmara bauxite deposit, Greece, (Maksimović and Pantó, 1980); 6, bastnäsite-(Ce), Milovici bauxite deposit, Montenegro, (Maksimović, 1987).

Table 10.2 Microprobe analyses (wt%) of authigenic hydroxylbastnäsites

	1	2	3	4	5	6
La_2O_3	31.7	23.0	27.1	17.3	17.3	27.7
Ce_2O_3	3.7	1.2	0.3	2.4	–	–
Pr_2O_3	5.5	5.7	8.5	6.0	8.0	7.3
Nd_2O_3	21.8	26.7	31.5	37.2	34.5	29.9
Sm_2O_3	3.6	4.0	4.4	6.5	5.3	4.7
Eu_2O_3	1.4	1.3	1.3	1.7	1.4	1.5
Gd_2O_3	2.7	3.1	1.4	3.8	2.1	2.6
Tb_2O_3	0.1	0.9	–	–	–	–
Dy_2O_3	1.1	1.5	–	–	0.3	0.7
Er_2O_3	0.1	0.3	–	0.8	–	0.1
Yb_2O_3	–	0.1	–	–	–	0.2
Y_2O_3	3.7	4.6	0.2	1.9	0.6	0.1
CaO	1.5	2.6	0.3	0.7	4.5	–
F	4.3	4.5	3.3	1.4	0.2	–
H_2O*	0.6	1.9	2.26†	2.0	3.8	4.2
CO_2*	19.5	20.5	20.63†	18.9	22.1	21.0
	102.1	101.9	101.19	100.6	100.1	100.0
$-O = F$	2.1	1.9	1.39	0.6	0.1	–
Total	100.0	100.0	99.80	100.0	100.0	100.0

Atomic ratios calculated to (O, OH, F) = 4

	1		2		3		4		5		6	
La	0.44		0.30		0.36		0.25		0.22		0.37	
Ce	0.05		0.01		<0.01		0.03		–		–	
Pr	0.08		0.07		0.11		0.08		0.10		0.10	
Nd	0.29		0.34		0.41		0.51		0.42		0.38	
Sm	0.05		0.05		0.06		0.09		0.06		0.08	
Eu	0.02		0.01		0.02		0.02		0.02		0.02	
Gd	0.03	1.12	0.04	1.02	0.02	0.99	0.05	1.12	0.02	1.02	0.03	0.99
Tb	<0.01		0.01		–		–		–		–	
Dy	0.01		0.02		–		–		<0.01		0.01	
Er	0.01		<0.01		–		–		–		<0.01	
Yb	–		<0.01		–		–		–		<0.01	
Y	0.07		0.07		<0.01		0.04		<0.01		<0.01	
Ca	0.07		0.10		0.01		0.03		0.17		0	
F	0.58	0.73	0.50	0.95	0.38	0.93	0.17	0.69	0.02	0.89	–	1.00
OH	0.15		0.45		0.55		0.52		0.87		1.00	
C	1.00		0.99		1.03		1.00		1.03		1.02	

*Carbon determined semiquantitatively by electron microprobe; H_2O is assumed to make up the remainder.
†CO_2 and H_2O determined by thermogravimetric and evolved gas analysis.
1, hydroxylbastnäsite-(La), Liverovici bauxite deposit, Montenegro (Maksimović and Pantó 1983); 2, hydroxylbastnäsite-(Nd), Marmeiko nickel deposit, Greece (Pantó and Maksimović, unpublished results); 3, hydroxylbastnäsite-(Nd), Zagrad bauxite deposit, Montenegro (Maksimović and Pantó, 1985a); 4, hydroxylbastnäsite-(Nd), Štitovo bauxite deposit, Montenegro (Maksimović and Pantó, 1983); 5, hydroxylbastnäsite-(Nd), Nisi bauxite deposit, Greece (Pantó and Maksimović, unpublished results); 6, hydroxylcarbonate-(Nd), Nisi bauxite deposit, Greece (Pantó and Maksimović, unpublished results).

CHEMICAL PROPERTIES

Table 10.3 Microprobe analyses (wt%) of authigenic monazites

	1	2		Atomic ratios calculated to O = 4	
				1	2
ThO_2	–	0.8	Th	–	0.01
La_2O_3	16.4	23.8	La	0.24	0.34
Ce_2O_3	18.3	8.6	Ce	0.27	0.12
Pr_2O_3	6.0	6.0	Pr	0.09	0.09
Nd_2O_3	21.5	20.6	Nd	0.30	0.29
Sm_2O_3	3.6	3.7	Sm	0.05	0.05
Eu_2O_3	–	1.3	Eu	– }1.07	0.02 }1.05
Gd_2O_3	2.2	2.8	Gd	0.03	0.04
Tb_2O_3	–	0.2	Tb	–	0.01
Dy_2O_3	0.5	1.2	Dy	0.01	0.01
Er_2O_3	–	0.4	Er	–	0.01
Y_2O_3	1.1	1.3	Y	0.02	0.03
CaO	1.7	1.2	Ca	0.07	0.05
SO_3	1.4	–	S	0.04 }0.95	– }0.95
P_2O_5	26.8	28.0	P	0.91	0.95
Total	99.5	99.4			

1, monazite-(Nd), Marmara bauxite deposit, Greece (Maksimović and Pantó, 1980); 2, Monazite-(La), Liverovici bauxite deposit, Montenegro, (Maksimović and Pantó 1983).

In monazite-(Nd) from Marmara bauxite deposit in Greece, the PO_4 in this mineral is partially replaced by SO_4. A sulphate-bearing monazite has been reported by Kukharenko et al. (1961), in which the double substitution of $Ca(SO_4)$ for $Ce(PO_4)$ was proposed. Pantó (1975) has found 1.8 wt% S in monazites of hydrothermal origin. Monazite-(La) from Montenegro is characterized by a very low thorium content and Ce relatively depleted, compared with that of La and Nd (Table 10.3).

A new variety of goyazite, neodymian-goyazite, was discovered in karstbauxites from Eastern Bosnia (Maksimović and Pantó, 1985b). According to the formula (Table 10.4) the mineral composition lies between goyazite, $SrAl_3(PO_4)_2 \cdot (OH)_5 \cdot H_2O$, florencite, $CeAl_3(PO_4)_2 \cdot (OH)_6$ and crandallite, $CaAl_3(PO_4)_2 \cdot (OH)_5 \cdot H_2O$, with some substitution of SO_4^{2-} for PO_4^{3-} (probably with corresponding substitution of OH^- for H_2O). Using abbreviations Gz = goyazite, Fc = florencite and Cn = crandallite for these members of the crandallite group, and because Nd is the most abundant of the REE, this mineral can be described as $Cn_{24}Fc_{33}Gz_{43}$, i.e. neodymiangoyazite.

Table 10.4 Microprobe analysis of authigenic neodymian-goyazite[b]

	wt%		Atomic ratios calculated to (O, OH) = 14	
La_2O_3	2.2	La	0.07	
Ce_2O_3	3.1	Ce	0.09	
Pr_2O_3	0.8	Pr	0.02	
Nd_2O_3	4.1	Nd	0.12	
Sm_2O_3	0.8	Sm	0.02	1.10
Gd_2O_3	0.9	Gd	0.02	
Tb_2O_3	0.1	Tb	0.01	
Dy_2O_3	0.6	Dy	0.02	
Y_2O_3	0.4	Y	0.02	
CaO	3.1	Ca	0.26	
SrO	9.9	Sr	0.46	
Al_2O_3	32.5	Al	3.05	3.05
P_2O_5	27.8	P	1.87	2.01
SO_3	2.3	S	0.14	
	88.6			
H_2O[a]	11.4	OH	6.05	6.05
	100.0			

[a] H_2O is assumed to make up the remainder; [b] Maksimović and Pantó (1985b).

10.4 X-ray powder diffraction study

In only eight of 15 analysed samples (Tables 10.1–10.4), could identification of the RE mineral be confirmed by X-ray powder diffraction (XRPD). In the other samples insufficient material was obtained, even after applying various separation techniques, to enable a characteristic diffraction pattern to be developed.

Bastnäsites were identified by the strongest reflections at 4.896 Å, 3.564 Å and especially at 2.877 Å (Maksimović, 1979; Maksimović and Pantó, 1980, 1985b). More data are available for hydroxylbastnäsite after the discovery of relatively high concentrations of this mineral in the Upper Jurassic bauxites of Montenegro (Maksimović and Pantó, 1985a). The diffraction pattern showed some reflections of contaminating boehmite and hematite phases, but the total amount of impurities was only a few per cent and the peaks belonging to them were readily distinguished from those of the hydroxylbastnäsite. The powder diffraction pattern was indexed by analogy with that of bastnäsite (Farkas, Maksimović and Pantó, 1985). The lattice parameters were refined by least squares, utilizing the powder data.

Table 10.5 Cell parameters, volume and the strongest reflection of bastnäsite and hydroxylbastnäsite

Mineral	Unit cell dimension (Å)	Volume (Å^3)	Strongest reflection (I = 100) d_{112} (Å)
Bastnäsite-(Ce) (Ni, Hughes and Mariano, 1993)	$a_0 = 7.1175$ $c_0 = 9.7619$	428.3	–
Bastnäsite-(Ce) (Glass et al., 1958)	$a_0 = 7.129$ $c_0 = 9.774$	430.2	2.879
Hydroxylbastnäsite-(Nd) (Maksimović and Pantó 1985a)	$a_0 = 7.191$ $c_0 = 9.921$	444.3	2.911
Hydroxylbastnäsite-(Ce) (Kirillov, 1964)	$a_0 = 7.23(2)$ $c_0 = 9.98(5)$	451.8	2.92
Hydroxylcarbonate-(Nd) Nisi deposit, Greece		455.5[a]	2.93

[a] Estimated on Figure 10.3.

The refined unit cell parameters of the hexagonal cell were calculated as $a = 7.191(1)$ Å; $c = 9.921(2)$ Å, which gave a cell volume of 444.3 Å^3.

An XRPD study of the samples from the Marmeiko and Nisi deposits in Greece have shown reflections due to bastnäsite (Rosenberg, 1984; Maksimović, Skarpelis and Pantó, 1993). In most of the samples, however, only the strongest reflection was observed, which varied also in its d-spacing from $d = 2.875$ Å to $d = 2.930$ Å. The reflection with the highest d_{112} value appeared in the sample from the Nisi deposit where fluorine was absent in the analysis of hydroxylbastnäsite-(Nd) (Table 10.2, no. 6). The substitution of OH for F in the structure of bastnäsite is likely to lead to an increase in the unit cell volume due to the different sizes of these ions, and in the value of the strongest d_{112} reflection (Table 10.5). The increase of the unit cell volume from bastnäsite to hydroxylbastnäsite is linear (Fig. 10.3), and the intersection of the line with the ordinate representing 1.00 atoms of OH per unit cell gives a unit cell volume of hydroxylcarbonate-(Nd) of 455.5 Å^3. It seems that relative changes to La, Ce and Nd contents (the major cations) do not significantly affect the size of the unit cell volume, whereas substitution of OH for F does.

The chemical composition and XRPD data together confirm that hydroxylbastnäsite without fluorine is a new mineral, it being the end member of the series bastnäsite – hydroxylcarbonate-RE (Pantó and Maksimović, in preparation).

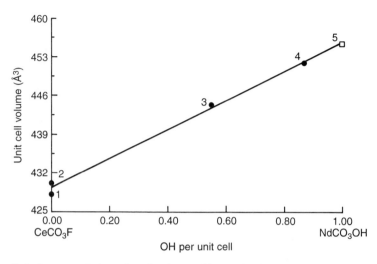

Figure 10.3 Increase of the unit cell volume of bastnäsite by substitution of OH for F. 1, bastnäsite-(Ce), Ni, Hughes and Mariano (1993); 2, bastnäsite-(Ce), Glass, *et al.* (1958); 3, hydroxylbastnäsite-(Nd), Maksimović and Pantó (1985a); 4, hydroxylbastnäsite-(Ce), Kirillov (1964); 5, hydroxylcarbonate-(Nd), Nisi deposit, Greece.

10.5 Rare earth minerals

The most frequent RE minerals in karstic deposits are members of the bastnäsite group: synchysite-(Nd), bastnäsite-(Ce) and -(Nd), and hydroxylbastnäsite. Relatively rare are RE phosphates and alkaline earth phosphates, such as monazite-(Nd), monazite-(La) and neodymian-goyazite.

10.5.1 Synchysite-(Nd)

Synchysite-(Nd) was described from the Cretaceous bauxite deposit in the Grebnik Mountain, Serbia (Maksimović and Pantó, 1978, 1980). It occurs between oolites as space fillings and micropore fillings in the cement (Figure 10.4a,b). Microprobe analysis gave the composition and empirical formula of this mineral (Table 10.1, no. 1), which is very close to the ideal formula of synchysite, but with Nd the most abundant REE. Synchysite-(Nd) occurs in association with Ni-chlorite which contains 5.1 wt% Ni (Maksimović, 1988).

10.5.2 Bastnäsite-(Ce)

Bastnäsite-(Ce) was found in several karst-bauxite deposits in Hungary, Greece, Bosnia and Montenegro, but mostly in only trace amounts.

RARE EARTH MINERALS

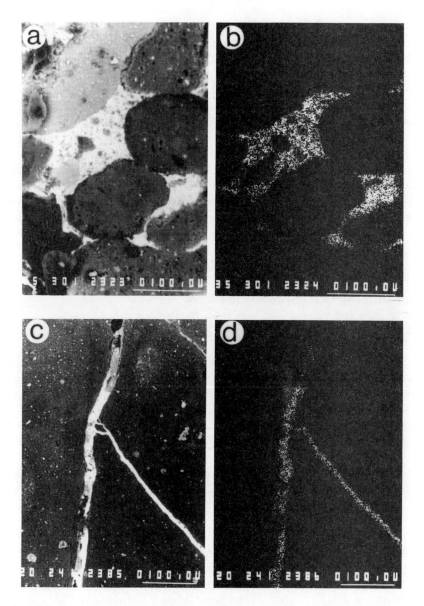

Figure 10.4 Synchysite-(Nd) in diasporic bauxite from Grebnik, Serbia: (a) backscattered electron image; (b) Nd Lα X-ray map. Bastnäsite-(Ce) microveins in a bauxite sample from Nagyharsány, Hungary: (c) backscattered electron image; (d) Ce Lα X-ray map.

Bastnäsite-(Ce) from the Lower Cretaceous deposit of Nagyharsány in Hungary occurs as irregular segregations, pore fillings, nebular impregnations and microveins (Figure 10.4c,d). Analysis of this mineral from segregations confirms Ce as the most abundant REE, although La and Nd also are present (Table 10.1, no. 2). Bastnäsite-(Ce) from the microveins, however, has a quite different composition with an extreme concentration of Ce (Table 10.1, no. 3), and with only minor amounts of the other REE present.

Trace amounts of bastnäsite have been found in several deposits of the bauxite-bearing area of Vlasenica in Eastern Bosnia. Bastnäsite-(Ce) was found in the Nazda deposit, about 2 m above the footwall limestone. It is interesting to note that in the same deposit, about 0.3 m above the footwall, Nd-rich goyazite was discovered (Maksimovic and Pantó, 1985b). In the Upper Jurassic bauxites of the Western Montenegro, with the Middle Jurassic footwall, a very rare occurrence of bastnäsite-(Ce) were observed, which contained a high concentration of fluorine (Table 10.1, no. 6). About 25 km to the east, in the Nikšićska Zupa, large deposits of Upper Jurassic bauxites occur on the Upper Triassic footwall. These bauxites contain more bastnäsite, though it is deficient in fluorine.

10.5.3 Bastnäsite-(La)

Bastnäsite-(La) was discovered in the Marmara bauxite deposit in Greece (Table 10.1, no. 4), together with bastnäsite-(Ce). In two samples of brindleyite (a hydrated Ni, Al silicate) collected from localities very close together in the deposit, the relative atomic ratios of the REE in the bastnäsite from the two samples were markedly different: La, 30.2 and 11.7; Ce, 23.7 and 72.2, and Nd, 28.5 and 7.5 atomic %. These extreme differences reveal, as in other deposits, a strong separation of individual REE in the formation of karst-bauxites within a very localized area (Maksimović and Pantó, 1980).

10.5.4 Hydroxylbastnäsite

Hydroxylbastnäsite is the most frequent and abundant mineral of the bastnäsite group. Most of the bastnäsites analysed show a deficiency of fluorine, replaced in the structure by OH groups, giving rise to hydroxylbastnäsite. Hydroxylbastnäsite is most abundant in Montenegro, in the Nikšićska Zupa, in the largest bauxite deposits in Europe. Both hydroxylbastnäsite-(Nd) and hydroxylbastnäsite-(La) (Table 10.2) occur here, the former a new mineral found in Montenegro.

Other occurrences of hydroxylbastnäsite-(Nd) are in Greece, in the Ni-rich bauxite deposit at Nisi, and in the Marmeiko karstic nickel deposit (Table 10.2, nos 2 and 5). In the bauxite deposit at Nisi, which in the lower part could be considered as a karstic nickel deposit, both hydroxylbastnäsite-

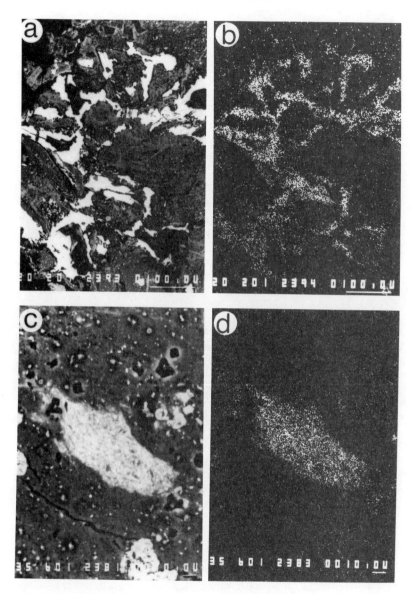

Figure 10.5 Hydroxylcarbonate-(Nd) from Nisi bauxite deposit, Greece. (a) Backscattered electron image; (b) Nd Lα X-ray map. Authigenic monazite-(Nd) from Marmara bauxite deposit, Greece: (c) backscattered electron image; (d) Nd Lα X-ray map.

(Nd) with fluorine, and hydroxylbastnäsite-(Nd) without fluorine (Table 10.2, no. 6), occur. The latter mineral represents a new phase, the end member of the series bastnäsite – hydroxylcarbonate-(RE) (Figure 10.5a,b). The distribution of the REE in the karstic nickel deposits is the same as in karst-bauxites, with the enrichment of the REE above the footwall limestone (Maksimović, Skarpelis and Pantó, 1993).

10.5.5 Monazite-(Nd)

Monazite-(Nd) was found in the Marmara bauxite deposit (Table 10.3, no. 1) associated with bastnäsite-(La), in accumulations of green brindleyite – $(Ni_{1.75}Al_{1.00})(Si_{1.5}Al_{0.5})O_5(OH)_4$. Here, it is commonly very fine grained and occurs as micropore fillings about $1-3\,\mu m$ in diameter, and also in fissure fillings up to $15\,\mu m$ in diameter. Larger grains (up to $45 \times 110\,\mu m$) have also been found (Figure 10.5c,d and Figure 10.6). The composition of this authigenic monazite is very variable in this deposit. In two adjacent samples examined, their relative atomic ratios of REE were markedly different: La, 23.6 and 11.4; Ce, 26.3 and 67.6; and Nd, 31.1 and 9.4 atomic % (Maksimović and Pantó 1980).

10.5.6 Monazite-(La)

Monazite-(La) occurs as rare space fillings and irregular grains $15-60\,\mu m$ in diameter in the Upper Jurassic bauxites of Montenegro (Table 10.3, no. 2). It is characterized by a very low thorium content and only minor amounts of cerium.

10.5.7 Neodymian-goyazite

Neodymian-goyazite forms micropore fillings and fissure fillings, as well as small grains $70 \times 30\,\mu m$ in diameter, occurring just above the footwall in the Nazda bauxite deposit in Bosnia (Figure 10.7). Bastnäsite-(Ce) occurs in the same deposit, but on a higher level above the footwall. The relative distribution of the REE in neodymian-goyazite and bastnäsite-(Ce) also reveals that separation of individual REE had occurred during formation of the karst-bauxites.

10.6 Genetic considerations

All the karst-bauxites and karstic nickel deposits investigated are composed of red bauxite, or red Ni–Fe ore, formed as continental products under oxidizing conditions in a karstic environment which had a good drainage system for the removal of surface water. The enrichment of the REE *per descendum* is a general process which occurred in all karstic deposits formed *in situ*. The overall concentration of the REE, however, varied over an extremely wide range: from 180 ppm total REE in the basal part of the Aghios Ioannis karstic nickel deposit in Greece (Maksimović, Skarpelis and Pantó, 1993) to 6.66 wt% for the sum of Nd, La and Pr in bauxite above the

GENETIC CONSIDERATIONS

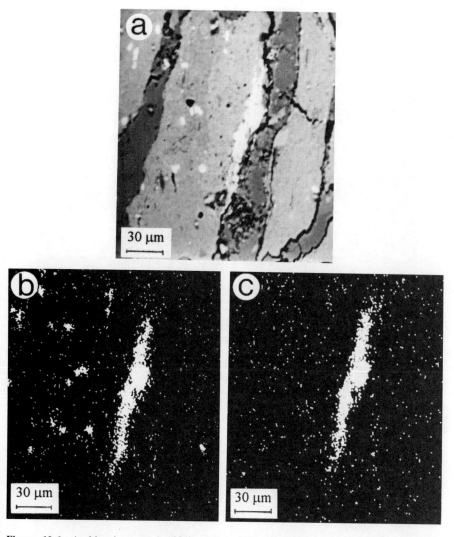

Figure 10.6 Authigenic monazite-(Ce) with bastnästite-(Ce) (white drops) in brindleyite sample from Marmara bauxite deposit, Greece. (a) Backscattered electron image; (b) Ce Lα map; (c) P Kα map.

footwall limestone in the Upper Jurassic bauxites of Montenegro. The observed concentrations are dependent on the original REE content of the initial clayey material which had collected in the karstic depression. The contribution of ultramafic material which, being extremely depleted in REE, greatly influenced the overall REE content, and also increased the nickel content. This was the situation at the Aghios Ioannis karstic nickel deposit,

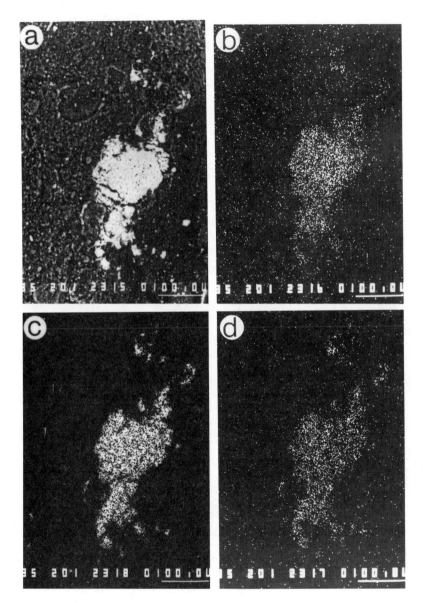

Figure 10.7 Nd-goyazite from Nazda bauxite deposit, Bosnia. (a) Backscattered electron image; (b) Nd Lα map; (c) Sr Kα map; (d) P Kα map.

where the source rocks were mainly weathered ultramafics. The Marmeiko karstic nickel deposit is the only one of this type containing authigenic RE minerals. Material for the formation of this deposit came from at least two different sources: (1) an ultramafic weathered rock rich in iron, nickel and chromium, and (2) a silicate rock rich in REE. The negative Eu anomaly observed in all samples from the Marmeiko deposit (Maksimović, Skarpelis and Pantó, 1993) could be used as an indicator of the source of the REE. The Nisi deposit is located only 3–4 km from the Marmeiko and Aghios Ioannis deposits. The bauxite deposit at Nisi is a karstic nickel deposit in the lower part, with high concentrations of nickel and REE, and associated authigenic Ni and RE minerals. These three deposits in the Lokris area in Greece provide excellent examples for understanding the genesis of karst-bauxites and karstic nickel deposits.

The authigenic RE minerals were formed during strong leaching of clayey material deposited in karstic depressions from surrounding weathered rocks. The available evidence indicates that in order to form authigenic RE minerals, the total REE content in the initial material should be greater than 1000 ppm (Table 10.6). The REE were transported into the karstic environments mainly as adsorbed ions on clay particles. A study of the relicts of the weathering crust on shales surrounding the Grebnik bauxite deposit in Serbia has shown that the amount of adsorbed REE in the red kaolinic clay ranged from 24 wt% to 79 wt%, with an average of 45.7 wt% of the toïl REE content (Maksimović, 1980). In the adsorbed form the REE were readily removed by percolating water and subsequently concentrated on the geochemical barrier of the footwall limestone, forming the authigenic RE minerals.

Table 10.6 Total REE of some karstic bauxites and karstic nickel deposits containing authigenic RE minerals

Deposit	Number of analyses (n)	Average REE content in the ore (ppm)	REE content in the mineral (%)[a]	Form of REE occurrence
Nazda, Vlasenica, Bosnia	59	1140	11.1	Nd-goyazite
			63.3	Bastnäsite-(Ce)
Marmeiko nickel deposit, Greece	9	1153	61.5	Hydroxylbastnäsite-(Nd)
Štitovo, Montenegro	9	1320	66.2	Hydroxylbastnäsite-(Nd)
Grebnik, Serbia	14	1450	44.6	Synchysite-(Nd)
Marmara, Greece	6	1582	59.3	Monazite-(Nd)
			64.6	Bastnäsite-(La)
Nisi, Greece	6	4576[b]	63.8	Hydroxylcarbonate-(Nd)

[a] Above the footwall limestone; [b] Average of the basal part (3 m) of the deposit.

Geochemical studies of the REE in karst-bauxites and karstic nickel deposits have shown that both light REE (LREE) and heavy REE (HREE) were 'mobile' during the formation of these deposits, and were both concentrated above the footwall limestone (Maksimović and Pantó, 1991; Maksimović, Skarpelis and Pantó, 1993). In six of the seven deposits, ratios of LREE/HREE and La/Y decrease downwards, showing an enrichment of HREE relative to LREE (Tables 10.7 and 10.8). However, an enrichment of LREE relative to HREE is observed in the authigenic RE minerals, where Nd-dominant RE minerals are often formed.

The cerium content of synchysite-(Nd) and of all hydroxylbastnäsites is surprisingly low, ranging from below detection limit to 3.7 wt% Ce_2O_3. This element behaves differently from the other REE: it may be oxidized in a relatively strong oxidizing environment. Therefore, in some karstic deposits Ce does not show the preferential enrichment *per descendum* of the other

Table 10.7 REE distribution along a vertical profile (7 m) in the Nazda karst-bauxite deposit, Vlasenica, Bosnia, including authigenic bastnäsite near the footwall

Bauxite samples	ΣREE^a (ppm)	$\Sigma LREE^b$ (ppm)	$\Sigma HREE^c$ (ppm)	$\Sigma LREE/\Sigma HREE$	La/Y
3827 (top)	737.6	644.2	3.4	6.9	2.0
3828	1 150.1	945.2	204.9	4.6	1.4
3829 (base)	1 653.4	1 161.1	492.3	2.3	0.7
Bastnäsite-(Ce)	633 300	546 600	87 600	6.2	3.0

$^a\Sigma REE$ = La–Lu, Y; $^b\Sigma LREE$ = La–Eu; $^c\Sigma HREE$ = Gd–Lu, Y.

Table 10.8 REE distribution along a vertical profile (10 m) in the Marmeiko karstic nickel deposit, Greece, including authigenic hydroxylbastnäsite-(Nd) above the footwall limestone

Samples	ΣREE^a (ppm)	$\Sigma LREE^b$ (ppm)	$\Sigma HREE^c$ (ppm)	$\Sigma LREE/\Sigma HREE$	La/Y
M9 Ferruginous clay (top)	375	279	96	2.9	1.5
M2 Ferruginous clay	363	272	111	2.4	1.2
M7 Ferruginous clay	385	266	99	2.7	1.3
M6 Ferruginous clay	517	393	124	3.2	1.7
M5 Ni–Fe ore	1 288	653	635	1.0	0.6
M4 Ni–Fe ore	1 444	819	625	1.3	0.7
M3 Ni–Fe ore	1 657	900	757	1.2	0.65
M2 Ni–Fe ore	2 432	1 302	1 150	1.1	0.64
M1 Ni–Fe ore (base)	1 918	972	946	1.0	0.62
Hydroxylbastnäsite-(Nd)	615 000	527 000	88 000	6.0	5.4

$^a\Sigma REE$ = La–Lu, Y
$^b\Sigma LREE$ = La–Eu
$^c\Sigma HREE$ = Gd–Lu, Y

GENETIC CONSIDERATIONS

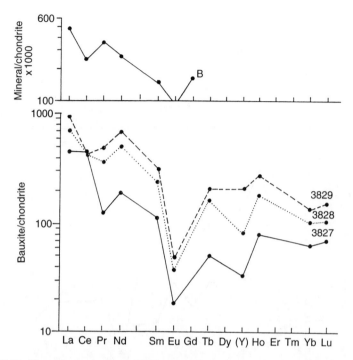

Figure 10.8 Chondrite-normalized REE patterns for red Cretaceous karst-bauxite and authigenic bastnäsite from Nazda deposit, Vlasenica, Bosnia; 3827–3829, bauxites; B = bastnäsite. (Chondrite abundances from Nakamura, 1974.)

REE (Maksimović and Roaldset, 1976; Koički, Koički and Maksimović, 1980; Maksimović, Skarpelis and Pantó, 1993).

Bauxites and Ni–Fe ores in the investigated profiles differ more in their absolute REE contents than in the relative distribution of the REE within them (Figures 10.8 and 10.9). A characteristic feature of some of the chondrite-normalized REE patterns is a well-marked negative Eu anomaly, as in the Vlasenica and Stitovo bauxite deposits. All the REE patterns, however, indicate a close genetic relation between the bauxites and Ni–Fe ores on one side, and the authigenic RE minerals in the same deposit on the other side (Maksimović and Pantó, 1991; Maksimović, Skarpelis and Pantó, 1993).

In many cases the compositions of authigenic RE minerals differ greatly within the same deposit, even over short distances, for example, bastnäsite-(Ce) from microveins and from segregations in the Nagyharsány bauxite deposit (compare analyses 2 and 3, Table 10.1); coexisting bastnäsite-(Ce) and bastnäsite-(La) in association with monazite-(Ce) and monazite-(Nd) in two samples of brindleyite from Marmara bauxite deposit; and

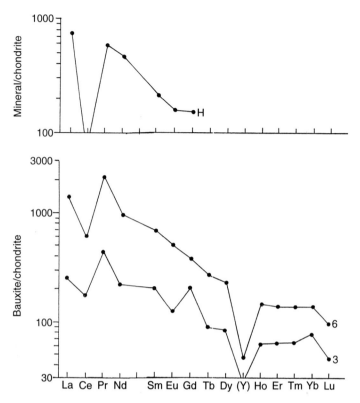

Figure 10.9 Chondrite-normalized REE pattern for red Jurassic karst-bauxite and authigenic hydroxylbastnäsite-(Nd) from Štitovo deposit, Montenegro; 3, 6 = bauxites; H = hydroxylbastnäsite-(Nd) (values ×1000).

hydroxylbastnäsite-(Nd) and hydroxylbastnäsite-(La) in deposits from Montenegro.

A remarkable feature of the Nazda bauxite deposit in Vlasenica is the occurrence of bastnäsite-(Ce) and Nd-rich goyazite in two different levels of the bauxite layer, Nd-rich goyazite being much closer to the footwall limestone. This phenomenon is explained by availability of CO_3^{2-} and PO_4^{3-} ions changing in the weathering solution in different parts of the deposit.

In many karstic deposits the percolating solutions were poor in fluorine, and consequently fluorine-deficient hydroxylbastnäsites were formed. This situation existed in the large bauxite deposits in Montenegro, and especially in the Nisi bauxite deposit in Greece, where hydroxylcarbonate-(Nd) was found. This mineral represents an end member of the isomorphic series bastnäsite–hydroxylbastnäsite–hydroxylcarbonate-(RE).

CONCLUSION

Separation and concentration of authigenic RE minerals near the footwall of karstic deposits were the result of several factors:

1. the initial content of the REE of the original weathered material which had collected in the karstic depressions;
2. the intensity of leaching of this material by percolating waters in a karstic environment;
3. the influence of the footwall limestone as an efficient geochemical barrier.

High-quality bauxites, with very low silica content, have a large enrichment of mobile trace elements, including REE, in the basal part of the deposit. This is seen in the Upper Jurassic bauxite deposits of Montenegro, where hydroxylbastnäsite concentrates to an extent that it could be a potential source of REE.

10.7 Conclusion

The most abundant authigenic RE minerals in karst-bauxites and karstic nickel deposits are members of the bastnäsite group. Relatively rare are RE phosphates and alkaline earth phosphates, such as monazite-(Nd), monazite-(La) and Nd-rich goyazite.

The compositional variability of all authigenic RE minerals in these deposits is now well established. It is caused basically by two types of substitution. Nd and La, but more frequently Nd, can replace Ce, forming Nd- or La-dominant minerals, and OH ions can substitute for F in the structure of bastnäsite, forming hydroxylbastnäsites. This substitution of OH for F has a significant effect on the unit cell volume, and a complete solid-solution series between bastnäsite and hydroxylbastnäsite is now established, which includes the fluorine-free end member – the hydroxylcarbonate-(RE).

The formation of authigenic RE minerals above the footwall limestone in karstic deposits was the result of several factors. These include (1) the relatively high REE content of the weathered parent rocks subsequently being washed down into karstic depressions; (2) the intensity of leaching of this material by surface water; and (3) the influence of the footwall limestone as an efficient geochemical barrier.

A large enrichment of the light REE relative to the heavy REE took place in authigenic RE minerals, where Nd most frequently played a leading role. A strong separation of individual REE in the RE minerals over small distances in the same deposit is also observed.

In general, these concentrations of the RE minerals in karst-bauxites and karstic nickel deposits have only a mineralogical, and not economic, importance. However, the large Upper Jurassic bauxite deposits in Montenegro, with high concentrations of hydroxylbastnäsite, could prove eventually to be a potential source of REE.

References

Bárdossy, Gy. and Pantó, Gy. (1973) Trace mineral and element investigation on bauxites by electron probe. 3éme Congres ICSOBA (Comité International pour l'Étude des Bauxites, de l'Alumine et de l'Aluminium), Nice, pp. 47–53.

Bárdossy, Gy., Pantó, Gy. and Várhegyi, Gy. (1976) Rare metals of Hungarian bauxites and conditions of their utilization. *Travaux ICSOBA*, **13**, 221–31.

Drake, M. J. and Weill, D. F. (1972) New rare earth element standards for electron microprobe analysis. *Chem. Geol.*, **10**, 179–81.

Farkas, L., Maksimović, Z. and Pantó, Gy. (1985) X-ray powder data and unit cell of natural hydroxyl-bastnaesite-(Nd). *N. Jb. Mineral. Mh.*, **H.7**, 298–304.

Glass, J. J., Evans, II. T. Jr, Carron, M. K. and Hildegrand, F. A. (1958) Cerite, from Mountain Pass, San Bernardino County, California. *Am. Mineral.*, **43**, 460–75.

Kirillov, A. S. (1964) New bastnaesite variety – hydroxyl-bastnaesite. *Dokl. Akad. Nauk USSR*, **159**, 1048–50, (in Russian).

Koički, S., Koički, A. and Maksimović, Z. (1980) Neutron activation analysis of lanthanides in domestic bauxites. *Glas 317 Acad. Serbe Sci. Arts*, **46**, 37–48.

Kukharenko, A. A., Bulakh, A. G. and Baklanova, K. A. (1961) A sulphate-bearing monazite in carbonatites from the Kola Peninsula, *Zap. Vses. Mineral. Obshch.*, **40**, 373–81 (in Russian).

Maksimović, Z. (1976) Genesis of some Mediterranean karstic bauxite deposits. *Travaux ICSOBA*, **13**, 1–14.

Maksimović, Z. (1978) Nickel in karstic environment: in bauxites and in karstic nickel deposits. *Bull. BRGM*, **2**, Sect. II, 173–83.

Maksimović, Z. (1979) Geochemical study of the Marmara bauxite deposit: implication for the genesis of brindleyite. *Travaux ICSOBA*, **15**, 121–31.

Maksimović, Z. (1980) The origin of yttrium and lanthanides in karstic bauxites of the Grebnik Mountain, Yugoslavia. *Bull. Acad. Serbe Sci. Arts*, **20**, 1–6.

Maksimović, Z. (1987) Trace elements in the Jurassic bauxites of Montenegro and their genetic significance. Special volume devoted to acad. Z. M. Bešić, *Acad. Montenegr. Sci. Arts*, 47–60, (in Serbian with English summary).

Maksimović, Z. (1988) Genesis of the Mediterranean karstic bauxites and karstic nickel deposits. *Glas 353 Acad. Serbe Sci. Arts*, **52**, 113–32, (in Serbian).

Maksimović, Z. (1993) Authigenic minerals of nickel and rare earth elements in karstic bauxites and karstic nickel deposits in Yugoslavia and Greece, *Bull. Geol. Soc. Greece*, **28/2**, 429–39.

Maksimović, Z. and Pantó, Gy. (1978) Minerals of the rare-earth elements in karstic bauxites: synchysite-(Nd), a new mineral from Grebnik deposit. Proceedings of 4th International Congress of ICSOBA, Athens, **1**, 540–52.

Maksimović, Z. and Pantó, Gy. (1980) Bastnaesite-(La) and monazite-(Nd), a new variety of monazite, from Marmara bauxite deposit (Greece). *Bull. Acad. Serbe Sci. Arts*, **20**, 35–42.

Maksimović, Z. and Pantó, Gy. (1981) Synchysite-(Nd) from Grebnik bauxite deposit (Yugoslavia). *Acta Geol. Acad. Sci. Hung.*, **24/2–4**, 217–22.

Maksimović, Z. and Pantó, Gy. (1983) Mineralogy of yttrium and lanthanide elements in karstic bauxite deposits. *Travaux ICSOBA*, **18**, 191–200.

Maksimović, Z. and Pantó, Gy. (1985a) Hydroxyl-bastnaesite-(Nd), a new mineral from Montenegro, Yugoslavia. *Mineral. Mag.*, **49**, 717–20.

Maksimović, Z. and Pantó, Gy. (1985b) Neodymiun goyazite in the bauxite deposit of Vlasenica (Yugoslavia). *Tschermaks Mineral. Petrog. Mitteilungen*, **34**, 159–65.

REFERENCES

Maksimović, Z. and Pantó, Gy. (1987) The occurrence and genesis of the hydroxyl-bastnaesites from Montenegro, Yugoslavia. *Bull. Acad. Serbe Sci. Arts*, **27**, 15–20.

Maksimović, Z. and Pantó, Gy. (1991) Contribution to the geochemistry of the rare earth elements in the karst-bauxite deposits of Yugoslavia and Greece. *Geoderma*, **51**, 93–109.

Maksimović, Z. and Roaldset, E. (1976) Lanthanide elements in some Mediterranean karstic bauxite deposits. *Travaux ICSOBA*, **13**, 199–220.

Maksimović, Z., Mindszenty, A. and Pantó, Gy. (1991) Contribution to the geochemistry of Hungarian karst bauxites and the allochthony/autochthony problem. *Acta Geol. Acad. Sci. Hung.*, **34/4**, 317–34.

Maksimović, Z., Skarpelis, N. and Pantó, Gy. (1993) Mineralogy and geochemistry of the rare earth elements in the karstic nickel deposits of Lokris area, Greece. *Acta Geol. Acad. Sci. Hung.*, **36/3**, 331–42.

Nagy, G. (1993) 'Quick' method for REE mineral analysis by EPMA. Conference on Rare Earth Minerals: Chemistry, Origin and Ore Deposits. The Natural History Museum, London. Abstracts, pp. 94–6.

Nakamura, N. (1974) Determination of REE, Ba, Fe, Mg, Na, and K in carbonaceous and ordinary chondrites. *Geochim. Cosmochim. Acta*, **38**, 757–75.

Ni, Y., Hughes, J. M. and Mariano, A. N. (1993) The atomic arrangements of bastnäsite-(Ce), $Ce(CO_3)F$, and structural elements of synchysite-(Ce), röntgenite-(Ce), and parisite-(Ce). *Am. Mineral.*, **78**, 415–18.

Pantó, Gy. (1975) Trace minerals of the granitic rocks of the Velence and Mecsek Mountains. *Acta Geol. Acad. Sci. Hung.*, **19/1–2**, 59–93.

Rosenberg, F. (1984) Geochemie und Mineralogie lateritischer Nickel- und Eisenerze in Lokris and Euböa, Griechenland. Unpublished Thesis, University of Hamburg, Hamburg, 169 pp.

CHAPTER ELEVEN
Rare earth deposits in China
C. Wu, Z. Yuan and G. Bai

11.1 Introduction

China is a country possessing abundant rare earth resources which are estimated at 40% of the total world reserves (US Bureau of Mines, 1993). Development of the Chinese rare earth industry began in 1927 when Dr Ding Daoheng made the first discovery of Bayan Obo iron ores (Obo means 'cairn' in Mongolian) during a field excursion to northwest China by a joint group of scientists from China, Sweden and Germany. Seven years later Prof. He Zuolin investigated one fluorite–iron ore specimen collected by Dr Ding, and identified two rare earth-containing minerals which were later recognized as bastnäsite and monazite, respectively (Zhao, 1983). Large-scale prospecting and exploration work on the Bayan Obo deposit commenced after 1950, initially with the help of Russian geologists. This work continued intermittently for more than 20 years, and has led to the confirmation of Bayan Obo as the largest rare earth ore deposit in the world.

During the 1950s and 1960s, geological surveys of rare earth resources in China were carried out by the central government and by the geological bureaus of local provinces. The main targets of those years were rare earth deposits hosted by carbonatitic rocks, alkali complexes, altered granites, pegmatites, migmatites and placers. In the early 1970s a new type of rare earth deposit, termed the ion-adsorption type, was discovered in southern Jiangxi Province. Numerous other deposits of this kind, where the REE are of low grade and easily extracted, were developed in south China during the 1980s, which had a significant influence on the world market.

In 1992 the rare earth production of China was reported to be about 21 000 tonnes, of which more than 60% were from Bayan Obo (bastnäsite, monazite and rare earth-rich iron concentrates), 18% from the Maoniuping mine (bastnäsite), 6% from the Weishan mine (bastnäsite), 14% from the ion-adsorption type (oxide) and only 0.3% from coastal placers. About half of the products were exported abroad (Rare Earth Office, 1993).

Rare Earth Minerals: Chemistry, origin and ore deposits. Edited by Adrian P. Jones, Frances Wall and C. Terry Williams. Published in 1996 by Chapman & Hall. ISBN 0 412 61030 2

11.2 Types of Chinese rare earth deposits

The major types of Chinese rare earth deposits, together with representative examples, are listed in Table 11.1, based on the revised classification of Zhang and Tao (1989) and a similar scheme from Möller (1989b). We prefer a classification that emphasizes the nature of the host rocks, rather than a genetic classification which might be controversial in many cases. For example, the genesis of the Bayan Obo deposit is the subject of great debate. Some of the occurrences quoted in Table 11.1 are of sub-economic significance at present and may be considered as potential resources in the future.

11.2.1 Carbonatitic rocks

Carbonatitic rocks refer, in a broad sense, to rocks that consist chiefly (>50%) of carbonate minerals, such as calcite, dolomite and ankerite, and show, regardless of their origins, mineralogical and chemical affinities to carbonatites defined by Le Bas (1977) and Heinrich (1980). They may occur as typical intrusive bodies associated with alkalic and ultrabasic complexes, or as stratiform layers and lenses in sedimentary successions. Both occurrences have similar REE and Nb mineralization characterized by very complex mineral assemblages of monazite, bastnäsite, pyrochlore, columbite, Nb-rutile, magnetite, fluorite, apatite and baryte. Examples of REE- and Nb-rich carbonatitic rocks are (Table 11.1):

1. stratiform: (a) Bayan Obo, Inner Mongolia; (b) Yangdun, Fujian; (c) Taohualashan, Inner Mongolia; (d) Yinachang, Yunnan;
2. intrusion-related: (a) Miaoya, Hubei and (b) Wajiertage, Xinjiang.

11.2.2 Quartz syenites

Quartz syenites and nordmarkites are host rocks of bastnäsite-barite-calcite veins which are generally developed along fracture zones. Bastnäsite is the only mineral of economic interest, although monazite, allanite, pyrochlore, aeschynite, chevkinite, columbite and thorite are also recognized in these systems. Typical examples include the Weishan mine, Shandong Province and the Maoniuping mine, Sichuan Province (Table 11.1).

11.2.3 Alkali granites

Gadolinite, hingganite, fergusonite and zircon occur as disseminations and veinlets in aegirine and riebeckite alkalic granites that are strongly altered to albite, quartz, fluorite and calcite-bearing rocks. Y and heavy REE abundances are relatively high in comparison with light REE. Zirconium,

Table 11.1 Characteristics of selected rare earth deposits in China

Localities	Economic remarks	Main features	References
1. Carbonatitic rock-hosted			
Bayan Obo, Inner Mongolia	REO 6%, Nb_2O_5 0.13%, Fe 35%, in production; Reserve: 35 million tonnes REO; about 600 million tonnes iron ores	Late Proterozoic dolomitic marble and slatehosted stratiform and lenticular REE–Nb–Fe ores with complex mineral assemblage of monazite, bastnäsite, columbite, etc.	Bai and Yuan, 1985; Institute of Geochemistry, 1988; Chao et al., 1992, 1993
Taohulashan, Inner Mongolia	REO 0.1–0.9%, Nb_2O_5 0.1–0.4%, potential	Late Proterozoic stratiform Nb–REE-bearing biotite–calcite marbles and calcite–biotite schists containing columbite, monazite, Nb-rutile and allanite	Bai and Yuan, 1985
Yinachang, Yunnan	Fe 50–60%; Cu 0.4–0.8%; REO 0.1–0.2%; P_2O_5 1–2%	Layer-like and lenticular Fe–REE–P ores hosted by Proterozoic dolomitic marbles intercalated with albite–biotite schists and tuffaceous slates	Qiu, Wang and Zhao, 1983; Bai and Yuan, 1985
Miaoya, Hubei	REO 1.7%, Nb_2O_5 0.12%, potential	REE–Nb-bearing syenite–carbonatite complex (231–278 Ma) intruding into late Proterozoic metavolcanic rocks and Devonian carbonaceous schist	Li, 1980; Bai and Yuan, 1985; Hubei Bureau of Geology and Mineral Resources, 1985
Wajiertage, Xinjiang	REO 0.15–4.3%; P_2O_5 1–8%, potential	REE-bearing carbonatite veins associated with Hercynian pyroxenite–gabbro complex intruding into Sinian metamorphic rocks	Bai and Yuan, 1985
2. Quartz syenites			
Maoniuping, Sichuan	REO 1–8%, in production	NNE-trending bastnäsite–baryte–carbonate veins associated with nordmarkite (12 Ma) intruding a granite batholith (78–134 Ma)	Pu, 1988; Chen and Pu, 1991; Yuan et al., 1993

Table 11.1 *Continued*

Localities	Economic remarks	Main features	References
Weishan, Shandong	REO ≥1.6%, in production	NW-trending bastnäsite–baryte–carbonate veins (110 Ma) associated with quartz syenite complex (140 Ma) emplaced in Archaean gneisses	Wang and Yan, 1980
3. Alkalic complexes			
Saima, Liaoning	REO 0.3–4.5% with significant U, Th and Nb, potential	U–REE-bearing alkali complex composed of trachytes, phonolites	Beijing Institute of Uranium Geology, 1974, 1975
4. Alkalic granites			
Baerzhe, Inner Mongolia	REO 0.2–0.7%, ZrO_2 0.5–3.4%, Nb_2O_5 0.02–0.3%, BeO 0.02–0.15%, potential	Veinlets and disseminations of zircon, columbite, hingganite, etc. in riebeckite granite (127 Ma) emplaced in Late Jurassic andesite tuffs	Bai *et al.*, 1980; Ding, Bal and Yuan, 1985
Cida, Sichuan	REO 0.05%, potential	Disseminated fergusonite in riebeckite granite (230 Ma) intruding into layered basic intrusion	Zhou *et al.*, 1985; Yang and Lu, 1988
5. Alkali pegmatites			
Qiganlaing, Inner Mongolia	REO 1.5%; P_2O_5 4–10%, potential	REE enrichments in pegmatitic diopsidites emplaced in Archaean gneisses	Zou *et al.*, 1985
Yousuobao, Hebei	REO 0.15–2.6%, P_2O_5 18–36%, potential	REE mineralization in Proterozoic pegmatite (1827 Ma) composed of diopside, phlogopite and apatite	Zou Tianren, 1993, personal communication

6. Metamorphic rocks Guangshui, Hubei	REO 0.2–0.3%, ZrO_2 0.5–1%, Nb_2O_5 0.027%, potential	Stratiform Zr–Y–HREE mineralization hosted by Late Proterozoic felsic schists	Li et al., 1981; Wu et al., 1993a
Shengtieling, Liaoning	REO 2%; Fe 34%; P_2O_5 1–3%, potential	Monazite–barite–magnetite segregations in Archaean gneisses	Yang, 1980
7. Phosphorites Zhijin, Guizhou and Kunyang, Yunnan	REO 0.05–0.n%; P_2O_5 8–25%, potential	Cambrian REE-bearing phosphorite beds	No. 6 Geological Party, 1974; Zhang, 1974
8. Bauxites Xiuwen, Guizhou	REO 0.1–0.2%, Al_2O_3 55–80%	REE-bearing layered and lenticular bauxites occurring at bottom of lower Carboniferous strata	No. 115 Geological Party, 1974
9. Lateritic weathering crusts Southern China	REO 0.05–0.2%, in production	REE leaching enrichment in clays of tropical–subtropical weathering crusts developed on various granites	Yang, Hu and Luo, 1981; Wu Huang and Gao, 1990; 1993b
10. Placers Guangdong, Hainan and Taiwan	Monazite 500–1000 g/m^2 associated with ilmenite, zircon and cassiterite, in production	Coastal placers occurring as lenses and layers	Clark and Zheng, 1991

Hf, Nb, Be and U are also considerably enriched. Examples of this type are Baerzhe, Inner Mongolia and Cida, Sichuan.

11.2.4 Alkali complexes

Alkali complexes are made up of a variety of rocks: monzonite, aegirine-syenite, nepheline-syenite, miaskite, lujavrite, trachyte and phonolite. Disseminated Na–Ca–REE titanosilicates such as mosandrite or rincolite may be elevated, up to 30% of the whole rock, in a certain phase associated with complex REE- and U-containing mineral assemblages, leading to potential economic value. A typical example is Saima, Liaoning Province (Table 11.1).

11.2.5 Alkali pegmatites

Alkali pegmatites emplaced into Archean to early Proterozoic gneisses may have high contents of REE. These pegmatites are composed mainly of giant crystals of diopside, phlogopite, anorthoclase, hyalophane, arfvedsonite, apatite, sphene and orthite, and are often associated with carbonatites (Zou et al., 1985). Monazite occurs as very fine-grained inclusions within apatite of which rare earth oxides (REO) may reach 3–8% and may be recovered as a by-product (Zou Tianren, 1993, personal communication). Typical examples of this group are Yousuobao, Hebei and Qiganliang, Inner Mongolia (Table 11.1).

11.2.6 Metamorphic rocks

In south China disseminated monazite, xenotime, zircon and ilmenite are enriched in migmatized gneisses and granitic migmatites. In northeast China elevated contents of monazite are found in apatite–barite–magnetite segregations in a Proterozoic metamorphic terrain consisting of biotite granulites, felsic schists, tourmaline–albite schists and some amphibolites (Yang, 1980).

Stratiform Zr–Y–HREE mineralization characterized by zircon, gadolinite, fergusonite, xenotime, monazite and allanite is found in strongly sheared Proterozoic felsic schists in the Dabie Mountains, central China (Li, 1980; Li et al., 1981; Wu et al., 1993a; Table 11.1).

11.2.7 Phosphorites

Cambrian sedimentary phosphorite beds in southwest China are enriched in REE (up to 0.2% REO) which are statistically positively correlated with phosphorus abundance and show the highest contents in collophane (colloidal apatite). Very fine-grained monazite are recognized in interstices among collophane, variscite, chalcedony, calcite and clay minerals; however, the mode of REE occurrence has not been clearly determined (Zhang, 1974; No. 6 Geological Party of Yunnan, 1974). Representative examples include Zhijin, Guizhou and Kunyang, Yunnan.

11.2.8 Bauxites

REE are also enriched in some Carboniferous bauxites in central Guizhou Province. The bauxites occur as layers and lenses at the bottom of Carboniferous strata uncomfortably overlaying Cambrian dolomites. The mode of occurrence of the REE is not well known in comparison to reported authigenic rare earth minerals from karst-bauxite deposits of Yugoslavia and Greece (Maksimović and Pantó, 1991; also Chapter 10), although trace amounts of fine-grained (0.03–0.2 mm) rutile, anatase, zircon, sphene, tourmaline, epidote and garnet are observed to be hosted by or associated with diaspore, gibbsite, kaolinite, chlorite and muscovite (No. 115 Geological Party of Guizhou, 1974).

11.2.9 Lateritic weathering crusts

REE are concentrated in weathering crusts developed on various granitic rocks in south China. The weathering crusts formed under tropical and subtropical conditions and belong to the 'monosiallitization type' of Samama (1986). They can be divided into three horizontal zones from top to bottom: (1) surface soil zone (soil cap), 0.5–2 m in thickness, consisting of kaolinite, halloysite-(1 Å), gibbsite, goethite and organic materials as the main components; (2) wholly weathered zone, about 3–20 m thick and composed mainly of halloysite-(7 Å) and kaolinite, with variable amounts of quartz and feldspar fragments; (3) sub-weathered zone, up to 5 m, having significant amounts of residual quartz and feldspar, along with halloysite-(7 Å), kaolinite, some montmorillonite and vermiculite (Song and Shen, 1982; Yang, 1987; Wu, Huang and Guo, 1990).

Rare earth orebodies, about 3–10 m in thickness, occur mainly in the wholly weathered zone, and sometimes include the lower part of the soil zone and/or the upper part of the sub-weathered zone. About 50–90% of the total REE can be extracted by an ion-exchange process employing electrolyte solutions (Wang, Shen and Song, 1980; Yang, Hu and Luo, 1981; Song and Shen, 1986; Wu, Huang and Guo 1990; Wu et al., 1993b).

11.2.10 Placers

Monazite and xenotime are the main rare earth minerals, in association with zircon, ilmenite and cassiterite, in alluvial and coastal placers which are widely distributed in the southeast coastal area of China, Hainan Island and the western coast of Taiwan. Fergusonite was mined from eluvial and alluvial placers in northern Guangxi during the 1960s.

11.3 Spatial distribution and age of formation

11.3.1 Distribution

Generally speaking, most of Chinese endogenic rare earth deposits, although different in age, are distributed in areas marginal to Precambrian cratons and/or adjacent orogenic belts, and show close spatial relationships to major fault zones.

The deposits of Saima, Yousuobao, Qiganliang, Bayan Obo, Taohualashan and Wajiertage occur from east to west, along the northern edge of the Sino-Korean and Tarim cratons. The Weishan deposit is located on the eastern margin of the Sino-Korean craton and very close to the Tancheng–Lujiang fault zone. In the Qinling–Dabieshan orogenic belt and on the northern edge of the Yangtze craton, the deposits of Guangshui, Miaoya, Lijiahe and Shuimo–Pinghe occur from east to west. The Maoniuping, Cida, Muluo and Yinachang deposits occur at the southwest margin of the Yangtze craton and adjacent to a north–south-trending ancient rift belt.

In regard to exogenetic deposits, on the other hand, their distribution is controlled by ancient and modern climate and sedimentary environments. Rare earth-bearing phosphorites and bauxites are clustered in the southwestern part of the Yangtze craton. Lateritic weathering crusts are restricted to the area south of latitude 27°N. Alluvial placers are found chiefly in two regions, i.e. southern Hubei–northern Hunan and southeast coastal areas where streams pass through large granite and migmatite terrains. Coastal placers are doubtless distributed along coastlines under favourable morphologic conditions.

11.3.2 Ages of formation

As shown in Table 11.1 and discussed in section 11.5, the Chinese rare earth deposits show a large range of ages of formation varying from Proterozoic to Cainozoic. However, for some of these deposits ages have not been clearly determined, and others are believed to involve multiple mineralization episodes, such as Bayan Obo (Chao *et al.*, 1992, 1993; section 11.5.1).

11.4 Geochemistry

11.4.1 REE and other element patterns

With regard to REE distribution patterns, rare earth deposits hosted by carbonatitic rocks, quartz syenites and alkalic complexes are characterized by distinctive light REE-enriched, steep chondrite-normalized patterns without apparent Eu anomalies (Figure 11.1a,b), similar to those of carbonatites and alkalic complexes elsewhere in the world (Möller, Morteani and Schley, 1980). Some dolomite samples from Bayan Obo display positive Eu anomalies (Qiu, Wang and Zhao, 1983; Gross, 1993). The La/Yb and Sm/Nd ratios are between 60 and 1000, mostly 100-400 and 0.08-0.18, respectively. These host rocks are also rich in Nb (500-2500 ppm), Ba (400-50 000 ppm), Sr (600-15 600 ppm), P and F, and depleted in Rb (<10-100 ppm) (Li, 1980; Wang and Yan, 1980; Bai and Yuan, 1985; Institute of Geochemistry, Academia Sinica, 1988; Gross, 1993). On the other hand, alkalic granites show normalized REE patterns with relatively elevated heavy REE and very strong Eu depletions (Figure 11.1c). The La/Yb and Sm/Nd ratios vary from 0.4 to 20 and from 0.17 to 0.44, repectively, combined with high Zr (700-25 000 ppm), Nb (200-3000 ppm) and Rb (200-1500 ppm), and low Ba (<100 ppm) and Sr (<50 ppm) contents (Zhang, 1986; Wang, 1987; Yang and Lu, 1988). The felsic schists from Guangshui show similar REE patterns to those of alkali granites (Figure 11.1c), and so do their Zr, Nb, Rb, Sr and Ba abundances (Wu *et al.*, 1993a, in preparation).

REE-rich clays have variable chondrite-normalized REE patterns (Figure 11.1d) which are inherited from their parent rocks (Yang, Hu and Luo, 1981; Wu, Huang and Guo, 1990). These chemical features of different types of rare earth deposits are attributed to different fractionation processes during which high pressure, high temperature, limited partial melting and CO_2-rich fluid extraction may be the main processes leading to strong separation between light REE and heavy REE of the former group (Möller, Morteani and Schley, 1980), and fractional crystallization may be responsible for heavy REE enrichment in residual melts for the latter group (Wu *et al.*, 1993c). These REE distribution patterns of rare earth deposits are markedly reflected by their products of monazite, bastnäsite and rare earth oxides (Table 11.2).

11.4.2 Isotope geochemistry

Isotopic data accumulated to date indicate without exception a mantle-related source for carbonatitic rocks and alkalic complexes. $^{87}Sr/^{86}Sr$ initial ratios were reported to be between 0.7036 and 0.7042 for apatite and dolomite from Bayan Obo (Institute of Geochemistry, Academia Sinica, 1988; Yuan *et al.*, 1992) and between 0.7043 and 0.7048 for siderites occurring as porphyrotopes in dolomite from the West ores of Bayan Obo (this work), which are in agreement with a high $\varepsilon_{Nd}(t)$ value of +6.1 ± 2.4 (Yuan *et al.*, 1992). $^{87}Sr/^{86}Sr$ initial ratios of other rare earth deposits were recorded as 0.7047-0.7052 for carbonates from Yangdun, 0.7029-0.7031 for calcites from Miaoya, 0.7050-0.7069 for calcites from Weishan (Institute of

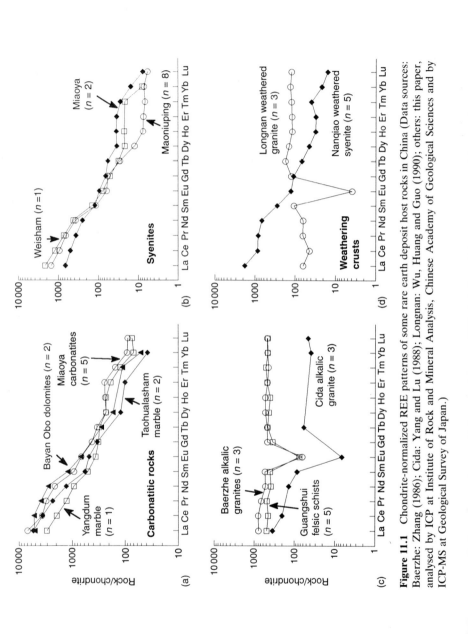

Figure 11.1 Chondrite-normalized REE patterns of some rare earth deposit host rocks in China (Data sources: Baerzhe: Zhang (1986); Cida: Yang and Lu (1988); Longnan: Wu, Huang and Guo (1990); others: this paper, analysed by ICP at Institute of Rock and Mineral Analysis, Chinese Academy of Geological Sciences and by ICP-MS at Geological Survey of Japan.)

Table 11.2 Rare earth distributions of minerals and products from the main rare earth deposits of China

Localities Province	Bayan Obo Inner Mongolia	Bayan Obo Inner Mongolia	Maoiuping Sichuan	Weishan Shandong	Xunwu Jiangxi	Longnan Jiangxi	Longyan Fujian
Products	Monazite[a] ($n = 23$)	Bastnäsite[a] ($n = 7$)	Bastnäsite[b] ($n = 22$)	Bastnäsite[c] ($n = 6$)	Oxide[d]	Oxide[d]	Oxide[d]
La_2O_3	25.7	26.0	37.4	34.2	29.84	2.18	23.16
Ce_2O_3	50.1	50.4	47.1	49.8	7.18	1.09	2.03
Pr_2O_3	5.4	5.4	3.61	4.0	7.41	1.08	5.11
Nd_2O_3	16.5	15.8	10.3	10.8	30.18	3.47	19.48
Sm_2O_3	1.11	1.07	0.82	0.71	6.32	2.34	4.42
Eu_2O_3	0.21	0.21	0.10	0.12	0.51	0.37	0.83
Gd_2O_3	0.52	0.52	0.26	0.21	4.21	5.69	4.89
Tb_2O_3	0.02	0.03	0.05	n.d.	0.46	1.13	0.75
Dy_2O_3	0.28	0.14	0.07	0.04	1.77	7.48	4.68
Ho_2O_3	n.d.	0.02	0.02	n.d.	0.27	1.60	0.53
Er_2O_3	n.d.	n.d.	0.04	n.d.	0.88	4.26	2.38
Tm_2O_3	n.d.	n.d.	0.01	n.d.	0.10	0.60	0.33
Yb_2O_3	0.05	0.06	0.01	0.02	0.62	3.34	1.60
Lu_2O_3	0.01	n.d.	0.00	n.d.	0.13	0.47	0.30
Y_2O_3	0.28	0.29	0.14	0.27	10.07	64.90	29.47
Total	100.18	99.94	99.93	100.17	99.95	100.00	99.96
REO(%)	70.5	70.4	74.3	67–75	>90	>90	>90

[a] Bai and Yuan (1985); [b] No. 109 Geological Party (1989); [c] Wang and Yan (1980); [d] Clark and Zheng (1991).
n.d.: not detected

Geochemistry, Academia Sinica, 1988), and 0.703–0.704 for apatites from alkalic pegmatites (Zou et al., 1985).

In contrast, alkalic granites possess variable $^{87}Sr/^{86}Sr$ initial ratios, between 0.698 and 0.707 for the Baerzhe aegirine granite (Wang, 1987; Zhang and Yuan, 1988) and 0.7089 for the Cida aegirine granite (Yang and Lu, 1988). This implies that alkaline granites may originate from the lower crust (Möller, 1989b) or may be derived from a mantle-related source and significantly contaminated by crustal materials, with slightly high $\varepsilon_{Nd}(t)$ values (+1.9 to +2.5) for the Baerzhe alkalic granite (Wang, 1987).

Most of granitic rocks in south China, which are the parent rocks of REE-rich clays, have high $^{87}Sr/^{86}Sr$ initial ratios varying from 0.708 to 0.732 and possess high radiogenic lead isotopic compositions for K-feldspars (Wu et al., 1990, 1993c) and low negative $\varepsilon_{Nd}(t)$ values (−2.0 to −26; Yuan and Zhang, 1992) indicating that large amounts of mature crustal materials were involved in the genesis of these highly evolved granites.

In respect of stable isotopic data, a large span of positively correlated $\delta^{13}C_{(PDB)}$ and $\delta^{18}O_{(SMOW)}$ values, between −6.6 and +1.4‰ and between +10.3 and +20.4‰, respectively, was documented for calcite and dolomite from Bayan Obo (Jiang et al., 1982; Wei and Shangguan, 1983; Liu, 1986). The variation was interpreted, according to two different ideas, as (1) a result of replacement of sedimentary calcite by dolomite in a closed sea basin (Wei and Shangguan, 1983); and (2) a result of mixing of deep-source carbon with sea waters (Bai and Yuan, 1985; Chen, 1985, 1993; Institute of Geochemistry, Academia Sinica, 1988). Chen (1985, 1993) reported very low $\delta D_{(SMOW)}$ values (−142 to −259‰) measured on riebeckite and on fluid inclusions of magnetite, hematite and dolomite from Bayan Obo mine, and proposed a mantle-derived source for the hydrogen. A range of $\delta^{34}S_{sulphide}$ −4 to +4‰ measured on 130 samples from Bayan Obo was obtained by Yang Fengjun, indicating a deep sulphur source for the deposit (Yuan et al., 1992).

Magmatic $\delta^{13}C_{(PDB)}$ values were recorded for marbles from Yangdun (−5.4 to −6.6‰), carbonatites from Miaoya (−6.2 to −7.3‰) and calcite from Weishan (−7.8‰) (Institute of Geochemistry, Academia Sinica, 1988). For the Maoniuping deposit, $\delta^{18}O_{H_2O(SMOW)}$ values are calculated to be between +3.3 and −6.9‰ from measured values of quartz and calcite from bastnäsite–calcite–barite veins, implying a meteoric water-dominated or a magmatic–meteoric water mixed origin for waters of the mineralizing fluids. This contrasts with $\delta^{13}C_{(PDB)}$ values (−6.3 to −6.5‰) for calcites which indicate a magmatic carbon source for the system (Wu et al., in preparation). Very low $\delta^{18}O_{whole\ rock\ (SMOW)}$ values (+2.9 to −5.6‰) were also recorded at Baerzhe alkalic granite, implying a meteoric water-dominated hydrothermal system interacting with the granite and wall rocks (Zhang and Yuan, 1988).

11.5 Geological examples

11.5.1 Bayan Obo Fe-Nb-REE deposit

Situated about 130 km north of Baotou city, Inner Mongolia, the Bayan Obo deposit is geotectonically located on the northern margin of the Sino-Korean craton and in the transitional zone between the craton margin and the Mongolian Hercynian foldbelt, with a deep-seated fracture separating the two tectonic units. East–west-trending early Proterozoic gneisses and schists of the Wutai (Erdaowa) group constitute the basement in the area. Unconformably overlying the basement is the mid-Proterozoic Bayan Obo group, which is composed of coarse- to medium-grained clastic and carbonate rocks.

The deposit is located in the Bayan Obo syncline which is made up, in ascending order, of three major lithologic units (Li, 1959): (1) greyish, fine- to medium-grained quartz–feldspar sandstone and quartzite (H1–H7); (2) dolomitic marble intercalated with feldspar rock, biotite rock and slates (H8); and (3) feldspar rock, biotite schist, slates and metavolcanics (H9) which occupy the central part of the syncline (Figure 11.2).

The dolomitic marble occurs as a spindle-shaped stratiform body and extends 18 km from east to west, with a width of tens to 1000 metres. The marble has a fine- to medium-grained texture, massive and banded structure, and consists chiefly of dolomite and calcite, together with some feldspar, quartz, Na-tremolite, magnesio-arfvedsonite, phlogopite, apatite, fluorite and baryte. Very fine-grained (0.02–0.1 mm), disseminated and aggregated

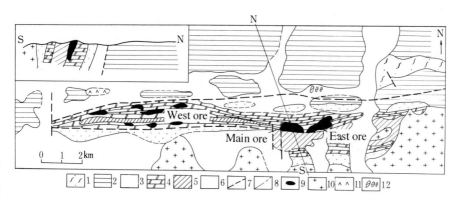

Figure 11.2 Schematic geological map of the Bayan Obo deposits, Inner Mongolia. 1, Schist and gneiss of Lower Proterozoic Wutai Group; 2, quartzite and limestones of the Bayan Obo Group; 3, quartz–feldspar sandstone and quartzite (H1–H7); 4, dolomitic marble (H8); 5, feldspar rock, biotite schists and slates (H9); 6, quaternary sediments; 7, faults; 8, geological boundary; 9, Fe–REE ore bodies; 10, granite; 11, gabbro and diorite; 12, carbonatite veins. (After Yuan et al., 1992).

bastnäsite, monazite, columbite and pyrochlore are abundant, and they are associated with magnetite and hematite and grade into massive iron ores. The ores consist of three major REE–Fe orebodies: Main ore body, East ore body and West ore body (Figure 11.2), mostly occurring in dolomite marble and partly extending upward into feldspar rock and slates. The average ore grades are 6% REO, 0.13% Nb and 35% Fe with total reserves of 35 million tonnes of REO and about 600 million tonnes of iron ores. The feldspar rock has light-coloured and dark-coloured varieties, and forms the hanging wall of the iron ore bodies. It is a finely recrystallized, slatey rock composed mainly of microcline, with minor biotite, quartz, calcite, riebeckite, aegirine, apatite and pyrite. The rock is also REE- and Nb-mineralized, containing monazite, bastnäsite, aeschynite, Nb-rutile and columbite.

Carbonatite dykes, with approximately east–west trends, thicknesses of 0.4–17 m and lengths of 30–150 m, intrude quartzites on both limbs of the syncline. Carbonatites with north–south trend also intrude the Kuangou gneiss and migmatite situated about 0.5 km north of the dolomitic marble. These carbonatite dykes show typical mineralogical and chemical features of igneous carbonatites (Le Bas *et al.*, 1992). Other igneous rocks in the area include gabbro, alkali gabbro, granite, quartz syenite and dykes ranging from basic to acidic in composition. The granite which occurs to the south of the deposit (Figure 11.2) has been dated at 255 Ma (Institute of Geochemistry, Academia Sinica, 1988) and is calc-alkalic in composition.

Numerous mineralogical and geochemical data have been accumulated since the 1960s. The new minerals of baotite $(Ba_4(Ti,Nb)_8O_{16}(Si_4O_{12})Cl)$, daqingshanite $((Sr,Ca,Ba)_3(Ce,La)(PO_4)(CO_3)_{3-x}(OH,F)_x)$, huanghoite $(BaCe(CO_3)_2F)$ and zhonghuacerite $(Ba_2(Ce)(CO_3)_3F)$ were found in the deposit (Zhang and Tao, 1986). The genesis of the deposit is still the subject of great debate. The following hypotheses were put forward by different researchers to interpret its origin.

1. The ore-forming materials, such as Fe, Nb and REE, came mainly from terrigenous deposits (Meng, 1982), but parts were introduced from a granitic magma (Wang, 1973).
2. The deposit is very similar to other iron formations and hydrolithic stratiform sediments formed by hydrothermal-effusive or volcanogenic processes, except for their high contents of REE, Nb and other minor elements (Qiu, Wang and Zhao, 1983; Tu, Zhao and Qiu, 1985; Gross, 1993).
3. The ore-hosting dolomite was sedimentary, and Fe, Nb and REE mineralization were brought about by multi-episode, epigenic, hydrothermal metasomatic processes during a period from Late Proterozoic (1260 Ma; $^{40}Ar/^{39}Ar$; Conrad and Mckee, 1992) to Hercynian with the Caledonian (596–421 Ma) as the peak period of mineralization (Chao

et al., 1990, 1992, 1993; Drew, Meng and Sun, 1990, Ren and Chao, 1990).
4. The dolomitic marble is considered as an intrusive carbonatite and the ore-forming materials are products of differentiation of the carbonatitic magma (Zhou *et al.*, 1980; Meng, 1981).
5. The deposit shows both apparent sedimentary occurrence and impressive mineralogical and geochemical affinities to carbonatite-related REE deposits in the world (Le Bas *et al.*, 1992). It is considered to be derived from a mantle-related carbonatite magmatic source through a volcanic exhalation and associated metasomatic process in a closed sea basin during the mid-Proterozoic period and subject to reactivation during later geological events (Bai and Yuan, 1983, 1985; Chen, 1985, 1993; Nakai *et al.*, 1989; Yuan *et al.*, 1992).

11.5.2 Maoniuping bastnäsite–carbonate–baryte vein deposit

Situated about 22 km southwest of Mianning, Sichuan Province, the Maoniuping deposit is located tectonically in the northern part of the north–south-trending Panzhihua–Xichan ancient rift belt occupying an area transitional between the Kangdian shield to the east and the Yanyuan-Lijiang depression to the west. The host nordmarkite, termed the Maoniuping stock, belongs to part of the Mianxi granite batholith, which is about 90 km in length south–north and 6–14 km in width east–west with an area of about 700 km^2 and is emplaced mainly into Permian basalt and Triassic coal-bearing strata. Having alkali-feldspar granite as the main composition, the Mianxi batholith is a multiphase granitic complex. In the area around the Maoniuping deposit, the following rock types are recognized: (1) purple-coloured medium- to fine-grained alkali-feldspar granite; (2) greyish, white-coloured, medium-grained alkali-feldspar granite; (3) graphic granite; (4) nordmarkite (ore-bearing), and (5) granitic porphyry (post-ore).

The Maoniuping nordmarkite stock, about 0.3 km^2 in surface area, outcrops on the east margin of the Mianxi batholith between alkali-feldspar granites and rhyolite (Figure 11.3). It is composed of microcline-microperthite, albite, quartz, aegirine-augite and aegirine. Minor minerals are biotite, calcite, baryte, fluorite and epidote, along with accessory magnetite, apatite, sphene, zircon, bastnäsite, parisite, monazite, allanites, pyrochlore and thorite. Zircons from the rock yield an apparent U–Pb age of 22 ± 8 Ma with a lower disconcordia intercept age of 12 ± 2 Ma (Yuan *et al.*, 1993), much younger than the major part of the Mianxi batholith (78–134 Ma; No. 109 Geological Party, 1992).

A group of bastnäsite-bearing pegmatite-like veins are developed along northeast-trending fractures and breccia zones which are secondary structures of the regional Haha deep-seated fault belt. Having variable compositions of

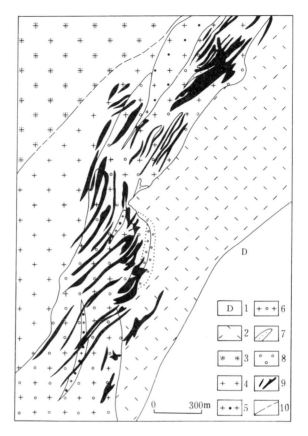

Figure 11.3 Schematic geological map of the Maoniuping bastnäsite–carbonate–baryte vein deposit, Sichuan Province. 1, Devonian clastic rocks; 2, rhyolites and felsites; 3, purple-coloured alkali-feldspar granite; 4, greyish-coloured alkali-feldspar granite; 5, graphic granite; 6, nordmarkite; 7, granite porphyry; 8, hybridized zone; 9, ore bodies; 10, geological boundaries. (After no. 109 Geological Party, Sichuan Bureau of Geology and Mineral Resources, 1992.)

aegirine-augite–fluorite–baryte, aegirine-augite–microcline and aegirine-augite–baryte–calcite, these veins constitute a 3 km long mineralized zone with a width of 200–600 m, and have a close spatial relationship to the Maoniuping nordmarkite stock (Figure 11.3). These veins and adjacent veinlets comprise a number of northeast-trending orebodies with variable sizes ranging from 200 to 1000 m long by 5 to 20 m wide, dipping northwesterly at 65–80° down to 300–500 m depth (Pu, 1988; Chen and Pu, 1991). The veins are intensively weathered and fractured down to 300 m producing black and dark-brown earthy ferromanganese oxides enriched in REE. Coarse-grained bastnäsite is the dominant economic ore mineral associated with chevkinite (Yang et al., 1991), and occasionally with britholite, xenotime, allanite, monazite, pyrochlore and betafite (Yang and Yan, 1991). Gangue minerals include baryte, calcite, fluorite, quartz, aegirine-augite, microcline, albite and the sulphide minerals pyrite, galena and molybdenite. Arfvedsonite and biotite from bastnäsite-bearing pegmatite-like veins give K–Ar dates of 27.8–40.3 Ma (No. 109 Geological Party, 1992).

Geological work on the area was started at the beginning of 1960s by the Sichuan Bureau of Geology and Mineral Resources. Since 1985 an exploration project on the deposit has being undertaken by the No. 109 Geological Party. The ore grade varies from 1 to 8% REO with an average of about 2% REO, without significant enrichment of other rare metals, and the proved reserve is about 400 000 tonnes of REO. The property has been mined since 1989 by several companies.

11.5.3 Weishan bastnäsite–baryte–carbonate vein deposit

The Weishan deposit, also known as the Chishan or '101' deposit, is located to the east of Weishan Lake district, about 200 km south of Jinan City, Shandong Province. The main exposure in the area is the Archaean Taishan Group which is composed mainly of biotite-plagioclase gneiss, amphibole–plagioclase gneiss, two-feldspar gneiss and amphibolites. These rocks are intruded by a group of small bosses and dykes including quartz syenite, syenite porphyry and aegirine syenite porphyry, as well as aegirine-augite granophyre, albite aplite and kersantite (Figure 11.4).

The quartz syenite porphyry (140 Ma; K–Ar; Wang and Yan, 1980) constitutes the major part of the complex and is the main host rock of bastnäsite–baryte–carbonate veins. Elongated roughly in a NW–SE direction with an surface area of about $0.4\,km^2$, this porphyritic rock body consists mainly of perthite orthoclase (70–80%), albite oligoclase (5% ±) and quartz (12%), along with small amounts of aegirine, aegirine-augite, diopside, alkali amphibole and biotite. Accessory minerals include sphene, zircon, apatite, monazite, baryte, fluorite, magnetite, occasionally, rutile, columbite and anatase. These rocks are partially altered and contain secondary chlorite, epidote and carbonates (Wang and Yan, 1980).

Rare earth mineralization occurs as veins and veinlets along northwest-trending fractures and breccia zones (Figure 11.4). The larger bastnäsite–quartz–baryte veins dip south westerly at 50–70° and are dated at 110 Ma (K–Ar; Wang and Yan, 1980). The largest vein (vein I) is several hundred metres long by 0.1 to 1.1 m wide. The average REO grade of vein I is about 1.6% with a maximum of 59.1%. Veinlets are of less economic importance. They are several centimetres in width and often comprise veinlet zones of up to 4–10 m in width. Mineralogically, these veins and veinlets are composed of baryte, quartz, calcite and dolomite, with small amounts of fluorite, feldspar, chlorite, muscovite, phlogopite, riebeckite, aegirine-augite, tremolite, diopside and epidote. The main rare earth mineral is bastnäsite

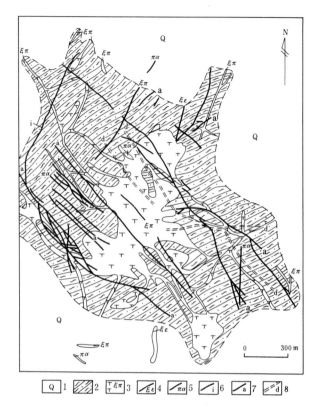

Figure 11.4 Schematic geological map of the Weishan bastnäsite–baryte–carbonate vein deposit, Shandong Province. 1, Quaternary sediments; 2, Archaean gneiss; 3, quartz syenite; 4, Aegirine syenite porphyry; 5, albitite prophyry and lamprophyre; 6, unidentified dykes; 7, bastnäsite–baryte–carbonate veins; 8, fracture zones. (After No. 6 Geological Party, Shandong Bureau of Geology and Mineral Resources.)

(Table 11.2), together with parisite, monazite, allanite, pyrochlore, britholite, calcian ancylite and carbocernaite (Qian, 1979). They occur as disseminations and/or aggregations in the veins, and are often brecciated and cemented by geothite, quartz, baryte and carbonates. Other recognized minerals include rutile, apatite, thorite, perovskite, columbite, anatase, pyrite, magnetite, hematite and wulfenite (Wang and Yan, 1980).

The deposit has been mined by the Weishan Rare Earth Mining Company since the 1970s; 3500 tonnes of bastnäsite concentrate (30–70% REO) were produced in 1992.

11.5.4 Miaoya syenite–carbonatite Nb–REE deposit

The Miaoya REE–Nb deposit was discovered by the Regional Geological Party, Northwest Bureau of Geology, in 1962, and was explored by the

GEOLOGICAL EXAMPLES

No. 5 Geological Party, Hubei Bureau of Geology and Mineral Resources during 1971 and 1980. The following descriptions are based on Li (1980), Hubei Bureau of Geology and Mineral Resources (1985) and Bai and Yuan (1985).

The deposit is located in the north west part of the Zhushan County, north-western Hubei Province. Tectonically, the area is situated on the western edge of the Wudang Block which belongs to the southeastern part of the Qinling Mountain fold belt, consisting mainly of Late Proterozoic metavolcanic rocks of the Yaolinghe Group. The REE–Nb-bearing body is a north westerly elongated lentiform syenite–carbonatite complex, about 3 km long by 560–820 m wide with a northeasterly dip of 60–80°. It intrudes the above metamorphic terrain, composed of Late Proterozoic metaspilites and keratophyres, together with corresponding agglomerates and tuffs to the south, and Devonian carbonaceous sericite schists interbedded with dolomitic limestone and marble to the north (Figure 11.5). The syenite–carbonatite complex consists of a variety of lithologies including syenite, 'contaminated' syenite, 'contaminated' albitite, syenite porphyry, albitite porphyry, calcite carbonatite, biotite–calcite carbonatite and ankerite carbonatite. These rocks are strongly stressed and deformed, showing orientation parallel to the strikes of the wall rocks (Figure 11.5). The so-called 'contaminated syenite' constitutes the major part (about 60%) of the complex and is made up of two components of contrasting colour, i.e. a light-coloured microcline-

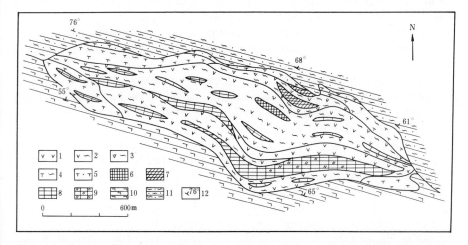

Figure 11.5 Schematic geological map of the Miaoya Nb-REE-bearing syenite–carbonatite complex, Hubei Province, 1, syenite; 2, contaminated syenite; 3, contaminated albitite; 4, syenite porphyry; 5, albitite porphyry; 6, fine-grained biotite–calcite carbonatite; 7, fine-grained ankerite carbonatite; 8, medium- to coarse-grained calcite carbonatite; 9, medium- to coarse-grained biotite–calcite carbonatite; 10, Proterozoic quartz keratophyre; 11, Devonian carbonaceous sericite schists; 12, dip and strike of strata. (After Bai and Yuan, 1985.)

dominated part (syenite) and a dark-coloured carbonaceous sericite-dominated part (schist). The latter occurs as schlieren and lenses (0.5 cm to several metres) in the former and accounts for 5–30% of the rock. Microcline from the syenite and biotite from the biotite–calcite carbonatite were dated at 251 Ma and 278 Ma, respectively (Li, 1980). The apparent U–Pb age of a zircon from the syenite was determined at 255 Ma (Bai and Yuan, 1985).

Niobium and rare earth mineralization occurs as fine-grained (0.01–0.6 mm) disseminated columbite, Nb-rutile, bastnäsite, parisite, monazite, allanite, zircon, pyrochlore and apatite. Monazite and bastnäsite are mostly concentrated in the calcite and ankerite carbonatite, while columbite is enriched in all phases of the syenites and carbonatites except the albitites and ankerite carbonatite. The dimensions of the ore bodies vary from hundreds to 2000 metres long by several metres to about 100 metres wide. Average grades are 1.7% REO and 0.12% Nb_2O_5.

11.5.5 Baerzhe alkali granite Zr–Nb–heavy REE–Be deposit

The Baerzhe deposit is situated in the southern part of the northeast-trending Daxinganling Mountain range, Inner Mongolia, northeastern China. The area is largely covered by Mesozoic volcanic basins developed along a NE- and NNE-trending deep-seated fault belt on basement of Variscan age. The Baerzhe alkali granite (127 Ma, Rb–Sr isochron, Zhang and Yuan, 1988), which intrudes late Jurassic andesite lavas and tuffs, consists of two parts, the West rock body and the East rock body (Figure 11.6).

The West rock body is circular at the surface (0.11 km^2) and enlarges downwards. The rock consists of perthite (50–55%), quartz (30–35%), riebeckite (10–15%), aegirine (2–7%), oligoclase–albite (1–3%), and magnetite and hematite (1–3%). Accessory minerals include synchysite, pyrochlore, ferrothorite, zircon, fluorite, calcite, chlorite, stilpnomelane, astrophyllite, electrum, galena, sphalerite, smithsonite and cerussite (Bai *et al.*, 1980; Ding, Bai and Yuan, 1985).

The East rock body crops out in a northeasterly elongated shape (0.3 km^2) and is composed mainly of quartz (40–45%), perthite (20–35%), albite (10–30%), riebeckite (5–8%), aegirine (0–3%), and magnetite and hematite (3–5%). Rare-metal mineralization occurs as veinlets and disseminations of zircon (2–8%) and hingganite (0–5%), associated with pyrochlore, ferrothorite, synchysite, monazite, genthelvite, Nb–Fe-bearing rutile, columbite, ilmenite, fluorite, calcite and geothite (Bai *et al.*, 1980; Ding, Bai and Yuan, 1985). The East body is intensively altered by albite, quartz, fluorite and aegirine, and five zones are recognized by drilling from the top down to 300 m depth: (1) pegmatitic granites; (2) strongly albitized alkali granite to which the mineralization is closely related; (3) intermediately albitized alkali granite; (4) weakly albitized alkali granite; and (5) porphyrytic riebeckite

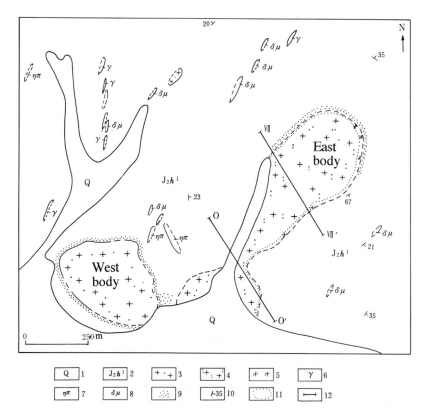

Figure 11.6 Schematic geological map of the Baerzhe alkaline granite, Inner Mongolia, northeastern China. 1, quaternary sediments; 2, late Jurassic andesite tuffs; 3, riebeckite granite; 4, albitized alkaline granite; 5, pegmatitic granite; 6, aplites; 7, monzanite porphyries; 8, diorite porphyries; 9, hornfels; 10, dip and strike of strata; 11, boundaries between different phases; 12, cross-section line. (After Prospecting Geological Party, Jilin Bureau of Geology and Mineral Resources.)

granite (Figure 11.7). Average grades are determined as REO 0.2–0.7%, ZrO_2 0.5–3.4%, Nb_2O_5 0.06–0.32% and BeO 0.02–0.15% (Bai et al., 1980).

11.5.6 Saima alkali complex REE–U deposit

The Saima deposit is located southeast of Shenyang City, Liaoning Province, northeastern China. The area is tectonically situated in the northeastern margin of the Sino-Korean craton with east–west-trending faults and folds as the main structures, and Proterozoic and Sinian metamorphic rocks being the main exposures.

The Saima alkali complex intrudes Proterozoic dolomitic marbles and

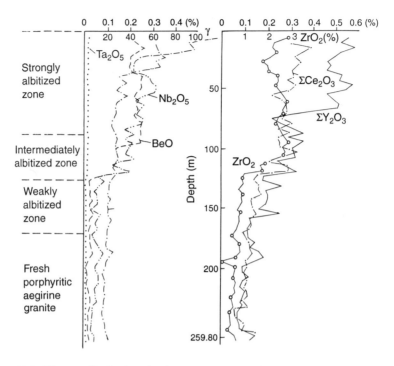

Figure 11.7 Element distributions in the cross-section of the East body of Baerzhe alkalic granite, Inner Mongolia. (Data after Prospecting Geological Party, Jilin Bureau of Geology and Mineral Resources.)

phyllites in the north and west, and Sinian–Cambrian quartzite and limestones in the northeast. It is unconformably overlain by Jurassic sandstones to the south and east (Figure 11.8). According to the Beijing Institute of Uranium Geology (1974), the complex has a surface exposure of about 20 km^2 and is made up of a variety of lithologies in the following sequence: (1) Extrusive rocks, mainly trachytes and phonolites; (2) intrusive rocks including (2a) miaskites and dark-coloured aegirine-nepheline syenites; and (2b) grass green-coloured aegirine-nepheline syenites; (3) dykes and porphyries, namely syenite aplite, nepheline syenite aplite and alkali-lamprophyres.

The phase 2a miaskites comprise the major part (60%) of the complex, and consist mainly of alkali feldspar (50–70%), nepheline (20–30%) and mafic minerals (biotite, melanite and aegirine; 10–30%), together with rare riebeckite, cancrinite and natrolite, and accessory sphene, apatite, ilmenite, zircon, lamprophyllite, eudialite and rincolite.

The phase 2b aegirine-nepheline syenite occurs as a small triangle-shaped stock at the northwestern edge of the complex (Figure 11.8) and is the main host rock of U and REE mineralization. The rock is grass green-coloured,

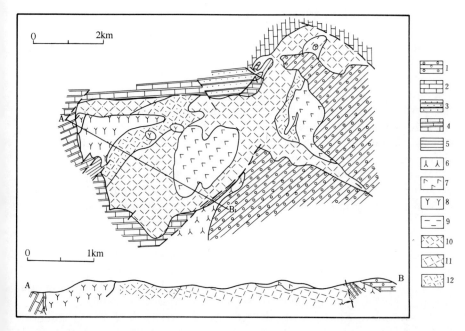

Figure 11.8 Generalized geological map and cross-section of the Saima alkalic complex, Liaoning Province 1, Jurassic sandstones; 2, Cambrian limestone; 3, sinian quartizte; 4–5, Proterozoic; 4, marble; 5, phyllite; 6, leucitophyre, pseudoleucite phonolite and tuffs; 7, phonolite and trachyte; 8, grass green-coloured aegirine–nepheline syenite (ore-bearing); 9, arfvedsonite-rich rock; 10, miaskite; 11, melanite miaskite; 12, dark-coloured aegirine–nepheline syenite. (Simplified after Beijing Institute of Uranium Geology, 1975.)

coarse-grained and composed of microcline (40–60%), nepheline (10–20%) and aegirine (10–40%), associated with rincolite (>2%), eudialyte, lamprophyllite (3%) and pectolite (up to 10%). Alteration including fenitization, carbonatization, silicification and kaolinitization is pervasively developed and skarns are formed at the western contact between the complex and dolomitic marbles. Ore bodies occur mostly at the northwestern contact of the complex, as parallel zones and lenses dipping northeasterly at 5–25° with a maximum thickness of 40 m. Rincolite is the dominant U- and REE-containing mineral (up to 30% of the rock), accompanied by loparite, catapleiite, nenadkevite, cancrinite, natrolite, strontianite, uranpyrochlore, betafite, pitchblende, uranothorianite, aldanite and ?britholite. REO contents vary from 0.3 to 4.5%, together with significant enrichment of U, Th and Nb (Beijing Institute of Uranium Geology, 1974, 1975).

Biotites from the miaskites are dated by the K–Ar method at 217–245-Ma, and rincolites from aegirine–nepheline syenite yield apparent U–Th–Pb ages between 223 and 240-Ma (Beijing Institute of Uranium Geology, 1975).

11.5.7 Metamorphic rock-hosted Zr–Y–heavy REE mineralization

The Guangshui stratiform Zr, Y and heavy REE occurrence was first found in 1978 in Late Proterozoic metamorphic rocks from the western Dabie Mountain area by the Hubei Regional Geological Party (Li et al., 1981). The area is located in northern Hubei Province and belongs tectonically to the east–west-trending Qinling–Dabie Mountain fold belt between the Sino-Korean and the Yangtze cratons. Highly metamorphosed hornblende-plagioclase gneiss, granitic gneiss and migmatites of the Late Archean Dabie Group constitute the basement, which is unconformably overlain by basal conglomerate of the Late Proterozoic Hongan Group (937–950 Ma; zircon Pb–Pb isochron; Tao et al., 1986). The Hongan Group comprises a sequence of schists, phosphate-bearing layers, dolomitic marbles, albite-muscovite schist, felsic schist and variety of greenschists. The mineralized area near Guangshui township may belong to the low greenschist facies belt of Kang et al. (1989). Distinctive paired ultra-high- and high-pressure metamorphic belts consisting of a coesite- and diamond-bearing eclogite belt in the north and a blueschist belt at the south were recognized in the area (Zhang and Kang, 1989; Wang et al., 1990; Wang, Liu and Maruyama, 1992).

The stratiform Zr, Y and heavy REE mineralization unit is composed of felsic schists and albite–muscovite schists, and occurs concordantly within the Tiantaishan Formation, the lower part of the Hongan Group. The unit was highly strained and deformed with strong schistosity developed in the direction of NW/70–80°SW or NW/70–80°NE (strike/dip) which is parallel to the strike of the metamorphic strata. The mineralization occurs in two areas. Near Guangshui township the mineralized zone crops out discontinuously in a core portion of a tightly overturned synclinorium along a NW–SE trend for more than 10 km, and near Dengjiawan village it crops out along a NE–SW trend for about 5 km (Figure 11.9). The mineralized zone is made up of several individual mineralized layers that vary from 560 to 2500 m in length and from 1 to 36 m in width with an average of about 10 m. Boundaries between mineralized and felsic schists are very difficult to distinguish macroscopically.

Felsic schists are the main host rocks of Zr, Y and heavy REE mineralization. They are light-coloured, fine-grained (0.02–0.5 mm), schistoid quartzofeldspathic rocks or muscovite-poor schists and consist mainly of quartz (25–35%), albite (An 0–2) (25–35%), microcline (15–20%), muscovite (5–15%) and magnetite (5–10%), with rare spessartine-almandine garnet. Rare-metal-bearing ore minerals including zircon, gadolinite, fergusonite, xenotime, monazite, allanite, epidote, gahnite and apatite occur as very fine grains (0.001–0.1 mm) contained by quartz, albite and microcline (Wu et al., 1993a, in preparation). Average rare metal contents are 0.086% Y_2O_3, 0.2–0.3% REO, 0.8–1% ZrO_2 and 0.027% Nb_2O_5, which are of only potential economic interest at present.

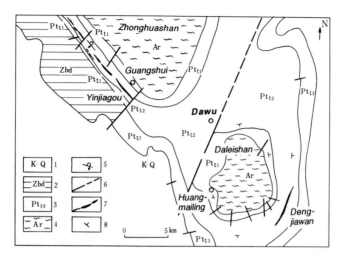

Figure 11.9 Schematic geological map of the late Proterozoic metamorphic rock-hosted Zr-Y-HREE mineralization, Guangshui area, Hubei Province. 1, Cretaceous-quaternary sediments; 2, sinian metasedimentary rocks; 3, late Proterozoic Hongan Group, mainly phosphorite and dolomitic marbles (Pt_{t1}), felsic schists (Pt_{t2}) and green schists (Pt_{t3}); 4, Archaean Dabie Group, gneisses and migmatites; 5, overturned synclinorium; 6, faults and fractures; 7, REE-bearing layers; 8, dip and strike of strata. (Simplified after Li *et al.*, 1981.)

11.6 Concluding remarks

Chinese rare earth deposits may be divided into 10 groups according to their host rocks (Table 11.1). Among them, those hosted by carbonatitic rocks, quartz syenites, weathering crusts (ion-absorption type) and placers are currently in production and are of greatest economic importance. Bayan Obo is the largest rare earth deposit in the world and produces more than 60% of the rare earth products of China. Ion-absorption type rare earth deposits are important in respect of their varieties with high concentrations of europium and heavy rare earths, including yttrium (Table 11.2).

Most of the Chinese endogenic rare earth deposits are distributed in marginal areas of older cratons and/or adjacent orogenic belts, and show close spatial relations to deep-seated fracture zones. The formation ages of the deposits vary from Proterozoic to Cainozoic. The distribution of exogenetic rare earth deposits, on the other hand, is largely controlled by ancient and modern sedimentary environments and weathering conditions.

Rare earth-hosting carbonatitic rocks, quartz syenites and alkalic complexes are characterized by distinctive light REE-enriched, steep right-dipping chondrite-normalized REE patterns, associated with high concentrations of Nb, Ba, Sr and P, and low contents of Rb. Alkalic granites show heavy REE-elevated patterns with strong negative Eu anomalies, accompanied by Zr, Nb and Rb enrichments, and Ba and Sr depletions.

Available isotopic geochemical data suggest a mantle-related source for the genesis of most rare earth deposits hosted by carbonatitic rocks, quartz syenites, alkalic complexes and alkalic pegmatites, a lower crustal or a contaminated mantle-derived source for alkalic granites, and a mature crust-derived source for the various granites which are protoliths of the REE-rich clays of south China.

Acknowledgements

The authors are very grateful to Liu Yuanjun, Cun Jie and Li Jiangzhou, and to Shi Zemin and Li Xiaoyu for their assistance during our field work in Guangshui in 1991, and in Maoniuping from 1990 to 1993. Thanks are also to Prof. Shunso Ishihara, Drs Terumasa Nakajima, Shigeko Togashi and Noboru Imai for their help to the first author in finishing part of the analytical work at the Geological Survey of Japan. We also thank Prof. Zou Tienren for allowing us to quote his unpublished data and for his review of the paper.

References

Bai, G. and Yuan, Z. (1983) On the genesis of the Bayan Obo ore deposit. *Bull. Inst. Mineral Depos., Chinese Acad. Geol. Sci.*, **4**, 1–15 (in Chinese with English abstract).

Bai, G. and Yuan, Z. (1985) Carbonatites and related mineral resources. *Bull. Inst. Mineral Depos., Chinese Acad. Geol. Sci.*, **1**, 99–195 (in Chinese with English abstract).

Bai, G., Yuan, Z., Ding, X. and Sun, L. (1980) Discussion on the petrogenesis and minerogenesis of the rare-metal-bearing alkaline granites in Baerzhe, Jilin Province. *Bull. Inst. Mineral Depos., Chinese Acad. Geol. Sci.*, **1**, 97–113 (in Chinese with English abstract).

Beijing Institute of Uranium Geology (1974) Characteristics of radioactive and rare elements mineralization in a certain alkalic rock complex, northeast China. Proceedings of the national symposium on geology of rare elements, Vol. 2, Geol. Publ. House, Beijing, pp. 260–271 (in Chinese).

Beijing Institute of Uranium Geology 1975. Basic characteristics of the Saima alkalic complex and related uranium deposit. Research Report, Beijing Institute of Uranium Geology, pp. 1–65, (in Chinese).

Chao, E. C. T., Minkin, J. A., Back, J. M. and Ren, Y. (1990) Field and petrographic textural evidence for the epigenetic hydrothermal metasomatic origin of the Bayan Obo rare earth ore deposit of Inner Mongolia, China. 15th General Meeting of the International Mineralogical Association, Abstracts, Vol. 2, June 28–July 3, 1990, pp. 930–1.

Chao, E. C. T., Back, J. M., Minkin, J. A. and Ren, Y. (1992) Host-rock controlled epigenetic, hydrothermal metasomatic origin of the Bayan Obo REE–Fe–Nb ore deposit, Inner Mongolia, P. R. C. *Appl. Geochem.*, **7**, 443–58.

Chao, E. C. T., Tatsumoto, M., Minkin, J. A. *et al.* (1993) Multiple lines of evidence for establishing the mineral paragenetic sequence of the Bayan Obo rare earth ore deposit of Inner Mongolia, China, in Proceedings of the Eighth Quadrennial IAGOD Symposium (ed. Yuon T. Maurice), Aug. 12–18, 1990, Ottawa, Canada, E. Schweizerbart'sche Verlagsbuchhandlung, Stuttgart, pp. 55–73.

Chen, C. and Pu, G. (1991) Geological features and genesis of the Maoniuping rare earth element deposit, Sichuan. *Geology Explor.*, **5**, 18–23 (in Chinese with English abstract).

REFERENCES

Chen, H. (1985) Hydrogen isotopic composition of the Bayan Obo Nb-REE-Fe deposit, Inner Mongolia. B. S. Unpublished thesis, Beijing University (in Chinese with English abstract).

Chen, H. (1993) Hydrogen, oxygen and carbon isotopes and their genetic significances of the Bayan Obo deposit. Proceedings of the Fifth All-China Symposium on Mineral Deposits, Geol. Publ. House, Beijing, pp. 561–2 (in Chinese).

Clark, A. L. and Zheng, S. (1991) China's rare earth potential, industry and policy. In *Rare Earths Future Prospects 1991*, pp. 40–48.

Conrad, J. E. and McKee, E. H. (1992) $^{40}Ar/^{39}Ar$ dating of vein amphibole from the Bayan Obo iron-rare earth element-niobium deposit, Inner Mongolia, China: constraints on mineralization and deposition of the Bayan Obo group. *Econ. Geol.*, **87**, 185–8.

Ding, X., Bai, G. and Yuan, Z. (1985) A study of mineralogical characteristics of some rare-metal alkaline granites in Inner Mongol Autonomous Region. *Bull. Inst. Mineral Depos, Chinese Acad. Geol. Sci.*, **2**, 97–113 (in Chinese with English abstract).

Drew, L. J., Meng, Q. and Sun, W. (1990) The Bayan Obo iron–rare-earth–niobium deposits, Inner Mongolia, China. *Lithos*, **26**, 43–65.

Gross, G. A. (1993) Rare earth elements and niobium in iron-formation at Bayan Obo, Inner Mongolia, China, in Proceedings of the Eighth Quadrennial IAGOD Symposium, Aug. 12–18, 1990, Ottawa, Canada, (ed. Y. T. Maurice), E. Schweizerbart'sche Verlagsbuchhandlung, Stuttgart, pp. 477–90.

Heinrich, E. W. (1980) *The Geology of Carbonatites*, Robert E. Krieger Publ. Company, Huntington, New York, 11–3.

Hubei Bureau of Geology and Mineral Resources (1985) *Study on the Miaoya carbonatite type Nb-REE deposit, Zhushan County, Hubei Province*. Monograph of Ore Deposits, Rare Metals, No. 1, Hubei Bureau of Geology and Mineral Resources, pp. 1–128 (in Chinese).

Institute of Geochemistry, Academia Sinica (1988) *Geochemistry of the Bayan Obo deposit*, Press Sci., Beijing, 554 pp. (in Chinese).

Jiang, C., Zhang, R., Pan, J. *et al.* (1982) Study on isotopic geology of the Bayan Obo deposit. In *Contributions to the Bayan Obo Deposit*, Institute of Geology, Academia Sinica, pp. 10–23 (in Chinese).

Kang, W., Zhang, S., Zhou, G. and Liu, J. (1989) The petrological characteristics of the high pressure metamorphic belt in the northern Hubei Province: geology of the Hubei-Anhui blueschist belt. *J. Changchun Univ. Earth Sci.*, Spec. Issue, 18–40 (in Chinese with English abstract).

Le Bas, M. J. (1977) *Carbonatite–nephelinite Volcanism*. John Wiley & Sons, pp. 37–8.

Le Bas, M. J., Keller, J., Tao, K. *et al.* (1992) Carbonatite dykes at Bayan Obo, Inner Mongolia, China. *Mineral. Petrol.*, **46**, 195–228.

Li, S. (1980) Geochemical features and petrogenesis of Miaoya carbonatites, Hubei. *Geochimica*, **4**, 345–55 (in Chinese with English abstract).

Li, S., Wu, G., Wang, T. *et al.* (1981) A new type of heavy rare earth mineralization, *Geology Explor.*, **12**, 47–8 (in Chinese).

Li, Y. (1959) *Geology and Exploration of the Bayan Obo Iron Ore Deposit*, Geol. Publ. House, Beijing (in Chinese).

Liu, T. (1986) A discussion on the genesis of dolomite in Bayan Obo, Inner Mongolia: with emphasis on the composition of oxygen and carbon isotopes. *Geol. Rev.*, **32**, 150–9 (in Chinese with English abstract).

Maksimović, Z. and Pantó, G. (1991) Contribution to the geochemistry of rare earth elements in the karst-bauxite deposits of Yugoslavia and Greece. *Geoderma*, **51**, 93–109.

Meng, Q. (1981) On the genesis of the Bayan Obo ore deposit related to carbonatite. *Geology Explor.* **3**, 10–7 (in Chinese).

Meng, Q. (1982) The genesis of the host rock dolomite of the Bayan Obo iron ore deposit and the analysis of its sedimentary environment. *Geol. Rev.*, **28**, 481–9 (in Chinese with English abstract).

Möller, P. (1989) Rare earth mineral deposits and their industrial importance, in *Lanthanides, Tantalum and Niobium* (eds P. Möller, P. Cerny and F. Saupé), Springer-Verlag, Berlin, pp. 171–88.

Möller, P., Morteani, G. and Schley, E. (1980) Discussion of REE distribution patterns of carbonatites and alkalic rocks. *Lithos*, **13**, 171–179.

Nakai, S., Masuda, A., Shimizu, H. and Qi, L. (1989) La–Ba dating and Nd and Sr isotope studies on the Bayan Obo rare earth element ore deposits, Inner Mongolia, China. *Econ. Geol.*, **84**, 2296–9.

No. 6 Geological Party of Yunnan (1974) Occurrence modes of associated rare earth elements in a certain phosphorite deposit, Yunnan Province. Proceedings of the national symposium on geology of rare elements, Vol. 2, Geol. Publ. House, Beijing, pp. 299–307 (in Chinese).

No. 109 Geological Party (1992) Report on investigation of REE occurrence mode and comprehensive utilization of the Maoniuping REE deposit, Mianning County, Sichuan Province. Sichuan Bureau of Geology and Mineral Resources, 178 pp. (in Chinese).

No. 115 Geological Party of Guizhou (1974) Rare and dispersed elements in bauxites, central Guizhou Province. Proceedings of the national symposium on geology of rare elements, Vol. 2, Geol. Publ. House, Beijing, pp. 315–8 (in Chinese).

Pu, G. (1988) Discovery of an alkali pegmatite–carbonatite complex zone in Maoniuping, southwestern Sichuan Province. *Geol. Rev.*, **34**, 88–92 (in Chinese with English abstract).

Qian, D. (1979) An occurrence of calcian ancylite and carbocernaite in China. *Geol. Rev.*, **25**, 29–35 (in Chinese).

Qiu, Y., Wang, Z. and Zhao, Z. (1983) Preliminary study of REE iron formations in China. *Geochemistry, Chinese Soc. Mineral. Petrol. Geochem., Acad. Sin.*, **2**, 186–200.

Rare Earth Office (1993) Rare earths in China – 1992. *Rare Earth Informations*, **3**, 1–5 (in Chinese).

Ren, Y. and Chao, E. C. T. (1990) The periods of mineralization and mineral assemblages of the Bayan Obo Fe–Nb–REE ore deposit of Inner Mongolia. 15th General Meeting of the International Mineralogical Association, Abstracts, Vol. 2, June 28–July 3, 1990, pp. 950–1.

Samama, J. C. (1986) *Ore Fields and Continental Weathering*. Von Nostrand Reinhold Company, New York.

Song, Y. and Shen, L. (1982) Discussion on clay minerals and their formation conditions occurring in the weathering crust of an acid volcanic rock in Jiangxi Province. *Acta Mineral. Sin.*, **3**, 207–21 (in Chinese with English abstract).

Song, Y. and Shen, L. (1986) REE geochemistry of the weathered crust of acid volcanic rocks – an experimental study. *Geochimica*, **3**, 225–34 (in Chinese with English abstract).

Tao, Q., Dong, J., Yuan, X. and Ying, J. (1986) Geochronology of Precambrian sequence in northeastern Hubei. *China. Bull. Tianjin Inst. Geol. Min. Res.*, **16**, 75–89 (in Chinese with English abstract).

Tu, G. Z., Zhao, Z H. and Qiu, Y. Z. (1985) Evolution of Precambrian REE mineralization. *Precambrian Res.*, **27**, 131–51.

REFERENCES

US Bureau of Mines (1993) Rare earths. In *Mineral Commodity Summaries*, U.S. Bureau of Mines, Washington DC, pp. 138–9.

Wang, H. and Yan, S. (1980) Characteristics of rare earth minerals and mineralization of a certain rare earth deposit, Shandong. *Geol. Informations of Shandong*, **2**, 38–50 (in Chinese).

Wang, X., Liu, J. G. and Maruyama, S. (1992) Coesite-bearing eclogites from the Dabie Mountains, central China: petrogenesis, P–T paths, and implication for regional tectonics J. *Geol.*, **100**, 231–50.

Wang, X., Shen, L. and Song, Y. (1980) An experimental investigation on the REE weathering crust of granites in South China, in *Scientific Papers on Geology for International Exchange (2)*, Prepared for 26th Internatl. Geol. Congr. Geol. Publ. House, Beijing, pp. 139–45 (in Chinese with English abstract).

Wang, X., Liu, J. G., Pan, G. *et al.* (1990) Field occurrences and petrology of eclogites from the Dabie Mountains, Anhui, central China. *Lithos*, **25**, 119–31.

Wang, Y. (1987) Evidence on the existence of a special enriched mantle. International Symposium on Petrogenesis and Mineralization of Granitoids, Abstract, Dec. 7–10, 1987, Guangzhou, China, pp. 329–330.

Wang, Z. (1973) The genetic features of the REE-Fe ore deposit related to sedimentation–metamorphism–hydrothermal metasomatism. *Geochimica*, **1**, 5–11.

Wei, J. and Shangguan Z. (1983) Oxygen and carbon isotope composition and genesis of dolomite in the Bayan Obo iron deposit, Inner Mongolia. *Petrological Research (2)*, Geol. Publ. House, Beijing, pp. 14–21, (in Chinese with English abstract).

Wu, C., Huang D. and Guo, Z. (1990) REE gochemistry in the weathered crust of granites, Longnan area, Jiangxi Province. *Acta Geol. Sin.* **3**, 194–209.

Wu, C., Huang, D., Bai, G. and Ding, X. (1990) Differentiation of rare earth elements and origin of granitic rocks, Nanling Mountain area. *Acta Petrol. Mineral.*, **9**, 106–16 (in Chinese with English abstract).

Wu, C., Lu H. Xu, L. and Hou, L. (1993b) A preliminary study on modes of occurrence of rare earth elements in the tropical–subtropical weathering crust of Nanling region. *Mineral Deposits*, **12**, 297–307, (in Chinese with English abstract).

Wu, C., Bai, G., Yuan, Z. *et al.* (1993a) Proterozoic metamorphic rock-hosted Zr, Y and heavy REE mineralization in the Dabie Mountain area, central China in *Rare Earth Minerals, Chemistry, Origin and Ore Deposits*, Abstracts, April 1–2, 1993, Natural History Museum London, pp. 160–2.

Wu, C., Huang, D., Yuan, Z. *et al.* (1993c) Trace element geochemistry and genetic implications for some highly evolved granitic rocks from south China. *J. Resource Geol*, Spec. Issue.

Yang, G. and Yan, C. (1991) Betafite from the Mianning aegirine alkaline granite, Sichuan Province. *Minerals and Rocks*, **11**, 9–13 (in Chinese with English abstract).

Yang, G., Pan, Z., Wu, X. and Zhao, W. (1991) Chevkinite from the Changbei (north of Xichan) rare earth deposit, Sichuan Province. *Acta Mineral.* **11**, 109–14 (in Chinese with English abstract).

Yang, R. and Lu, D. (1988) Trace element geochemistry of granitoids in the Panxi area. *Sci. Sin. Ser.* B, **31**, 1109–21.

Yang, Y., Hu, Z. and Luo, Z. (1981) Geological characteristics of the ion-absorption type rare earth deposits and their prospecting direction. *Bull. Inst. Mineral Depos., Chinese Acad. Geol. Sci.*, **1**, 102–18 (in Chinese with English abstract).

Yang, Z. (1980) Geological characteristics of the Shengtieling monazite type rare earth deposit. Report of the Regional Geological Party, Liaoning Bureau of Geology, pp. 1–11, (in Chinese).

Yang, Z. (1987) A study on clay minerals from the REE-rich weathered crust developed on the

Longnan granite, Jiangxi Province. *Sci. Geol. Sin.*, **1**, 70–80 (in Chinese with English abstract).

Yuan, Z. and Zhang, Z. (1992) Sm–Nd isotopic characteristics of granitoids in the Nanling region and their petrogenetic analysis. *Geol. Rev.*, **38**, 1–15 (in Chinese with English abstract).

Yuan, Z., Bai, G., Wu, C. *et al.* (1992) Geological features and genesis of the Bayan Obo REE ore deposit, Inner Mongolia, China. *Appl. Geochem.*, **7**, 429–42.

Yuan, Z., Bai, G., Ding, X. *et al.* (1993) U–Pb isotopic age of zircon from the Maoniuping alkali granite, Sichuan Province, and its geological significance. *Mineral Deposits*, **12**, 189–92 (in Chinese with English abstract).

Zhang, M. (1986) Geochemistry of rare and rare earth elements and oxygen isotopics of the '801' alkalic granite, Inner Mongolia. Unpublished BS thesis, Chinese Acad. Geol. Sci, 63 pp. (in Chinese).

Zhang, P. and Tao, K. (1986) *Bayan Obo Mineralogy*. Sci. Press, Beijing, pp. 3–36, (in Chinese).

Zhang, P. and Tao, K. (1989) A study on the genetic classification of rare earth mineral deposits of China. *Sci. Geol. Sin.*, **1**, 26–32, (in Chinese with English abstract).

Zhang, M. and Yuan, Z. (1988) Oxygen isotopic geochemistry of the '801' (Baerzhe) rare element alkaline granite, Inner Mongolia. *Bull. Inst. Mineral Depos., Chinese Acad. Geol. Sci.*, **1**, 139–46, (in Chinese with English abstract).

Zhang, S. and Kang, W. (1989) The character of the blueschist belt on northern edge of the Yangtze platform. In *Geology of the Hubei–Anhui Blueschist Belt*, J. Changchun Univ. Earth Sci., Spec. Issue, 1–9, (in Chinese with English abstract).

Zhang, Y. (1974) Geologic characteristics and ore processing experiment of a certain REE-bearing phosphorite deposit, Guizhou Province. Proceedings of the National Symposium on Geology of Rare Elements, Vol.2, Geol. Publ. House, Beijing, pp. 308–14 (in Chinese).

Zhao, C. (1983) Chronicle of the Bayan Obo rare earths (1927–1982). 1st Symposium on Geology and Mining of Rare Earths, Geological Society of China Papers, (in Chinese).

Zhou, Z., Li, G., Song, T. and Liu, Y. (1980) Geological features of the Bayan Obo dolomite carbonatite, Inner Mongolia and its genetic discussion. *Geol. Rev.*, **26**, 35–41 (in Chinese with English abstract).

Zhou, B., Shi, Z., Zhang, Y. and Li, X. (1985) A-type granites of the Panxi rift zone, in *Contribution to Panzhihua–Xichang Rift Belt, China, No. 1*, Geol. Publ. House, Beijing, pp. 201–26, (in Chinese with English abstract).

Zou, T., Yang Y., Guo, Y. and Ni, Y. (1985) China's crust- and mantle-source pegmatites and their discriminating criteria. *Geochemistry, Chinese Soc. Mineral. Petrol. Geochem., Acad. Sin.*, **4**, 1–17.

CHAPTER TWELVE
Yttrium and rare earth element minerals of the Kola Peninsula, Russia

A. P. Belolipetskii and A. V. Voloshin

12.1 Introduction

The Kola Peninsula, which forms the northeastern part of the Baltic Shield in Russia, is renowned for its rare element mineralogy. In addition to the classic agpaitic nepheline syenites of Khibina and Lovozero, there are many other complexes which host significant Li, Cs, Be, Nb, Ta, Zr, Y, REE and Sc mineralization. This chapter summarizes the features of the main stages of mineralization and presents a more detailed example of an extraordinary heavy REE pegmatite deposit. There are three main stages of intensive rare-metal ore genesis on the Kola Peninsula (Belolipetskii and Kozyreva, 1989; Belolipetskii, 1990; Belolipetskii and Gordienko, 1991; Belolipetskii et al., 1992):

1. Late Archaean – Kola stage, 2800–3000 Ma;
2. Early Proterozoic – Karelian stage, 1700–2400 Ma;
3. Palaeozoic – Caledonian–Hercynian stage, 360–385 Ma.

There are distinct differences between the three stages and the greatest amounts of RE mineralization are found in stages (2) and (3). Although RE minerals occur in various geological formations on the Kola Peninsula, they are mostly confined to alkaline rock complexes and granite and alkaline pegmatites (Figure 12.1). Over seventy mineral species of Y and REE have been distinguished. They belong to the following groups: halides (3), oxides and hydroxides (18), silicates (30), phosphates (7), and carbonates (15) (Table 12.1).

12.2 Late Archaean – Kola stage, 2800–3000 Ma

Amongst the geological complexes of Archaean age on the Kola Peninsula, the highest concentrations of rare elements occur in the various types of

Rare Earth Minerals: Chemistry, origin and ore deposits. Edited by Adrian P. Jones, Frances Wall and C. Terry Williams. Published in 1996 by Chapman & Hall. ISBN 0 412 61030 2

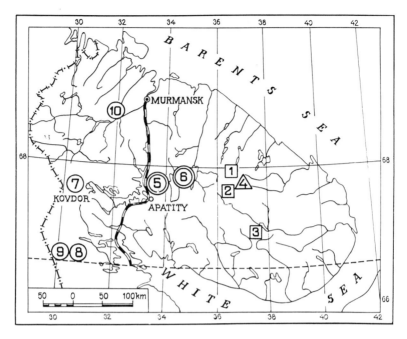

Figure 12.1 The occurrence of the main types of REE mineralization on the Kola Peninsula. Complexes of peralkaline granites and associated silexites (1, West Keivy: Rovozero, Platongora, Yumperuaiv), metasomatites (2, Eljozero; 3, Lavrent'evskii massif), rare earth pegmatites and amazonitic rand-pegmatites (4, Ploskaya, Parusnaya, Serpovidny). Complexes of nepheline syenites (5, Khibina; 6, Lovozero). Complexes of alkaline–ultrabasic rocks and associated carbonatites (7, Kovdor; 8, Vuoriyarvi; 9, Sallanlatvi; 10, Seblyavr)

granitoids and, particularly, in granite pegmatites. Large concentrations of spodumene, pollucite, beryl and columbite–tantalite are associated with granite pegmatities of Voronji tundry, Kolmozero, Polmos and other localities of the Kola Peninsula (Sosedko, 1961; Gordienko, 1970). There are also high concentrations of holmquistite in the metasomatically altered gabbro-labradorites (apobasic metasomatites) situated in the deep fault zones (Gordienko, Krivovitshev and Syritso, 1987).

However, the RE minerals, in contrast to the Be, Li, Nb and Ta minerals, do not form local concentrations in these rocks. Instead, they are typomorphic accessory minerals and are represented usually only by allanite and monazite. Other RE minerals and RE-bearing minerals, including pyrochlore, euxenite and xenotime occur in metasomatically altered granites (Belkov, 1979). Xenotime, fergusonite and euxenite are common in mica-bearing and rare-metal pegmatites where processes of albitization and

LATE ARCHAEAN–KOLA STAGE

Table 12.1 RE minerals and RE-bearing minerals found on the Kola Peninsula and their geological occurrence

Mineral	AP	ALG	ALP	AMZ	ALS	CRB	NFP
Halides							
Ce-Fluorite				AMZ	ALS		
Fluocerite				AMZ			
Y-Fluorite				AMZ			
Oxides							
Pyrochlore	AP	ALG	ALP			CRB	NFP
Uranpyrochlore						CRB	
Pyrochlore-(Ce)	AP		ALP			CRB	NFP
Plumbopyrochlore				AMZ			
Strontiopyrochlore							NFP
Uranmicrolite	AP						
Betafite	AP						
Aeschynite		ALG	ALP	AMZ			
Euxenite	AP	ALG	ALP	AMZ			
Samarskite	AP						
Fergusonite	AP	ALG	ALP	AMZ			NFP
Formanite				AMZ			
Fersmite	AP					CRB	NFP
Loparite							NFP
Lueshite						CRB	
Natroniobite						CRB	
Vigezzite						CRB	
Zirconolite						CRB	
Silicates							
Allanite-(Ce)	AP	ALG	ALP		ALS		
Britholite		ALG	ALP		ALS		NFP
Caysichite				AMZ			
Cerite							NFP
Chevkinite		ALG	ALP				NFP
Eudialyte							NFP
Gadolinite-(Ce)			ALP				
Gadolinite-(Y)		ALG	ALP	AMZ			NFP
Hingganite-(Yb)				AMZ			
Hingganite-(Y)				AMZ			
Iimoriite				AMZ			
Ilimaussite							NFP
Ilmajokite							NFP
Kainosite				AMZ			
Karnasurtite							NFP
Keivyite-(Yb)				AMZ			
Keivyite-(Y)				AMZ			
Kuliokite				AMZ			
Laplandite							NFP
Mosandrite					ALS		NFP
Nordite-(Ce)							NFP
Nordite-(La)							NFP
Perrierite			ALP				
Rowlandite			ALP				
Sazhinite							NFP

313

Table 12.1 *Continued*

Steenstrupine						NFP
Thalenite		ALG	ALP	AMZ		
Tundrite-(Ce)						NFP
Vyuntspakhkite-(Y)				AMZ		
Yttrialite		ALG	ALP			
Phosphates						
Belovite					CRB	NFP
Churchite				AMZ		
Monazite	AP	ALG	ALP	AMZ	CRB	NFP
Phosinaite						NFP
Rhabdophane		ALG	ALP			NFP
Vitusite						NFP
Xenotime	AP	ALG	ALP	AMZ		NFP
Carbonates						
Ancylite					CRB	NFP
Bastnäsite		ALG	ALP	AMZ	CRB	NFP
Burbankite					CRB	NFP
Carbocernaite					CRB	NFP
Cordylite					CRB	
Donnayite					CRB	NFP
Ewaldite					CRB	NFP
Hydroxylbastnäsite					CRB	
Huanghoite					CRB	
Kamphaugite				AMZ		
Mckelveyite					CRB	NFP
Synchysite-(Ce)						NFP
Synchysite-(Y)				AMZ		
Tengerite		ALG	ALP	AMZ		
Zhonghuacerite					CRB	NFP

AP = granites and pegmatites of Archaean age.
Complexes of peralkaline granites and syenites: ALG = granites and silexites; ALP = pegmatites and metasomatites; AMZ = amazonitic rand-pegmatites; ALS = syenites and nephelinic syenites.
Complexes of ultrabasic alkaline rocks: CRB = carbonatites.
Complexes of nephelinic syenites: NFP = pegmatites, metasomatites and hydrothermalites.

silicification have been sufficiently important (Kalita, 1961). Thus the characteristic features of the rare earth minerals in the Archaean geological complexes are the narrow range of species and the predominance of light REE over the heavy REE and Y. Other minerals, including those enriched with Y and heavy REE, are found only in metasomatic rocks. Overall, the rare earth mineralization in the Archaean complexes is much less significant than that of the later phases of Kola mineralization.

12.3 Early Proterozoic – Karelian stage, 1700–2400 Ma

In contrast to the Kola stage, geological complexes belonging to the Karelian stage of the Early Proterozoic contain RE mineralization which, although still associated with other rare metals, is on a much larger scale.

The Karelian stage is characterized by active intrusion of peralkaline granites and intensive postmagmatic hydrothermal and metasomatic processes. Peralkaline granites are widespread on the Kola Peninsula, occupying an area of about 3000 km. There are 14 large massifs and many veins and dykes which are usually situated in deep fault zones. The largest massifs are West Keivy (West Keiv), Ponoj and Lavrent'evskii; all are situated in the central part of the Kola Peninsula within the Keivy structural zone. Amongst the granites, rocks containing aegirine-arfvedsonite, lepidomelane–ferrohastingsite and augite–lepidomelane have been distinguished (Batieva, 1976). These peralkaline granites typically have a high content of rare elements and they are notable for being saturated with rare element accessory minerals and, in particular, rare earths and yttrium.

According to Belkov (1979), accessory RE minerals are represented by allanite, chevkinite, pyrochlore, fergusonite, aeshynite, monazite, britholite and bastnäsite. A considerable increase in the amounts of these minerals and in the variety of mineral species is observed in the final products of peralkaline granite intrusion, namely silexites (at West Keivy), and in metasomatically altered granites and host rocks (at the Eljozero and Lavrent'evskii massifs) (Belkov et al., 1988). Specifically, the rare-metal silexites contain increased contents of britholite-(Y), yttrialite and fergusonite (at Rovozero, Platongora and Yumperuaiv). Yttrium and heavy REE predominate in all these minerals. Spatially and genetically, RE mineralization is associated closely with zirconium (zircon) and tantalum–niobium mineralization (fergusonite and aeschynite).

Local concentrations of RE minerals are observed widely in metasomatically altered granites and their host rocks and, especially, in the metasomatites. A metasomatite is a mineral derivative of the action of metasomatic solutions on the original rocks. Thus, it is a secondary mineral assemblage which is a result of the adaptation of these rocks to new thermodynamic and chemical conditions. Taking into consideration the intensity of metasomatic process, we can distinguish three varieties of metasomatite: poorly revealed (5–20% vol. of the rock consists of the new mineral assemblages), markedly revealed (20–80%) and fully revealed (>80%). We also differentiate between apogranite metasomatite evolved after granite rocks and apobasic metasomatites formed after gabbro-labradorites, amphibolites and other basic rocks. Fully revealed apogranite metasomatites are characterized by the most diverse RE-bearing mineral species. Abundant aeschynite-(Y), fergusonite, yttrotitanite, thalenite, britholite-(Y), gadolinite, allanite, chevkinite, pyrochlore, euxenite, monazite and bastnäsite have been found (at Rovozero, Eljozero and Lavrent'evskii). Quantitatively, Y- and heavy RE-bearing minerals prevail, although Ce-rich minerals are developed locally. Increased contents of RE-bearing minerals suggest a direct dependence on the intensity of albitization and silicification. A characteristic feature of apobasic metasomatites derived from gabbroid rocks is the abundance of an uncommon variety of REE-epidote, which contains mostly Y and heavy REE (Belolipetskii *et al.*, 1967).

Microcline and amazonite pegmatites containing complex Zr–Nb–Be–Y–REE mineralization are spatially and genetically associated with peralkaline granites. This mineralization arises due to metasomatic processes such as albitization and silicification (Lunz, 1972; Kalita, 1974). The main RE minerals are gadolinite, fergusonite, chevkinite, britholite and yttrialite. Of greatest interest are the amazonite rand-pegmatities, of which a typical example is the pegmatite of Ploskaya Mountain (West Keivy). Their rare element mineralization is characterized by an extraordinary diversity and, essentially, no data of this kind have previously been reported. Thus, later in this chapter particular attention is paid to the description of these rare minerals.

A rich and diverse RE mineralization is associated with alkaline, alkaline–ultrabasic and carbonatite complexes of the Kola Peninsula. The earliest alkaline magmatism in the region occurs in the Karelian stage in the form of the metasomatically altered nepheline syenites of the Sacharijok massif. These syenites have high concentrations of britholite and pyrochlore. Allanite and bastnäsite are also reported to occur in minor amounts.

12.4 Palaeozoic – Caledonian–Hercynian stage, 300–400 Ma

The major part of the alkaline and alkaline–ultrabasic magmatism on the Kola Peninsula dates from the Palaeozoic stage of autonomous tectonic-magmatic activation of the Baltic Shield (300–400 Ma). This stage is significant in terms of the rare-metal specialization of the Kola metallogenic province, due to the enrichment of rare elements in these alkaline magmas. The Palaeozoic intrusions were formed over a long period of time, and this caused varied accompanying autometasomatic and hydrothermal–metasomatic processes, which have caused the redistribution and concentration of rare elements. The cumulative effect of this alkaline igneous activity is an extraordinary diversity of RE mineral species.

The most widespread mineralization is primary loparite and eudialite in agpaitic nepheline syenites of the Lovozero alkaline massif. The mineralized nepheline syenites were formed by normal magmatic differentiation. The widespread occurrence of magmatic stage rare-metal mineralization in alkaline rocks is a typical feature of the Lovozero massif. In the Khibina massif this type of occurrence is more scarce, being limited to an accessory mineralization only. Here instead, the highest concentrations of rare metal minerals appear in the alkaline metasomatites (albitites, microclinites, aegirine–feldspar rocks) which are widespread in both massifs. In these complexes pyrochlore is predominant, in association with loparite, eudialite, mosandrite and zircon. All of these minerals are light REE-enriched. The best-known rare-metal mineralization occurs in the alkaline pegmatites and late hydrothermal veins (Semenov, 1972; Khomyakov, 1990). It is characterized by an extremely wide variety of RE mineral species (Table 12.1). Whereas the RE-bearing minerals are represented mainly by oxides and silicates of aluminium, titanium and zirconium in the early magmatic formations, in the alkaline pegmatites and late hydrothermal veins the characteristic minerals are widespread light REE-enriched RE phosphates and carbonates (e.g. monazite, ancylite, burbankite, bastnäsite, huanghoite, rhabdophane and synchysite).

Rare-metal mineralization including diverse rare earth mineralization is common in the carbonatites of alkaline–ultrabasic massifs such as Kovdor, Vuoriyarvi, Seblyavr and Sallanlatvi (Kapustin, 1971). A typical RE mineral assemblage in the early tetraphlogopite–calcite carbonatites would be light REE-enriched pyrochlore, hatchettolite (uranpyrochlore) and lueshite (natroniobite; see also Mitchell, Chapter 3). In these carbonatites the rare earth minerals are closely associated with baddeleyite, zircon and calzirtite. Lueshite predominates in phlogopite–calcite carbonatites of the Sallanlatvi and Seblyavr massifs and late calcite and dolomite carbonatites contain a wide range of RE minerals; for example burbankite, bastnäsite, carbocernaite, cerite, monazite, huanghoite, rhabdophane, cordylite and ancylite. These latter minerals are commonly associated with Sr and Ba minerals and are light REE-enriched, but some Y-bearing minerals, namely mckelveyite, ewaldite and donnayite, have now also been reported.

12.5 Y–REE mineralization in amazonitic rand-pegmatites

Many new mineral species and rare minerals of Y and Yb have been discovered in the amazonitic rand-pegmatites of the Kola Peninsula (Voloshin *et al.*, 1983a, 1983b, 1986; Voloshin, Pakhomovskii and Tyusheva, 1983, 1985; Voloshin, Pakhomovskii and Zezulina, 1986; Voloshin and Pakhomovskii, 1987). These pegmatites appear to be unique and are therefore given special attention here. A rand-pegmatite is an area (vein or body) which occurs in the endo- and exocontact zone of a metasomatically altered granite. It consists of very large crystals or blocks of quartz and feldspar, which are intergrown or 'cemented' by fine-grained aggregates of quartz–albite–mica. There are two stages in the evolution of amazonitic rand-pegmatites. The first conforms to a classic magmatic scheme, the second is postmagmatic and is characterized by multiphase processes of metasomatism and recrystallization. In the postmagmatic stage two phases of albitization and three phases of fluoritization, silicification and dissolution have been traced (Voloshin and Pakhomovskii, 1986; Voloshin, 1989). It is possible to distinguish some of the postmagmatic processes at various localities but they are best developed in the amazonitic rand-pegmatites. Postmagmatic processes are best seen in differentiated amazonitic rand-pegmatites.

The Y and REE mineralization comprises silicates and phosphates and is believed to have formed as the result of a second-stage fluoritization process, termed fluoritization-II, which has produced yttrofluorite. All the Y and REE minerals are characterized by several generations and crystals with narrow zones, caused by varying Y/REE ratios (Figure 12.2). There is a negative correlation between the Y and Yb contents of specific zones: as the Y abundance increases from the central to the marginal zones, so the levels of Yb and the other REE decrease (Voloshin and Pakhomovskii, 1986).

Early mineral generations are commonly selectively Yb or heavy REE-enriched. The following generations crystallized under conditions of high Y potential and thus are Y-enriched, and eventually the crystallization of selective Y phases begins. The chemical composition of minerals with the highest contents of Y and Yb (and other REE) are shown in Tables 12.2–12.4. The relationships between Y, Yb and REE in the formulae of all analysed minerals are shown in Figures 12.3–12.6.

At present only two Yb mineral species, namely keivyite-(Yb) and hingganite-(Yb) are known (Figures 12.3–12.6), but according to the available data (Tables 12.2–12.4), there are other possible varieties with the following formulae.

Yb-xenotime-(Y)	$(Y_{0.6}Yb_{0.4})PO_4$
Yb-vyuntspakhkite-(Y)	$(Y_{2.4}Yb_{1.6})_4Al_5Si_5O_{18}(OH)_5$
Yb-kuliokite-(Y)	$(Y_{2.4}Yb_{1.6})_4Al(SiO_4)_2(OH)_2F_5$
Yb-thalenite-(Y)	$(Y_{1.8}Yb_{1.2})_3Si_3O_{10}(F,OH)$

A chondrite-normalized plot (Figure 12.7) serves to highlight the extreme enrichment in heavy REE and the difference between the (Y) and (Yb) species of keivyite. Keivyite-(Yb) is not only extremely heavy REE-enriched, it also has a negative Y anomaly.

A general scheme for the RE mineralization needs to have two compositional branches; an Yb branch and an Y branch. The minerals change their composition and pass from one branch to another gradually. All Y and RE minerals in the amazonitic rand-pegmatites occur in bodies of yttrofluorite and crystallize separately. Therefore, it is not possible to determine the temporal relationship between those minerals enriched in Yb and the other varieties. It is possible, however, to observe the relationships between the different generations of one mineral species, and also the late growth of tengerite and churchite on early Y–Yb-minerals.

The distribution of Y, Yb and REE is controlled by crystallochemical factors, due to the capacity of the elements to substitute into isomorphic

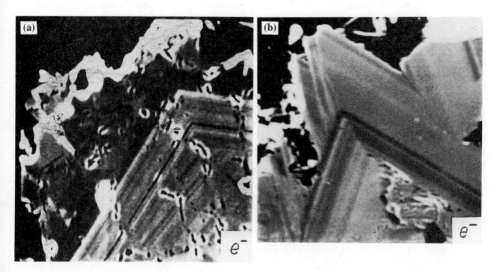

Figure 12.2 Zoned crystals of (a) thalenite and (b) kuliokite. Backscattered electron image; magnification = 200.

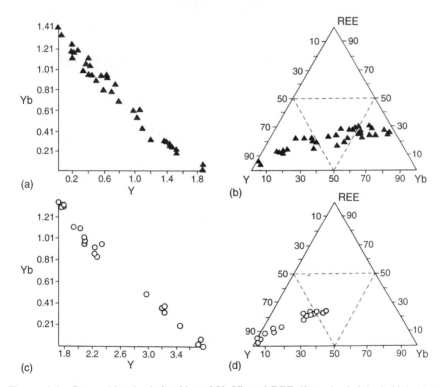

Figure 12.3 Compositional relationships of Y, Yb and REE (formula site) in (a,b) keivyite and (c,d) vyuntspakhkite.

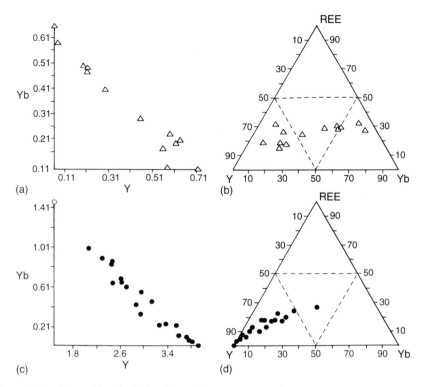

Figure 12.4 Compositional relationships of Y, Yb and REE (formula site) in (a,b) hingganite and (c,d) kuliokite.

Table 12.2 Chemical composition of Y and REE silicates from amazonitic rand-pegmatites, Kola Peninsula

Oxide (wt%)	Keivyite		Hingganite		Vyuntspakhkite-(Y)		Kuliokite-(Y)	
	1	2	3	4	5	6	7	8
Y_2O_3	1.64	57.92	8.56	26.91	17.89	40.98	31.01	55.11
Yb_2O_3	55.14	1.24	34.07	15.31	22.99	1.14	21.06	1.40
Lu_2O_3	7.36	0.06	4.50	1.69	3.13	0.47	2.28	0.14
Tm_2O_3	3.21	0.34	3.10	1.38	1.83	0.34	1.90	0.16
Er_2O_3	5.79	2.42	8.22	4.42	6.57	1.23	8.16	1.21
Ho_2O_3	0.72	0.70	1.03	0.21	0.26	0.21	1.23	0.36
Dy_2O_3	2.10	1.86	2.47	2.39	2.84	0.73	4.18	0.91
Tb_2O_3			0.05	0.33	0.27	0.12		
Gd_2O_3	0.05	0.78	0.11	0.53	0.24	0.09	0.30	0.10
Sm_2O_3		0.47						
CaO		0.27	1.14	2.42				1.02
Al_2O_3					13.41	14.17	6.51	7.08
SiO_2	24.15	33.03	22.11	27.59	26.07	32.14	14.56	17.94
Total	100.16	99.09	85.36	83.18	95.50	91.62	91.19	85.43

1 = keivyite-(Yb); 2 = keivyite-(Y); 3 = hingganite-(Yb); 4 = hingganite-(Y).

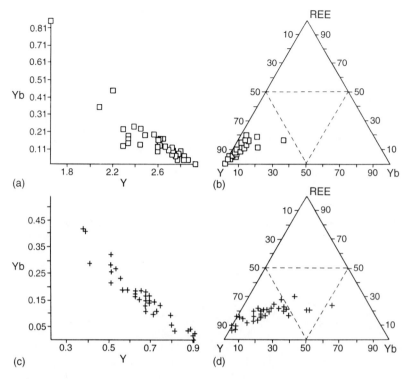

Figure 12.5 Compositional relationships of Y, Yb and REE (formula site) in (a,b) thalenite and (c,d) xenotime.

Table 12.3 Chemical composition of Y and REE silicates and carbonate-silicates from amazonitic rand-pegmatites, Kola Peninsula

Oxide (wt%)	Thalenite-(Y)		Iimoriite-(Y)		Kainosite-(Y)		Caysichite-(Y)	
	1	2	3	4	5	6	7	8
Y_2O_3	29.07	57.62	55.16	60.19	12.12	31.34	13.76	30.56
Yb_2O_3	26.18	1.57	4.75	1.07	16.31	0.76	16.61	2.18
Lu_2O_3	3.42	0.33	0.20	0.19	2.62	0.09	2.15	0.12
Tm_2O_3	2.05	0.57	0.64	0.22	1.55	0.08	1.33	
Er_2O_3	6.35	1.66	2.92	1.56	3.82	0.97	4.50	1.29
Ho_2O_3	0.78	0.03	0.58	0.16	0.51		0.57	0.22
Dy_2O_3	2.13	1.27	1.24	1.87	1.09	0.59	1.64	0.57
Tb_2O_3		0.12	0.15	0.14				0.18
Gd_2O_3	0.13	0.23	0.04	0.58			0.08	0.20
Sm_2O_3				0.16				
CaO					14.87	16.23	7.54	7.36
SiO_2	28.54	35.32	18.73	19.41	32.34	36.73	25.22	30.55
Total	98.65	98.72	84.41	85.55	85.23	86.79	73.40	73.23

CO_2, H_2O not analysed.

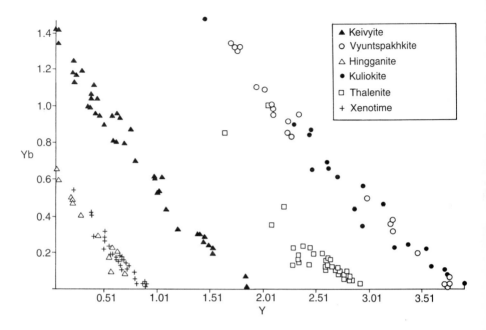

Figure 12.6 Comparative diagram of Yb and Y (formula site) in keivyite, vyuntspakhkite, hingganite, kuliokite, thalenite and xenotime.

Table 12.4 Chemical composition of Y and REE phosphates from amazonitic rand-pegmatites, Kola Peninsula

Oxide (wt%)	Xenotime-(Y)		Churchite-(Y)	
	1	2	3	4
Y_2O_3	10.32	51.00	32.26	42.24
Yb_2O_3	43.38	1.47	13.80	4.11
Lu_2O_3	8.48	0.27	2.01	0.45
Tm_2O_3	3.33	0.18	1.33	0.59
Er_2O_3	5.87	2.65	5.57	6.40
Ho_2O_3		1.13	0.90	1.57
Dy_2O_3	0.86	1.90	3.50	4.25
Tb_2O_3		0.39	0.38	0.39
Gd_2O_3	0.09	0.90	0.77	0.77
Sm_2O_3		0.04		
Nd_2O_3		0.02		
CaO	0.24			
P_2O_5	24.47	35.47	28.98	29.70
Total	97.04	95.42	89.50	90.65

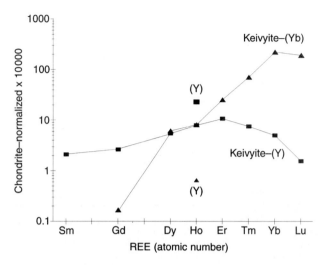

Figure 12.7 Chondrite-normalized plot of heavy REE in keivyite-(Y) and keivyite-(Yb) from amazonitic rand-pegmatites. In addition to the extreme heavy REE-enrichment in keivyite-(Yb) there is a negative Y anomaly compared with the Ho and Dy values. (Chondrite values from Wakita, Rey and Schmitt, 1971.)

crystal structures. For example, although there is no synthetic $Y_2Si_2O_7$ analogue of the thortveitite-type structure ($Yb_2Si_2O_7$), keivyite-(Y), $Y_2Si_2O_7$, always contains a small amount of Yb which defines its structural stability. In contrast, the other REE (excluding Y) which are significant in the composition of keivyite-(Yb) have no compounds isostructural with $Yb_2Si_2O_7$ (thortveitite) and therefore form substructural units (Yakubovich et al., 1986).

The Y branch of mineralization is characterized by an evolutionary series of minerals. Early phosphates such as monazite and xenotime are replaced by later bastnäsite and churchite respectively, whereas silicate minerals evolve to silicate–carbonates, carbonates and then to a second phase of phosphates such as churchite (Figure 12.8).

12.6 Economic REE mineralization

It is possible to make the following points regarding the economic evaluation of RE mineralization on the Kola Peninsula. At present, the most important source of light REE is that obtained from loparite concentrate mined from the Lovozero alkaline massif. In the very near future the output of REE is likely to increase if eudialite ores are also worked from Lovozero. Second in importance is the pyrochlore (or lueshite) concentrate extracted from the rare-metal carbonatite ores in the alkaline–ultrabasic massifs.

Rare-metal ores associated with peralkaline granites, silexites, metasoma-

YTTRIUM AND REE MINERALS OF THE KOLA PENINSULA

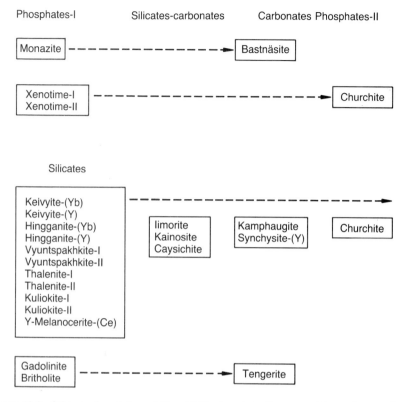

Figure 12.8 Scheme of evolution of Y and RE minerals in the amazonitic rand-pegmatites.

tites and amazonite rand-pegmatites are very promising, their advantage being the high content of Y and heavy REE. However, the alkaline pegmatites and hydrothermal veins which are characterized by the greatest variety of RE minerals are most likely to be of mineralogical interest only.

References

Batieva, I. D. (1976) *Petrology of Peralkaline Granites of the Kola Peninsula*, Science Press, Leningrad, 224 pp. (in Russian).

Belkov, I. V. (1979) *The Accessory Minerals of Granitoids of the Kola Peninsula*. Science Press, Leningrad, 184 pp (in Russian).

Belkov, I. V., Batieva, I. D., Vinogradova, G. V. and Vinogradov, A. N. (1988) *The Mineralization and Fluid Activity in the Contact Zones of Peralkaline Granites Intrusions*. Kola Branch, Academy of Science USSR Press, Apatity, 110 pp. (in Russian).

Belolipetskii, A. P. (1990) Metasomatic rare-metal ore-forming in early Precambrian history of the Kola Peninsula. In *New Data in the Study of Mineral Resources of Murman Region. 1989. (The models of typical metallogenic systems)*. Kola Science Centre, Russian Academy of Science Press, Apatity, pp. 29–33 (in Russian).

REFERENCES

Belolipetskii, A. P. and Gordienko, V. V. (1991) Mineralogical and geochemical peculiarities of rare-metal ore-forming in the Precambrian complexes of Kola Peninsula, in *New Data on the Rare Element Mineralogy of the Kola Peninsula*, Kola Branch, USSR Academy of Science Press, Apatity, 20–7 (in Russian).

Belolipetskii, A. P. and Kozyreva, L. V. (1989) Rare-metal ore-forming in the deep fault zones of the north-eastern part of the Baltic Shield. In *Endogenic Processes in the Deep Fault Zones*, Irkutsk, pp. 215–7 (in Russian).

Belolipetskii, A. P., Denisov, A. P., Elina, N. A. and Kultshitskaya, E. A. (1967) The results of chemical and X-ray investigation of mineral group epidote–allanite, in *The Materials on Mineralogy of the Kola Peninsula*, **5**, Science Press, Leningrad, pp. 129–37. (in Russian).

Belolipetskii, A. P. Voloshin, A. V., Britvin, S. N. *et al.* (1992) *Rare-metal Mineralization of Kola Metallogenic Province*, Kola Science Centre, Russian Academy of Science Press, Apatity, 50 pp. (in Russian, English).

Gordienko, V. V. (1970) *Mineralogy, Geochemistry and Genesis of Spodumene Pegmatites*, Nedra Press, Moscow, 176 pp. (in Russian).

Gordienko, V. V., Krivovitshev, V. G. and Syritso, L. F. (1987) *Metasomatites of Pegmatite Fields*, Leningrad University Press, Leningrad, 222 pp. (in Russian).

Kalita, A. P. (1961) *Rare-earth Pegmatites of Alakurtti and Ladoga Coast*, USSR Academy of Science Press, Moscow, 120 pp. (in Russian).

Kalita, A. P. (1974) *Pegmatities and Hydrothermalites of Peralkaline Granites of the Kola Peninsula*, Nedra Press, Moscow, 139 pp. (in Russian).

Kapustin, Yu. L. (1971) *Mineralogy of Carbonatites.*, Science Press, Moscow, 288 pp. (in Russian). Translated into English as: Kapustin, Yu. L. (1980) *Mineralogy of Carbonatites*. Amerind Publishing, New Dehli, 259 pp.

Khomyakov, A. P. (1990) *Mineralogy of Hyperagpaitic Alkaline Rocks*, Science Press, Moscow, 196 pp (in Russian). Translated into English as Khomyakov, A. P. (1995) *Mineralogy of Hyperagpaitic Alkaline Rocks*, Clarendon Press, Oxford, 223 pp.

Lunz, A. Ya. (1972) *Mineralogy, Geochemistry and Genesis of Rare-earth Pegmatites of Northwest USSR*, Nedra Press, Moscow, 176 pp. (in Russian).

Semenov, E. I. (1972) *Mineralogy of Lovozero Alkaline Massif*, Science Press, Moscow, 308 pp. (in Russian).

Sosedko, A. F. (1961). *The Materials on Mineralogy and Geochemistry of Granite Pegmatites*, Gosgeoltechizdat Press, Moscow, 154 pp. (in Russian).

Voloshin, A. V. (1989) Mineralization of rare metals and rare earths in Proterozoic randpegmatites. *Precambrian Granitoids*. Abstr. 141, Helsinki.

Voloshin, A. V. and Pakhomovskii, Ya. A. (1986) *Minerals and Evolution of Mineral Formation in the Amazonitic Pegmatites of the Kola Peninsula*. Science Press, Leningrad, 168 pp. (in Russian).

Voloshin, A. V. and Pakhomovskii, Ya. A. (1987) Iimoriite from amazonitic pegmatites from Kola Peninsula (the first occurrence in USSR). In *Mineralogical associations and minerals in the magmatic complexes* (ed. I. V. Belkov), Kola Branch, USSR Academy of Science Press, Apatity, pp. 34–42 (in Russian).

Voloshin, A. V., Pakhomovskii, Ya. A. and Tyusheva, F. N. (1983) Keivyite–a new ytterbium silicate from the amazonitic pegmatites from the Kola Peninsula. *Mineral. J.*, **5**, 94–9 (in Russian).

Voloshin, A. V., Pakhomovskii, Ya. A. and Tyusheva, F. N. (1985) Keivyite-(Y)–a new diorthosilicate and thalenite from the amazonitic pegmatites of the Kola Peninsula. Diortho- and triorthosilicates of yttrium. *Mineral. J.*, **7**, 79–94 (in Russian).

Voloshin, A. V., Pakhomovskii, Ya. A. and Zezulina E. P. (1986) Caysichite from the amazonitic pegmatites of the Kola Peninsula. *Mineral. J.*, **8**, 90–3 (in Russian).

Voloshin, A. V., Pakhomovskii, Ya. A., Men'shikov, Yu. P. *et al.* (1983a) Vyuntspakhkite–a new Y–Al silicate from the amazonitic pegmatites of the Kola Peninsula. *Mineral. J.*, **5**, 89–94 (in Russian).

Voloshin, A. V., Pakhomovskii, Ya. A., Men'shikov, Yu. P. *et al.* (1983b) Hingganite-(Yb)–a new mineral species from the amazonitic pegmatites of the Kola Peninsula. *Dokl. Acad. Sci. USSR*, **270**, 1188–92 (in Russian).

Voloshin, A. V., Pakhomovskii, Ya. A., Tyusheva, F. N. *et al.* (1986) Kuliokite-(Y)–a new Y–Al fluorsilicate from the amazonitic pegmatites of the Kola Peninsula. *Mineral. J.*, **8**, 94–9. (in Russian).

Wakita, H., Rey, P. and Schmitt, R. A. (1971) Abundances of the 14 rare earth elements and 12 other elements in Apollo 12 samples: five igneous and one breccia rocks and four soils. *Proc. 2nd Lunar Sci. Conf., Geochim. Cosmochim. Acta Suppl.*, **2**, 1319–29.

Yakubovich, O. V., Simonov, M. A., Voloshin, A. V. and Pakhomovskii, Ya. A. (1986) The crystal structure of the keivyite $Yb_2[Si_2O_7]$. *Dokl. Acad. Sci. USSR*, **291**, 863–7 (in Russian).

Further reading

Further information and references, in Russian and English, on some of the localities in this chapter are given in:

Kogarko, L. N., Kononova, V. A., Orlova, M. P. and Woolley, A. R. (1995) *Alkaline Rocks and Carbonatites of the World. Part Two: Former USSR*, Chapman & Hall, London, 226 pp.

CHAPTER THIRTEEN
Analysis of rare earth minerals
C. T. Williams

13.1 Introduction

Fundamental to the characterization of RE minerals, to the understanding of their formation and to many geochemical and petrological problems involving the REE, is accurate chemical analysis of the minerals themselves. For routine chemical analysis, microanalytical techniques have largely superseded the need for mineral separation procedures. With modern electron microprobes, RE minerals can be analysed quantitatively for 20 to 25 elements at detection limits down to 0.05 wt% in 10–15 minutes. Fast and accurate analyses of RE minerals on the micron scale have enabled much new information to be obtained on compositional changes during the crystallization of the minerals, on subtle differences in element substitutions, and on the role these minerals can play in providing insights into changes in fluid composition during evolution of the fluid. Some recent microanalytical studies have also shown that differences in mineral composition can be related to structural variations within individual crystals.

The objectives of this chapter are to outline those techniques used in the analysis of RE minerals, concentrating more specifically on the microanalytical techniques currently available, and to provide examples of their applications in mineralogy.

13.2 Historical background

The chemical analysis of RE minerals is of course inextricably related to the discovery of the REE themselves. A new mineral from Ytterby in Sweden, subsequently named *gadolinite* after the Finnish chemist Johann Gadolin, promoted the initial recognition of previously unknown 'rare earths' yttria (1794) and ceria (1803). After Davy in the early 19th century showed that the 'earths' were in fact compounds of metallic elements with oxygen, the elements were renamed yttrium and cerium. During the early part of the 19th century yttria and ceria were observed in many other rare minerals, many of the earliest analyses being undertaken by the Swedish chemist Berzelius and his students such as F. Wöhler (*wöhlerite*) and C. G.

Rare Earth Minerals: Chemistry, origin and ore deposits. Edited by Adrian P. Jones, Frances Wall and C. Terry Williams. Published in 1996 by Chapman & Hall. ISBN 0 412 61030 2

Mosander (*mosandrite*), based on the gravimetric and titrimetric analysis techniques available at that time. Mosander indeed showed by fractional precipitation that both ceria and yttria could each be separated into two more oxides named lanthana and didymia (from ceria) and erbia and terbia (from yttria). The invention of the spectroscope in 1859 was instrumental in proving that didymia was itself a complex oxide comprising oxides of samarium (1879), neodymium (1885), praesydimium (1885) and europium (1901). Terbia and erbia were themselves resolved into oxides of holmium (1878), thulium (1879), dysprosium (1886), ytterbium (1878) and lutetium (1907). Thus by the time Moseley, during 1913–14, was able to confirm the positions of the REE within the modern version of the Periodic Table of Elements, all of the REE (with the exception of promethium, for which no stable isotope exists) had been discovered.

From that period in the early part of the 20th century up to the present day, traditional 'wet chemical' analyses were augmented with spectroscopic techniques such as colorimetry, emission, absorption, X-ray and gamma-ray spectrometry. More recently spectroscopic techniques such as inductively coupled plasma emission spectrometry (ICPES) and ICP mass spectrometry (ICPMS) are increasingly being used to determine the REE. For analysis of RE minerals, however, all of these methods suffer the disadvantages of requiring pure mineral concentrates, often followed by dissolution of the mineral and a chemical separation procedure prior to analysis. Although these may not pose serious problems as only small amounts of material may be necessary for the analysis, and procedures are well established, much potential information on chemical zonation within the mineral itself and on the relationships between compositional variations and textural features is lost.

The advent of *in situ* microchemical techniques, such as electron probe microanalysis (EPMA) in the late 1950s, obviated the need for mineral separations and revolutionized the analysis of all minerals. However, it was not until the early 1970s that analyses of RE minerals by EPMA started to enter the scientific literature. Since then most of the published analytical data have been by this technique, although the analysis of RE minerals is far from routine. More recently, new microchemical techniques such as ion probe, proton probe, synchrotron XRF and laser-ablated ICPMS have improved detection limits for the REE. However, access to these techniques is at present limited because of the small number of instruments available; thus the predominant method of analysis of RE minerals, at least for the next decade, is likely to be EPMA.

13.3 Techniques of analysis

There are now many different techniques capable of providing accurate quantitative analysis of RE minerals. These can be broadly categorized into

BULK ANALYTICAL TECHNIQUES

'bulk' and 'microanalytical' techniques described below, with one or two specialized techniques providing additional chemical and structural information. In addition to these techniques, recent improvements in scanning electron microscopes and image processing software have enabled aspects of compositional, textural and structural information to be combined into 'compositional imaging'.

13.3.1 Compositional imaging

Scanning electron microscopes (SEMs) now routinely have backscatter electron detectors which provide compositional information based on differences in mean atomic number of the mineral matrix. Backscatter electron images (BEIs) have now become invaluable aids in mineralogy (e.g. Paterson, Stephens and Herd, 1989), particularly in locating small mineral phases of high mean atomic number, such as RE minerals, which may occur only at very low modal abundances and be only a few microns in size. Textural relationships, fine-scale mineral intergrowths and subtle compositional zoning can all be revealed in BEIs, where some examples are illustrated in Figure 13.1.

Recent developments in high-pressure (low-vacuum) specimen chambers on SEMs have enabled high-resolution BEIs to be obtained from uncoated samples (e.g. Taylor and Jones, in press). Thus, 'photographic-quality' images of individual crystals, which may be located within vugs and surrounded by matrix material, can be obtained (Figure 13.2a,b). With an energy-dispersive detector fitted to an SEM, chemical information can additionally be acquired from uncoated samples (such as those in Figure 13.2) without detaching the crystal from its surrounding matrix, and can assist in identification of the mineral. Figures 13.3a and 13.3b are energy-dispersive spectra obtained from the parisite crystals shown in Figures 13.2a and 13.2b respectively, and indicate that in Figure 13.2a the mineral contains La, Ce and Nd, whereas in Figure 13.2b Ce (and Nd) are depleted relative to La. Both samples are from Langesundfjord, Norway (Larsen, Chapter 6).

Elemental X-ray maps can be obtained from several of the microanalytical techniques outlined below (e.g. EPMA, SIMS and PIXE), and if in a digitized format can be enhanced with image-analysis software to provide information on mineral textures and structures not apparent from optical, or even backscatter, images. An example of sector zoning in monazite-(Ce) from Kangankunde, Malawi, (Wall and Mariano, Chapter 8) revealed by a La X-ray map is shown in Figure 13.4.

13.4 Bulk analytical techniques

Bulk techniques require an initial mechanical separation procedure to obtain pure mineral separates prior to analysis, and some require a further chemical

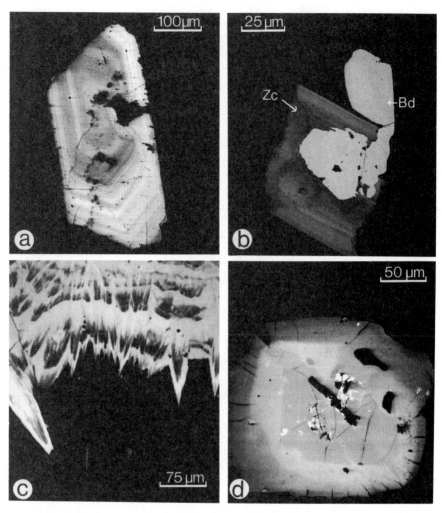

Figure 13.1 Examples of compositional zoning and intergrowths of RE minerals revealed by backscatter electron images. (a) Oscillatory zoned monazite-(Ce) from Western Carpathians (Broska and Williams, in preparation); (b) Sector-zoned zirconolite-(Nd) from Schryburt Lake, British Columbia, Canada (Platt and Williams, in preparation); (c) bästnastite-(Ce)/parisite-(Ce) intergrowth from Langesundfjord, Norway (Larsen, Chapter 6); (d) zirconolite from Zomba, Malawi (Platt *et al.*, 1987). All BEIs were obtained using a KE Developments solid state backscatter electron detector attached to a Hitachi S2500 scanning electron microscope operated at 15 kV accelerating potential and 1 nA specimen current.

separation procedure to isolate the REE as a group. Once these procedures have been completed, however, the REE can be determined, often to sub-ppm concentration levels. Classical gravimetric procedures of analysis have now generally been replaced by a range of instrumental techniques.

BULK ANALYTICAL TECHNIQUES

Figure 13.2 (a) and (b) Backscatter electron images of isolated parisite crystals in vugs in rock matrix from two samples from Langesundfjord, Norway. Both BEIs were obtained using a Robinson scintillation backscatter electron detector attached to an ISI-ABT55 scanning electron microscope adapted for high-pressure operation. The SEM was operated at 20 kV and the specimen current was optimized for imaging. (Samples kindly provided by R. Bevins, National Museum of Wales, UK.)

Examples of bulk instrumental techniques are atomic absorption spectroscopy (AAS), ICPES, ICPMS, X-ray fluorescence (XRF), instrumental neutron activation analysis (INAA), ion chromatography (IC) and isotope dilution mass spectrometry (IDMS). For a full chemical analysis of the minerals, additional techniques may be required to measure concentrations of elements such as fluorine, carbon and hydrogen. Commercial laboratories involved in assaying RE mineral concentrates in the mining and extraction industries usually employ a combination of AAS and XRF.

Below are very brief outlines of some of the more important techniques used in the analysis of 'bulk' specimens. Details on theory, instrumentation, etc. can be found in any recent textbook of the subject (e.g. Potts, 1987 comprehensively covers these techniques). Where possible, I have cited references which relate specifically to analysis of RE minerals or to REE in geological samples.

13.4.1 Instrumental neutron activation analysis (INAA)

In neutron activation analysis, samples are irradiated in a thermal neutron flux to create unstable isotopes which decay at known rates according to their half-lives. The decay products include gamma-rays, the energy of which is characteristic of a particular isotope. Instrumental neutron activation analysis measures simultaneously the intensities of peaks from the gamma-ray spectrum of the samples, without radiochemical separation, by means of solid state intrinsic Ge or lithium-drifted Ge detectors. Several analytical schemes have been published for analysis of rocks and minerals

ANALYSIS OF RARE EARTH MINERALS

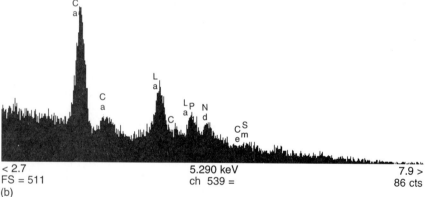

Figure 13.3 Energy-dispersive spectra from the unprepared (and uncoated) samples of those crystals in Figure 13.2. (a), Parisite with Ce and Nd-dominant (from Figure 13.2a); (b), parisite with La-dominant (from Figure 13.2b; note that Ce Lα peaks are absent from this spectrum. Spectra were obtained using a LINK AN10 000 energy-dispersive system attached to a Hitachi S2500 SEM operated at 15 kV and 1 nA specimen current.

which include details for the REE (e.g. Henderson and Pankhurst, 1984; Williams and Wall, 1991). Not all the REE have isotopes which are favourable for analysis by INAA (e.g. Pr, Dy and Er cannot be determined using systems situated away from the reactor site because of the very short half-lives of their unstable isotopes), and detection limits vary for different

BULK ANALYTICAL TECHNIQUES

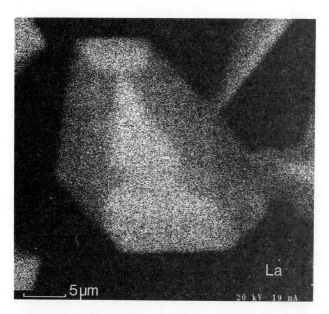

Figure 13.4 La X-ray distribution map revealing sector zoning in monazite-(Ce) from Kangankunde, Malawi (see Wall and Mariano, Chapter 8; Wall and Cressey, in preparation). Elemental map obtained using a Cameca SX50 electron microprobe operated at 20 kV and 20 nA current measured on a Faraday cup. Image processing performed using VISILOG software.

REE (e.g. La typically 0.5 ppm, Eu 0.02 ppm, Gd 2 ppm, Lu 0.05 ppm). For lower detection limits, separation procedures need to be undertaken after irradiation, this technique being termed radiochemical neutron activation analysis (RNAA), for which several analytical schemes have been developed for the REE (see Potts, 1987).

Advantages of INAA are that separated mineral grains need not be comminuted to a powder prior to analysis and that sample dissolution is not required, i.e. the technique is non-destructive. After analysis the mineral grains therefore may be subjected to further chemical or structural analysis necessary for complete characterization, or be used as standards in other microanalytical techniques such as EPMA or ion probe.

13.4.2 Inductively coupled plasma emission spectroscopy (ICPES)

In ICPES a solution, usually of 3–5% total dissolved solids, is introduced as an aerosol into a high-temperature plasma and heated to temperatures up to 8000 K. Characteristic atomic and ionic emission lines are detected and measured using a photomultiplier. Because of the large number of emission lines generated by many elements, a procedure to separate the REE from potentially interfering elements is generally required. Additional problems arise in that mutual spectral interferences of some REE may occur. There are many published protocols and procedures for analysis of REE in geological samples including Cantagrel and Pin (1994) and Watkins and Nolan (1992), and those cited in Potts (1987). All the REE can be analysed quantitatively, with detection limits typically 0.1 to 5 ppm, from 500 mg of sample.

13.4.3 Inductively coupled plasma mass spectroscopy (ICPMS)

In ICPMS, ions generated from the solution introduced into a plasma are separated in a quadrupole mass spectrometer on the basis of their mass to charge ratios. Unlike ICPES, preconcentration or separation of the REE is not required for analysis by ICPMS. Additionally, ICPMS has a much better sensitivity and fewer interferences than ICPES, although the total dissolved solids burden of the sample solution must be restricted to $\approx 0.2\%$. In the analysis of RE minerals, the extremely good sensitivity, and the fact that the detectors cannot measure accurately high concentrations, means that large dilution factors of 10^4 to 10^5 are necessary, but all the REE can be quantified over a large concentration range. Jarvis (1989) gives examples of analyses of bastnäsite, monazite and rhabdophane where concentrations range from 120 000 ppm La to 3 ppm Lu. Rivoldini and Fadda (1994) describe a procedure based on the use of potassium-based fluxes to ensure that refractory RE-bearing phases are completely dissolved.

13.4.4 Isotope dilution mass spectrometry (IDMS)

IDMS is generally regarded as the most precise and sensitive technique for many elements, including the REE. This technique relies on the fact that many elements have more than one naturally occurring isotope. In outline, the procedure involves addition of an aliquot of an element enriched in one these isotopes to a solution of the sample, followed by chemical separation and subsequent measurement of the isotope ratio by mass spectrometry.

Thirwell (1982), Henderson and Pankhurst (1984) and Potts (1987) detail several procedures for the determination of very low concentrations of REE in bulk rock samples.

13.4.5 X-ray fluorescence (XRF)

In XRF powdered bulk samples are compressed into a pellet, or fused with a flux into a glass bead which is then irradiated using monochromatic X-rays. The resulting X-rays emitted from the sample have energies characteristic of the elements in the sample, the intensities being proportional to their concentration, which can be detected either sequentially using wavelength-dispersive spectrometry, or simultaneously by energy-dispersive spectrometry. Matrix correction procedures are necessary for quantitative analysis. The large amount of material required makes this technique unsuitable for analysis of RE minerals unless a mineral concentrate is available.

XRF is a mature technique and there is a large amount of literature related to instrumentation and applications. Van Grieken and Markowicz (1993) provide a recent review of the different aspects of XRF and related techniques.

13.4.6 *Ion chromatography (IC)*

IC is potentially a very sensitive solution-based technique for the analysis of RE minerals since it is capable of separating parts-per-million concentrations of the REE in $\approx 100\,\mu L$ of sample. Modern equipment can be fitted with electrochemical and/or spectrophotometric detectors, the latter being used for REE quantification after reaction of the components with a chromophore. For rock analysis REE preconcentration combined with matrix elimination using chelating ion exchange resins is necessary before separation on an ion exchange column. There are no publications which refer specifically to RE minerals, but a procedure for the determination of REE in rock samples is given in a Dionex Technical Note (Anon., 1992).

13.5 Microanalytical techniques

Microanalytical techniques can provide quantitative analysis of RE minerals on a micron-sized scale without the need for mineral separation. Although detection limits generally are not as favourable as for bulk techniques, the ability to analyse micron-sized crystals, or characterize compositional zoning within individual crystals, renders microanalytical techniques as the preferred analysis option. The techniques are often non-destructive, and chemical data can readily be combined with textural information obtained from optical or backscatter electron images.

Of the techniques available, electron probe microanalysis (EPMA) is the most well-established, and a procedure for the analysis of RE minerals is described in detail below. Other microanalytical techniques employed in the analysis of RE minerals outlined below are secondary-ion mass spectrometry (SIMS) or 'ion probe', proton-induced X-ray emission spectroscopy (PIXE), synchrotron X-ray fluorescence, analytical transmission electron microscope (ATEM) and laser ablation microprobe inductively coupled plasma mass spectrometry (LAMP-ICPMS). For recent reviews on these and other microanalytical techniques see relevant chapters in Potts *et al.* (1995).

13.5.1 Electron probe microanalysis (EPMA)

Although EPMA is a well-established technique for the microanalysis of many minerals (e.g. Reed, 1993), quantitative analyses of the REE in minerals is not routine and poses specific analytical problems. Several published procedures refer in general to analysis of REE in minerals (e.g. Åmli and Griffin, 1975; Exley, 1980; Williams and Gieré, 1988; Dawson, Smith and Steele, 1994), and specifically to the overlap and interference problems (e.g. Roeder, 1985; Roeder et al., 1987; Potts, 1987). The preferred method is to use a wavelength-dispersive (WD) rather than energy-dispersive (ED) microprobe, because of the interference problems outlined below and because detection limits by WD are one to two orders of magnitude better than by ED (e.g. Potts, 1987). Smith and Reed (1982) consider some of the additional problems of energy-dispersive analysis of RE minerals. However, WD microprobes are fewer in number and not always generally available, and with careful standardization ED analysis can be used successfully in analysis of RE minerals (e.g. Platt, 1994), and on occasions may be a better choice than WD analysis with beam-sensitive (or hydrated) minerals (e.g. Nagy, 1993; see also Maksimović and Pantó, Chapter 10), or when analysing sub-micron intergrowths or inclusions (e.g. Wall and Mariano, Chapter 8).

Paramount to the accurate microprobe analysis of RE minerals is the choice of standard. Drake and Weill (1972) initially produced a series of four glasses, each containing four of the REE selected so as not to have mutually interfering X-ray lines, and which were used in many of the earlier published microprobe analyses. However, these glasses are no longer available. Roeder (1985) and Roeder et al. (1987) manufactured glasses with different concentrations of REE (see Taylor and Pollard, Chapter 7) which were used primarily to test overlap correction procedures and to assess detection limits. Jarosewich and Boatner (1991) reported the existence of a series of synthetic RE phosphates for use as potential reference standards for the electron microprobe. Other laboratories manufacture REE glasses for microanalysis, some of which can be purchased (e.g. P&H Developments, Derbyshire, England), and there is also available commercially a selection of synthetic compounds, such as light REE fluorides, cerium oxides and pure metals for the mid and heavy REE. In the procedure described below, single REE–Ca–Al silicate glasses prepared by Peter Hill (University of Edinburgh) are the primary standards used, where concentrations of each REE vary from 11 to 22 wt%. Some secondary standards are additionally used to provide an internal assessment of the quality of the data, these being well-characterized RE minerals.

Two major problems in EPMA analysis of RE minerals arise from the presence of a large number of peaks from excitation of L lines of the REE present in the sample: (1) the presence of interfering peaks, both from the

REE themselves, and from other commonly occurring elements such as Ti, Mn, Fe and Ba, and (2) the choice of background positions in a crowded region of the energy spectrum. Examples of spectra for light, mid and heavy RE mineral phosphates (Figure 13.5) illustrate the extent of these problems.

An analytical procedure has been developed based on a Cameca SX50 four spectrometer wavelength-dispersive microprobe located at the Natural

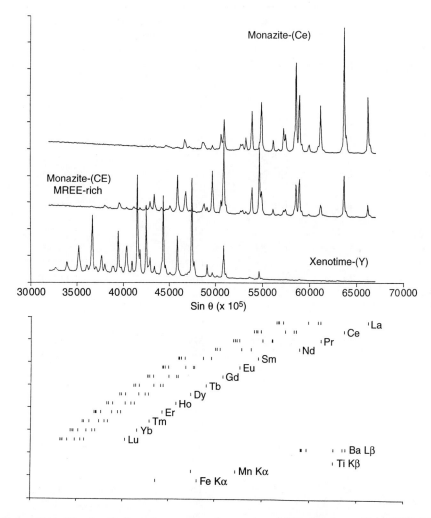

Figure 13.5 Wavelength-dispersive spectra of examples of RE phosphates (light REE-rich monazite-(Ce); mid REE-rich monazite-(Ce); heavy REE-rich xenotime-(Y) using a LIF 200 dispersing crystal. Microprobe used was a Cameca SX50; operating conditions were 20 kV accelerating voltage, 20 nA beam current measured on a Faraday cup.

History Museum, London. Table 13.1 details the elements, lines and crystals used in the procedure, together with empirically determined values for the mutually interfering REE, and values for 16 other elements often associated with RE-bearing minerals. The interference values given in Table 13.1

Table 13.1 Elements analysed and interferences observed in microprobe analysis of rare earth minerals

Element	Line	Crystal	Interfering element	% Interference
La	Lα	PET 002	No interferences observed	
Ce	Lα	PET 002	Ba	29.6
			Ti	0.10
			Sr	0.22
			Ta	0.18
Pr	Lα	LIF 200	La	8.9
Pr	Lβ	LIF 200	No interferences observed	
Nd	Lα	LIF 200	Ce	0.40
Sm	Lα	LIF 200	Ce	1.65
			Ta	0.11
Eu	Lα	LIF 200	Pr	14.9
			Nd	1.27
			Mn	0.51
Gd	Lα	LIF 200	Ce	6.95
			La	0.15
Gd	Lβ	LIF 200	Ho	145
Tb	Lα	LIF 200	Sm	0.75
			Fe	0.25
			Ce	0.08
Dy	Lα	LIF 200	Mn	31.4
			Th	1.34
			Eu	0.81
			Fe	0.30
Ho	Lβ	LIF 200	Dy	0.46
Er	Lα	LIF 200	Tb	7.00
			Fe	0.35
Tm	Lα	LIF 200	Sm	8.69
			Ta	2.48
			Dy	1.09
Yb	Lα	LIF 200	Tb	0.44
			Dy	0.42
			Sm	0.28
Lu	Lα	LIF 200	Dy	4.48

Elements measured for interferences: Y, REE as Ca, Al silicate glasses; Mn, Fe, Zr, Nb, Ta as pure metals; Na, Mg, Al, Si, P, Sr, Ba, Th, U as synthetic oxides or compounds.

should be used only as a guideline, as absolute values will depend primarily on the resolution of individual analysing crystals, and will differ for different microprobes.

The choice of background positions used is critical for accurate analysis, particularly at low concentration levels, and Table 13.2 details positions which for most RE minerals are interference-free. It is always good practice, however, to check whether any unsuspected lower-order lines interfere with the selected background positions when analysing minerals of unusual major element compositions.

Practical (achievable) detection limits for the REE depend on several factors: the spectral interferences and overlap corrections necessary; the operating conditions, such as accelerating voltage and specimen current used; the counting times for peak and background positions, and the ability of the mineral itself to stand up to high probe currents without specimen damage occurring.

In Table 13.3 detection limits for the REE are presented for some RE phosphate minerals based on operating conditions of 25 kV accelerating voltage and 100 nA current measured on a Faraday cup, and with counting times of 50 s on both peak and background. These values have been empirically determined by calculating the background values at the peak positions from interpolation of the overall background spectra. Under these operating conditions, detection limits as low as 0.02 wt% are achievable for some elements, although more typically values range from 0.03 to 0.13 wt%. Where large interference corrections are necessary, for example in the measurement of Eu and Gd using $L\alpha$ lines in light RE minerals, detection limits (and standard errors) are much larger and will depend on the concentration of the interfering element. Thus, the values given in Table 13.3 are estimates only, and are examples of what can be achieved under favourable operating conditions.

In practice RE-bearing oxides (e.g. perovskite, aeschynite and zirco-

Table 13.2 Background positions used for the REE in microprobe analysis of rare earth minerals

Wavelength (λ)	Crystal	$\sin\theta$ ($\times 10^5$)	θ
2.68931	PET 002	30 735	17.90
2.60531	PET 002	29 775	17.32
2.70473	LIF 200	67 170	42.20
2.60728	LIF 200	64 750	40.35
2.24086	LIF 200	55 650	33.81
2.08180	LIF 200	51 700	31.13
1.82812	LIF 200	45 400	27.00
1.55028	LIF 200	38 500	22.64

Table 13.3 Typical detection limits of REE (wt% oxide) calculated for some rare earth phosphates

Element	Line	Monazite-(Ce)	Xenotime-(Y)	Apatite
Y	$L\alpha$	0.046	0.058	0.038
La	$L\alpha$	0.036	0.030	0.020
Ce	$L\alpha$	0.034	0.028	0.018
Pr	$L\alpha$	0.052	0.048	0.032
Pr	$L\beta$	0.126	0.114	0.074
Nd	$L\alpha$	0.046	0.048	0.032
Sm	$L\alpha$	0.041	0.040	0.026
Eu	$L\alpha$	0.040	0.040	0.026
Gd	$L\alpha$	0.040	0.040	0.026
Gd	$L\beta$	0.070	0.070	0.050
Tb	$L\alpha$	0.040	0.040	0.026
Dy	$L\alpha$	0.038	0.042	0.026
Ho	$L\beta$	0.072	0.080	0.050
Er	$L\alpha$	0.036	0.044	0.026
Tm	$L\alpha$	0.036	0.044	0.026
Yb	$L\alpha$	0.038	0.046	0.028
Lu	$L\alpha$	0.038	0.046	0.028

Detection limits are the 6σ 'limit of determination' (Potts, 1987) and are based on 50 s count time on peak and backgrounds; gun operating potential of 25 kV, and 100 nA probe current. Y, La and Ce for PET 002 crystal, other REE for LIF 200 crystal.

nolite), silicates (e.g. chevkinite and allanite) and phosphates (e.g. monazite, xenotime and RE-rich fluorapatites) show least specimen damage and can withstand probe conditions of 25 kV and 100 nA, and detection limits similar to those in Table 13.3 can be achieved. In contrast, RE carbonates (e.g. lanthanite and burbankite), fluocarbonates (e.g. bastnäsite and parisite) and their hydrated analogues become unstable under high probe currents. Consequently, analysis of these minerals usually requires lower kV values and a defocused electron beam for which typical operating conditions would be 15 kV, 10–20 nA and a spot size of 5–25 μm. Detection limits then could be two to five times worse than those given in Table 13.3. Alternatively, these minerals could be analysed using ED analysis at much lower specimen current settings, but with lower accuracy, precision and poorer detection limits (10 to 20 times worse than those for WD).

Analyses of several RE-bearing minerals are presented in Table 13.4; the data were obtained using the WD procedure outlined above, and matrix corrections were made using a PAP ϕpz model. Chondrite-normalized REE plots are illustrated in Figure 13.6. With four wavelength spectrometers on the Cameca SX50 at the Natural History Museum, London, and with the operating conditions outlined above, a typical analysis for 25–30 elements takes 10–15 min. The SX50 has a stage reproducibility to within $\pm 1.5\,\mu$m,

Table 13.4 Rare earth minerals analysed using procedure described in the text (wt%)

	1	2	3	4	5
Na_2O	<0.05	<0.05	<0.05	<0.05	1.47
MgO	<0.05	<0.05	<0.05	<0.05	0.30
Al_2O_3	<0.05	<0.05	<0.05	<0.05	0.50
SiO_2	0.12	1.07	0.28	0.19	<0.05
P_2O_5	28.74	28.25	29.10	39.38	<0.05
CaO	0.16	3.93	<0.05	55.55	34.20
TiO_2	0.05	<0.05	<0.05	<0.05	33.77
MnO	<0.05	<0.05	<0.05	<0.05	0.07
FeO	<0.05	<0.05	<0.05	<0.05	5.45
SrO	1.00	<0.05	<0.05	<0.05	0.22
Y_2O_3	0.08	1.25	25.62	<0.038	<0.038
ZrO_2	<0.05	<0.05	<0.05	<0.05	0.88
Nb_2O_5	<0.05	<0.05	<0.05	<0.05	17.88
BaO	0.06	<0.05	<0.05	0.21	0.47
La_2O_3	20.39	3.94	<0.03	0.15	0.78
Ce_2O_3	33.48	12.53	<0.028	0.22	1.79
Pr_2O_3	3.21	1.54	<0.048	0.03	0.20
Nd_2O_3	7.68	7.64	0.28	0.07	0.58
Sm_2O_3	0.78	10.30	0.77	<0.026	0.04
Eu_2O_3	0.15	0.07	0.07	<0.026	<0.026
Gd_2O_3	0.32	8.78	4.76	<0.05	<0.05
Tb_2O_3	<0.040	0.71	1.38	<0.026	<0.026
Dy_2O_3	<0.038	0.83	11.74	<0.026	<0.026
Ho_2O_3	<0.072	<0.072	2.53	<0.050	<0.050
Er_2O_3	<0.036	<0.036	8.47	<0.026	<0.026
Tm_2O_3	<0.036	<0.036	1.51	<0.026	<0.026
Yb_2O_3	<0.038	<0.038	11.23	<0.028	<0.028
Lu_2O_3	<0.038	<0.038	1.75	<0.028	<0.028
Ta_2O_5	<0.1	<0.1	<0.1	<0.1	0.80
ThO_2	0.09	17.88	0.94	<0.05	<0.05
UO_2	<0.05	1.05	0.56	<0.05	0.05
F	<0.1	<0.1	<0.1	3.57	<0.1
Total	96.30	99.77	101.00	99.37	99.45
$-O = F$	–	–	–	1.50	–
Total	96.30	99.77	101.00	97.87	99.45

Analyses: 1, monazite-(Ce) from Kakangkunde carbonatite, Malawi; 2, monazite-(Ce) from Mozambique; 3, xenotime-(Y) BM1995,100; 4, apatite from Adamello, Italy; 5, perovskite from Magnet Cove.

automatic focusing of the specimen, and an extremely stable probe current; thus unattended overnight operation can routinely produce >50 analyses from up to six polished thin sections. The large amount of information acquired, which necessitates the use of modern computer spreadsheets to collate and manipulate the data, could not have been possible even 5

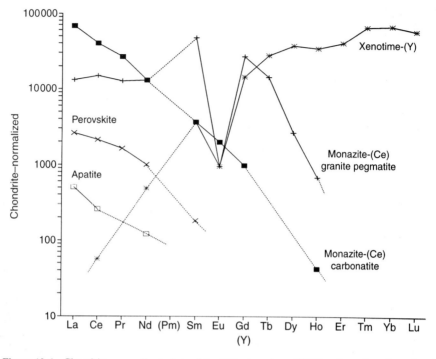

Figure 13.6 Chondrite-normalized plots of the RE minerals and RE-bearing minerals analysed by electron microprobe in the procedure described in the text, and given as examples in Figure 13.5 and Table 13.4. Xenotime-(Y) BM1995,100; monazite-(Ce) from granite-pegmatite at Muiane, Mozambique; monazite-(Ce) from carbonatite at Kangankunde, Malawi; perovskite from Magnet Cove, USA; apatite from Adamello, Italy.

years ago. With the extent of information now possible on a micron scale, paragenetic models will become more detailed, and crystal chemical trends within RE mineral species can be further refined.

13.5.2 Analytical transmission electron microscopy (ATEM)

As with EPMA, the transmission electron microscope uses an electron beam as the main source of excitation, the beam spot size however being much smaller, typically $0.02-0.05\,\mu m$, compared with the 0.2 to $2\,\mu m$ spot of electron microprobes and SEMs. The TEM provides primarily structural information from ultra-thin sections of minerals over a very small sample area, but with an energy-dispersive detector attached to the TEM quantitative analysis is possible (Champness, Cliff and Lorimer, 1982; Lorimer, 1987). Problems arise in quantitative analyses, however, from unavailability of suitable homogeneous standards (on the sub-micron scale), and appropriate data correction procedures for thin targets (Potts, 1987). Sample preparation using ion-thinning procedures is time consuming and unpredictable as obtaining suitable thinned sections of small mineral grains located within rock sections can prove difficult. However, a recent study on some RE-bearing minerals previously analysed by EPMA has shown a good comparison between analyses obtained by EPMA and those by ATEM (Lumpkin et al., 1994).

13.5.3 Proton-induced X-ray emission spectroscopy (PIXE)

Instead of an electron beam generating characterisitic X-ray lines, as in EPMA, a beam of high-intensity protons, accelerated to 2–3 MeV in a Van de Graaff generator and focused down to $\approx 1\,\mu m$, is used to induce X-rays from within the sample. The emitted X-ray spectrum is then detected and peak intensities are measured in a similar way to EPMA. The high primary energy of the proton beam (typically 3 MeV) potentially enables the measurement of the REE K X-ray lines (compared to their L lines in EPMA) with suitable solid state Ge detectors (as in SXRF, below), thus eliminating many of the spectral interferences outlined above for EPMA. However, very few PIXE instruments are equipped with high-energy detectors, and generally the spectra are measured using lithium-drifted Si detectors similar to those by ED analysis in EPMA.

A potential advantage of PIXE over EPMA is an improvement in sensitivity, particularly for Y and the heavy REE, where the detection limit for Y can be as low as a few ppm, although a relatively long analysis time (typically tens of minutes) is required to achieve such low detection limits. A further disadvantage of the technique is the relatively deep penetration of the proton beam below the sample surface, typically tens of microns, compared with $3-5\,\mu m$ with EPMA, giving a relatively poor spatial resolution. Thus X-rays can be emitted from inclusions at depth within the mineral, and any fine-scale compositional zoning will not be resolved. However, good agreement for analyses of the REE can be achieved between PIXE, EPMA and INAA (Roeder et al., 1987).

Currently, the number of PIXE probes operating worldwide is less than 50, so that accessibility to this technique is limited.

13.5.4 Synchrotron X-ray fluorescence (SXRF)

With a synchrotron source, high-energy photons are generated which, when focused onto the surface of a sample, will produce characteristic X-rays in a manner similar to EPMA and PIXE. SXRF is analogous to the 'bulk' XRF technique outlined above but, with a $25\,\mu m$ diameter beam now achievable, it is can be described as a microanalytical technique. Chen et al. (1993) have analysed quantitatively the REE, using their K lines, in monazite and apatite from the Bayan Obo rare earth deposit, China (see Wu, Yuan and Bai, Chapter 11) and obtained a good correspondence between values obtained from SXRF, EPMA and INAA. Chen et al. (1993) made their measurements on doubly polished thin sections of accurately known thickness and, for an analysis time of 60 min, minimum detection limits of 6 ppm (for La) and 26 ppm (for Lu) were calculated.

SXRF suffers similar disadvantages as those for PIXE; indeed, accessibility to a synchrotron source is even more restricted than to a proton source, with probably only two or three instruments wouldwide capable of the microanalysis of RE minerals.

13.5.5 Secondary-ion mass spectrometry ion probe (SIMS)

In SIMS a beam of primary ions is generated in a duoplasmatron, the ions are accelerated by a high voltage (10–30 kV) and bombard the sample, usually a polished thin section. Atoms sputtered from the sample surface are ionized and detected using a mass spectrometer. For microanalysis the ion beam can be focused, typically down to 5 μm in diameter, when the technique is more usually known as ion microprobe analysis (Reed, 1989). Although the spatial resolution of the ion beam is not as good as that of an electron beam in EPMA, the depth resolution is superior as only the surface few atoms are sputtered off. Detection limits for the REE are two orders of magnitude better than for EPMA, which enables a full chondrite-normalized profile to be obtained. Reed (1984, 1985) and Mitchell and Reed (1988) measured several RE minerals, including allanite, apatite and perovskite, using the ion probe.

13.5.6 Laser ablation microprobe (LAMP-ICPMS)

Conventional analysis by ICPMS described above (section 13.4.3) involves mineral separation followed by dissolution of the sample. The use of a high-intensity laser to ablate a small volume of the solid sample and then to introduce the ablated material as a vapour into a conventional ICP mass spectrometer has enormous potential in the analysis of RE minerals. Recent publications related to the quantitative analysis of REE using this technique (e.g. Fedorowich et al., 1993; Jarvis and Williams, 1993) indicate that initial problems relating to the stability of the laser and to mass yield and fractionation effects may now be being resolved. However, Morrison et al. (1995) report that the light REE show some variation of elemental response to laser ablation and/or plasma decomposition of powdered samples, which may relate to different minerals having different responses.

Recent work by Chenery and Cook (1993) on zoned monazite grains has shown that ablation craters less than 5 μm in depth and diameter can be achieved, and that the data obtained are comparable with EPMA, but that detection limits for the REE are superior. Increasing the crater size to 25 μm in depth and diameter enables an improvement in detection potentially to sub-ppm concentration levels.

13.6 Other techniques

Other techniques such Fourier-transform infrared spectroscopy (FTIR) and cathodoluminescence (CL) can provide additional information on chemical bonding, elemental mapping and textural aspects relating to RE minerals.

FTIR spectra in favourable circumstances can be used as a diagnostic identification tool, and examples of spectra for a range of RE carbonates and fluocarbonates are given by Jones and Jackson (1994).

CL spectra, resulting from interaction of REE with an electron beam, can provide information on mineral overgrowths or the paragenesis of RE-hosted phases. For an outline of the theory and practical aspects see, for example, Marshall (1988), and for example of CL spectra of some RE minerals see Mariano (1989).

β-autoradiography has also been used to analyse and map the distribution of specific β-emitting REE isotopes, either from experimental studies using radioactive isotopes (e.g. Sm partitioning between mineral and melt; Mysen, 1978), or in natural systems by neutron activation-induced β-autoradiography (e.g. distribution of Eu + Tb + Tm in fossilized bone samples; Williams and Potts, 1989).

13.7 Concluding statement

Microanalytical techniques will continue primarily to be used in the analysis of RE minerals, with bulk techniques retained for trace analysis of the REE in whole-rock samples. In EPMA advances in instrument stability and reliability have enabled unattended overnight operation to become routine, and advances made in computational power have been essential in assisting the collation and interpretation of the large amounts of data generated. Future trends in EPMA will utilize this improved instrument stability and, with higher kV equipment, will enable this technique to enter the field of 'trace' microanalysis, where improvements in detection limits to 20–50 ppm are foreseen. Image analysis, coupled with modern backscatter detectors, have highlighted many new textural features of RE minerals.

Techniques such as PIXE, synchrotron XRF (and, to a lesser extent, ion probe), although providing significant improvements in detection limits for the REE, are relatively inaccessible, and are unlikely to be used for routine analysis of RE minerals. The relatively lower cost of LAMP-ICPMS should ensure that this technique will become increasingly available in the near future, and the technique will mature to become complimentary to EPMA. Other more specialized techniques will continue to play an important supporting role in the analysis of RE minerals and will feature in specific aspects of their chemistry and structure.

Echoing the opening sentences of this chapter, accurate chemical analysis of RE minerals will continue to be essential to our understanding of the minerals themselves. Analysis of RE minerals even today still poses challenges, both for 'mature' techniques such as EPMA and for the newer techniques now becoming available. However, recent developments in microanalysis suggest that future trends will bring about improvements in both instrumentation and analytical procedures, and that much new information will be forthcoming. Thus, we will continue to increase our knowledge of these exciting minerals, and be able to understand better their chemistry, origin and their ore deposits.

Acknowledgements

I am extremely grateful to the following for providing specimens and advice: Richard Bevins, Linda Campbell, Andrew Clark, Vic Din, Akin Fadipe, Alf Olav Larsen, Garth Platt and Frances Wall.

References

Åmli, R. and Griffin, W. L. (1975) Microprobe analysis of REE minerals using empirical correction factors. *Am. Mineral.*, **60**, 599–600.

Anonymous (1992) Determination of lanthanide metals in digested rock samples by chelation ion chromatography. *Dionex Technical Note* **27**, 12 pp.

Cantagrel, F. and Pin, C. (1994) Major, minor and rare earth element determinations in 25 rock standards by ICP–atomic emission. *Geostand. Newsl.*, **18**, 123–38.

Champness, P. E., Cliff, G. and Lorimer, G. W. (1982) Quantitative analytical electron microscopy of metals and minerals. *J. Microsc.*, **108**, 231–49.

Chen, J. R., Chao, E. C. T., Back, J. M. *et al.* (1993) Rare earth element concentrations in geological and synthetic samples using synchrotron X-ray fluorescence analysis. *Nucl. Instr. Meth. Phys. Res.*, **B75**, 576–81.

Chenery, S. and Cook, J. (1993) Determination of rare earth elements in single mineral grains by laser ablation microprobe–inductively coupled plasma mass spectrometry–prelimary study. *J. Anal. Atom. Spectr.*, **8**, 299–303.

Drake, M. J. and Weill, D. F. (1972) New rare earth element standards for electron microprobe analysis. *Chem. Geol.* **10**, 179–81.

Dawson, J. B., Smith, J. V. and Steele, I. M. (1994) Trace element distribution between coexisting perovskite, apatite and titanite from Oldoinya Lengai, Tanzania. *Chem. Geol.*, **117**, 285–90.

Exley, R. A. (1980) Microprobe studies of REE-rich accessory minerals: implications for Skye granite petrogenesis and REE mobility in hydrothermal systems. *Earth Planet. Sci. Lett.*, **48**, 97–110.

Fedorowich, J. A., Richards, J. P., Jain, J. C. *et al.* (1993) A rapid method for REE and trace-

REFERENCES

element analysis using laser sampling ICP-MS on direct fusion whole-rock glasses. *Chem. Geol.*, **106**, 229–249.

Henderson, P. and Pankhurst, R. J. (1984) Analytical chemistry, in *Rare Earth Element Geochemistry* (ed. P. Henderson), Elsevier, pp. 467–99.

Jarosewich, E. and Boatner, L. A. (1991) Rare earth element reference samples for electron microprobe analysis. *Geostand. Newsl.*, **15**, 397–9.

Jarvis, K. E. (1989) Determination of rare earth elements in geological samples by inductively coupled plasma mass spectrometry. *J. Anal. Atom. Spectr.*, **4**, 563–70.

Jarvis, K. E. and Williams, J. G. (1993) Laser ablation inductively coupled plasma mass spectrometry (LA-ICP-MS): a rapid technique for the direct, quantitative determination of major, trace and rare-earth elements in geological samples. *Chem. Geol.*, **106**, 251–62.

Jones, G. C. and Jackson, B. (1994) *Infrared Transmission Spectra of Carbonate Minerals*, Chapman & Hall, London, 256 pp.

Lorimer, G. W. (1987) Quantitative X-ray microanalysis of thin specimens in the transmission electron microscope; a review. *Mineral. Mag.*, **51**, 49–60.

Lumpkin, G. R., Smith K. L., Blackford, M. G. *et al.* (1994) Determination of 25 elements in the complex oxide mineral zirconolite by analytical electron microscopy. *Micron*, **25**, 581–87.

Mariano, A. N. (1989) Cathodoluminescence emission spectra of rare earth element activators in minerals, in *Geochemistry and Mineralogy of Rare Earth Elements* (eds B. R. Lipin and G. A. McKay), Mineralogical Society of America, Reviews in Mineralogy, **21**, 339–348.

Marshall, D. J. (1988) *Cathodoluminescence of Geological Materials*, Unwin Hyman, London, 146 pp.

Mitchell, R. H. and Reed S. J. B. (1988) Ion microprobe determination of rare earth elements in perovskites from kimberlites and alnoites. *Mineral. Mag.*, **52**, 331–9.

Morrison, C. A., Lambert, D. D., Morrison, R. J. S., Ahlers, W. W. and Nicholls, I. A. (1995) Laser ablation–inductively coupled plasma–mass spectrometry: an investigation of elemental responses and matrix effects in the analysis of geostandard materials. *Chem. Geol.*, **119**, 13–29.

Mysen, B. O. (1978) Experimental determination of rare earth element partitioning between hydrous silicate melt, amphibole and garnet peridotite minerals at upper mantle pressures and temperatures. *Geochim. Cosmochim. Acta*, **42**, 1253–63.

Nagy, G. (1993) 'Quick' method for REE mineral analysis by EPMA. *Rare Earth Minerals: Chemistry, Origin and Ore Deposits*, Mineralogical Society Spring Meeting, Natural History Museum, London, UK. Conference Extended Abstracts, pp. 94–6.

Paterson, B. A., Stephens, W. E. and Herd, D. A. (1989) Zoning in granitoid accessory minerals as revealed by backscattered electron imagery. *Mineral. Mag.*, **53**, 55–61.

Platt, R. G. (1994) Perovskite, loparite and Ba–Fe hollandite from the Schryburt Lake carbonatite complex, nothwestern Ontario, Canada. *Mineral. Mag.*, **58**, 49–57.

Platt, R. G., Wall, F., Williams C. T. and Woolley A. R. (1987) Zirconolite, chevkinite and other rare earth minerals from the nepheline syenites and peralkaline granites and syenites of the Chilwa Alkaline Province, Malawi. *Mineral. Mag.*, **51**, 253–63.

Potts, P. J. (1987) *A Handbook of Silicate Analysis*, Blackie, Glasgow, 622 pp.

Potts, P. J., Bowles, J. F. W., Reed, S. J. B and Cave, M. R. (eds) (1995) *Microprobe Techniques in the Earth Sciences*, Chapman & Hall, 432 pp.

Reed, S. J. B. (1984) Secondary-ion mass spectrometry ion probe analysis for rare earths. *Scanning Electron Microscopy*, **II**, 592–35.

Reed, S. J. B. (1985) Ion-probe determination of rare earths in allanite. *Chem. Geol.*, **48**, 137–43.

Reed, S. J. B. (1989) Ion microprobe analysis – a review of geological applications. *Mineral. Mag.*, **53**, 3–24.

Reed, S. J. B. (1993) *Electron Microprobe Analysis* (2nd edn), Cambridge University Press, 326 pp.

Rivoldini, A. and Fadda, S. (1994) Inductively coupled plasma mass spectrometric determination of low-level rare earth elements in rocks using potassium-based fluxes for sample decomposition. *J. Anal Atom. Spectr.*, **9**, 519–24.

Roeder, P. L. (1985) Electron-microprobe analysis of minerals for rare-earth elements: use of calculated peak-overlap corrections. *Can. Mineral.*, **23**, 263–71.

Roeder, P. L., MacArthur, D., Ma, X.-P. *et al.* (1987) Cathodoluminescence and microprobe study of rare-earth elements in apatite. *Am. Mineral.*, **72**, 801–11.

Smith, D. G. W. and Reed, S. J. B. (1982) Rare earth element determinations by energy-dispersive electron microprobe techniques, in *Proceedings of the Institution of Physics Conference Series 61: Electron Microscopy and Analysis Group Conference 1981*, (ed. M. J. Gorinje), University of Cambridge, pp. 159–62.

Taylor, P. D. and Jones, C. G. (in press) Use of the environmental chamber in uncoated SEM of recent and fossil bryozoans. *Microscopy and Analysis*.

Thirwell, M. F. (1982) A triple filament method for rapid and precise analysis of rare earth elements by isotope dilution. *Chem. Geol.*, **35**, 155–66.

Van Grieken, R. E. and Markowicz, A. A. (1993) *Handbook of X-Ray Spectrometry*, Practical Spectroscopic Series, **14**, Marcel Dekker, New York.

Watkins, P. J. and Nolan, J. (1992) Determination of rare earth elements, yttrium, scandium and hafnium using cation-exchange separation and inductively coupled plasma-atomic emission spectrometry. *Chem. Geol.*, **95**, 131–92.

Williams, C. T. and Gieré, R. (1988) Metasomatic zonation of REE in zirconolite from a marble skarn at the Bergell contact aureole (Switzerland/Italy). *Schweiz. Mineral. Petrogr. Mitt.*, **68**, 133–40.

Williams, C. T. and Potts, P. J. (1989) Element distribution maps in fossil bones. *Archaeometry*, **30**, 237–47.

Williams, C. T. and Wall, F. (1991) An INAA scheme for the routine determination of 27 elements in geological and archaeological samples, in *Neutron Activation and Plasma Emission Spectrometric Analysis in Archaeology* (eds M. J. Hughes, M. R. Cowell and D. R. Hook), British Museum Occasional Paper, **82**, 105–19.

APPENDIX A
Glossary of rare earth minerals

A.1 Borates

Braitschite-(Ce)	$(Ca,Na_2)_7(Ce,La)_2B_{22}O_{43} \cdot 7H_2O$
Moydite-(Y)	$(Y,REE)[B(OH)]CO_3$
Peprossite-(Ce)	$(Ce,La)Al_2B_3O_9$

A.2 Carbonates, fluocarbonates and hydroxylcarbonates

*Ambatoarinite	$Sr_5(La,Ce)(CO_3)_{17}O_3$ (?)
Ancylite-(Ce)	$SrCe(CO_3)_2OH \cdot H_2O$
Baiyuneboite-(Ce)	$NaBaCe_2(CO_3)_4F$
Bastnäsite-(Ce)	$(Ce,La)(CO_3)F$
Bastnäsite-(La)	$(La,Ce)(CO_3)F$
Bastnäsite-(Y)	$(Y,Ce)(CO_3)F$
Burbankite	$(Na,Ca)_3(Sr,Ba,Ce)_3(CO_3)_5$
Calcio-ancylite-(Ce)	$(Ca,Sr)Ce(CO_3)_2OH \cdot H_2O$
Calcio-ancylite-(Nd)	$(Nd,Ce)_3Ca(CO_3)_4(OH)_3 \cdot H_2O$
Calkinsite-(Ce)	$(Ce,La)_2(CO_3)_3 \cdot 4H_2O$
Carbocernaite	$(Ca,Na)(Sr,Ce,Ba)(CO_3)_2$
Cebaite-(Ce)	$Ba_3Ce_2(CO_3)_5F_2$
Cordylite-(Ce)	$Ba(Ce,La)_2(CO_3)_3F_2$
Daqingshanite-(Ce)	$(Sr,Ca,Ba)_3(Ce,La)PO_4(CO_3)_{3-x}(OH,F)$ $x \approx 0.8$
Donnayite-(Y)	$Sr_3NaCeY(CO_3)_6 \cdot 3H_2O$
*Eisenkalkancylite	Carbonate of REE, Fe and Ca
Ewaldite	$Ba(Ca,Y,Na,K)(CO_3)_2$
Gysinite-(Nd)	$Pb(Nd,La)(CO_3)_2(OH) \cdot H_2O$
Huanghoite-(Ce)	$BaCe(CO_3)_2F$
Hydroxylbastnäsite-(Ce)	$(Ce,La)CO_3(OH,F)$
Hydroxylbastnäsite-(La)	$(La,Ce)CO_3(OH,F)$
Hydroxylbastnäsite-(Nd)	$(Nd,Ce,La)CO_3(OH,F)$
Khanneshite	$(Na,Ca)_3(Ba,Sr,Ce,Ca)_3(CO_3)_5$
Kimuraite-(Y)	$CaY_2(CO_3)_4 \cdot H_2O$
Lanthanite-(Ce)	$(Ce,La)_2(CO_3)_3 \cdot 8H_2O$
Lanthanite-(La)	$(La,Ce)_2(CO_3)_3 \cdot 8H_2O$
Lanthanite-(Nd)	$(Nd,Ce,La)_2(CO_3)_3 \cdot 8H_2O$

GLOSSARY OF RARE EARTH MINERALS

Lokkaite-(Y)	$CaY_4(CO_3)_7 \cdot 9H_2O$
Mckelveyite-(Y)	$Ba_3Na(Ca,U)Y(CO_3)_6 \cdot 3H_2O$
Parisite-(Ce)	$Ca(Ce,La)_2(CO_3)_3F_2$
Parisite-(Nd)	$Ca(Nd,Ce,La)_2(CO_3)_3F_2$
Remondite-(Ce)	$Na_3(Ce,La,Ca,Na,Sr)_3(CO_3)_5$
Röntgenite-(Ce)	$Ca_2(Ce,La)_3(CO_3)_5F_3$
Sahamalite-(Ce)	$(Mg,Fe^{2+})(Ce,La,Nd)_2(CO_3)_4$
Schuilingite-(Nd)	$PbCu(Nd,Gd,Sm,Y)(CO_3)_3OH \cdot 1.5H_2O$
†Shomiokite-(Y)	$Na_3Y(CO_3)_3 \cdot 3H_2O$
Synchysite-(Ce)	$Ca(Ce,La)(CO_3)_2F$
Synchysite-(Nd)	$Ca(Nd,Y,Gd)(CO_3)_2F$
Synchysite-(Y)	$Ca(Y,Ce)(CO_3)_2F$
Tengerite-(Y)	$CaY_3(CO_3)_4(OH)_3 \cdot 3H_2O$
Thorbastnäsite	$Th(Ca,Ce)(CO_3)_2F_2 \cdot 3H_2O$
Zhonghuacerite-(Ce)	$Ba_2Ce(CO_3)_3F$

A.3 Oxides

Aeschynite-(Ce)	$(Ce,Ca,Fe,Th)(Ti,Nb)_2(O,OH)_6$
Aeschynite-(Nd)	$(Nd,Ce,Ca,Th)(Ti,Nb)_2(O,OH)_6$
Aeschynite-(Y)	$(Y,Ca,Fe,Th)(Ti,Nb)_2(O,OH)_6$
Brannerite	$(U,Ca,Y,Ce)(Ti,Fe)_2O_6$
Calciobetafite	$(Ca,REE,Th,U)_2(Nb,Ta,Ti)_2O_7$
*Calciosamarskite	Niobate of REE (Y), U, Fe, Ca and Th
*Cerhomilite	Borosilicate of Ca, Be, Fe, Th and REE
Cerianite-(Ce)	$(Ce^{4+},Th)O_2$
Ceriopyrochlore-(Ce)	$(Ce,Ca,Y)_2(Nb,Ta)_2O_6(OH,F)$
Cerotungstite-(Ce)	$CeW_2O_6(OH)_3$
Crichtonite	$(Sr,La,Ce,Y)(Ti,Fe^{3+},Mn)_{21}O_{36}$
*Cuprovudyavrite	Silicate of Cu,Ti and REE
Davidite-(Ce)	$(Ce,La)(Y,U,Fe^{2+})(Ti,Fe^{3+})_{20}(O,OH)_{38}$
Davidite-(La)	$(La,Ce)(Y,U,Fe^{2+})(Ti,Fe^{3+})_{20}(O,OH)_{38}$
*Dysanalyte	$(Ca,Ce,Na)(Ti,Nb,Ta)O_3$
Euxenite-(Y)	$(Y,Ca,Ce,U,Th)(Nb,Ta,Ti)_2O_6$
Fergusonite-beta-(Ce)	$(Ce,La,Nd)NbO_4$
Fergusonite-beta-(Nd)	$(Nd,Ce)NbO_4$
Fergusonite-beta-(Y)	$YNbO_4$
Fergusonite-(Ce)	$(Ce,La,Nd)NbO_4$
Fergusonite-(Nd)	$(Nd,Ce)NbO_4$
Fergusonite-(Y)	$YNbO_4$
Fersmite	$(Ca,Ce,Na)(Nb,Ta,Ti)_2(O,OH,F)_6$
Formanite-(Y)	$YTaO_4$
Hibonite	$(Ca,Ce)(Al,Fe,Ti,Si,Mg)_{12}O_{19}$

HALIDES

*Irinite	$(Na,Ce,Th)_{1-x}(Ti,Nb)(O,OH)_3$
Ishikawaite	near $(U,Fe,Y,Ce)(Nb,Ta)O_4$
*Knopite	$(Ca,Ti,Ce_2)O_3$
Kobeite-(Y)	$(Y,U)(Ti,Nb)_2(O,OH)_6$
*Lessingite	$Ca_2Ce_4Si_3O_{13}(OH)_2$ (?)
Loparite-(Ce)	$(Ce,La,Na,Ca,Sr)(Ti,Nb)O_3$
*Loranskite-(Y)	$(Y,Ce,Ca)ZrTaO_6$ (?)
Loveringite	$(Ca,Ce)(Ti,Fe^{3+},Cr,Mg)_{21}O_{38}$
Lucasite-(Ce)	$CeTi_2(O,OH)_6$
Murataite	$(Na,Y)_4(Zn,Fe)_3(Ti,Nb)_6O_{18}(F,OH)_4$
Niobo-aeschynite-(Ce)	$(Ce,Ca,Th)(Nb,Ti)_2(O,OH)_6$
Niobo-aeschynite-(Nd)	$(Nd,Ce)(Nb,Ti)_2(O,OH)_6$
*Pisekite	$(Y,As,Ca,Fe,U)(Nb,Ti,Ta)O_4$
Plumbopyrochlore	$(Pb,Y,U,Ca)_{2-x}Nb_2O_6(OH)$
Polycrase-(Y)	$(Y,Ca,Ce,U,Th)(Ti,Nb,Ta)_2O_6$
*Polymignite	$\equiv$ zirconolite
*Priorite	$\equiv$ aeschynite-(Y)
*Risörite	$(Y,Er)(Nb,Ti,Ta)(O,OH)_4$
Samarskite-(Y)	$(Y,Ce,U,Fe^{3+})_3(Nb,Ta,Ti)_5O_{16}$
*Scheteligite	$(Ca,Fe,Mn,Sb,Bi,Y)_2(Ti,Ta,Nb,W)_2(O,OH)_7$
Tantalaeschynite-(Y)	$(Y,Ce,Ca)(Ta,Ti,Nb)_2O_6$
Tanteuxenite-(Y)	$(Y,Ce,Ca)(Ta,Ti,Nb)_2(O,OH)_6$
Uranmicrolite	$(U,Ca,Ce)_2(Ta,Nb)_2O_6(OH,F)$
†Uranopolycrase	$(U,Y)(Ti,Nb)_2O_6$
Uranpyrochlore	$(U,Ca,Ce)_2(Nb,Ta)_2O_6(OH,F)$
Vigezzite	$(Ca,Ce)(Nb,Ta,Ti)_2O_6$
Yttrobetafite-(Y)	$(Y,U,Ce)_2(Ti,Nb,Ta)_2O_6OH$
Yttrocolumbite-(Y)	$(Y,U,Fe^{2+})(Nb,Ta)O_4$
Yttrocrasite-(Y)	$(Y,Th,Ca,U)(Ti,Fe^{3+})_2(O,OH)_6$
Yttropyrochlore-(Y)	$(Y,Na,Ca,U)_{1-2}(Nb,Ta,Ti)_2(O,OH)_7$
Yttrotantalite-(Y)	$(Y,U,Fe^{2+})(Ta,Nb)O_4$
Yttrotungstite-(Y)	$YW_2O_6(OH)_3$
Zirconolite	$(Ca,Th,U,REE)Zr(Ti,Nb,Fe)_2O_7$
Zirkelite	$(Ti,Ca,Zr)O_{2-x}$ (see also zirconolite)

A.4 Halides

Fluocerite-(Ce)	$(Ce,La)F_3$
Fluocerite-(La)	$(La,Ce)F_3$
Gagarinite-(Y)	$NaCaY(F,Cl)_6$
Tveitite-(Y)	$Ca_{1-x}Y_xF_{2+x}$
*Yttrocerite	$(Ca,Ce,Y,La)F_3 \cdot nH_2O$
*Yttrofluorite	$(Ca,Y)F_{2-3}$

GLOSSARY OF RARE EARTH MINERALS

A.5 Silicates

Agrellite	$Na(Ca,REE)_2Si_4O_{10}F$
Allanite-(Ce)	$(Ce,Ca,Y)_2(Al,Fe^{3+})_3(SiO_4)_3OH$
Allanite-(Y)	$(Y,Ce,Ca)_2(Al,Fe^{3+})_3(SiO_4)_3OH$
*Alumobritholite	$(Ca,Ce,Y)_3(Al,Fe)_2[(Si,Al,P)_4]_3(F,O)$
*Anderbergite	Hydrous silicate of Zr, Ca and REE
Ashcroftine-(Y)	$KNaCaY_2Si_6O_{12}(OH) \cdot 4H_2O$
*Beckelite	$Ca_3(Ce,La,Y)_4(Si,Zr)_3O_{15}$
*Bilibinite	Hydrated silicate of U, Pb and REE
*Bodenite	Aluminosilicate of Ca, Fe^{2+} and REE
Britholite-(Ce)	$(Ce,Ca)_5(SiO_4,PO_4)_3(OH,F)$
Britholite-(Y)	$(Y,Ca)_5(SiO_4,PO_4)_3(OII,F)$
Byelorussite-(Ce)	$NaMnBa_2Ce_2Ti_2Si_8O_{26}(F,OH) \cdot H_2O$
Calciogadolinite	$(Y,Ca)_2(Fe^{2+},Fe^{3+})Be_2Si_2(O,OH)_{10}$
*Calcybeborosilite	$(REE,Ca)_2(B,Be)_2Si_2O_8(OH)_2$
Cappelenite-(Y)	$Ba(Y,Ce)_6Si_3B_6O_{24}F_2$
Caysichite-(Y)	$Y_2(Ca,Gd)_2Si_4O_{10}(CO_3)_3 \cdot 4H_2O$
Cerite	$(Ce,Ca)_9(Mg,Fe^{2+})Si_7(O,OH,F)_{28}$
Cervandonite-(Ce)	$(Ce,Nd,La)(Fe,Ti,Al)_3(Si,As)_3O_{13}$
Chevkinite	$(Ca,Ce,Th)_4(Fe^{2+},Mg)_2(Ti,Fe^{3+})_3Si_4O_{22}$
Dissakisite-(Ce)	$Ca(Ce,Y)MgAl_2Si_3O_{12}(OH)$
Dollaseite-(Ce)	$CaCeMg_2AlSi_3O_{16}(OH)F$
*Erdmannite	Aluminosilicate of REE, Fe and Ca
*Eucrasite	Silicate of Th, Ca, Na and REE
*Eucolite	≡ eudialyte
Eudialyte	$Na_4(Ca,Ce)_2(Fe^{2+},Mn,Y)ZrSi_8O_{12}(O,Cl)$ (?)
*Freyalite	$(Th,Ce)SiO_4$ (?)
Gadolinite-(Ce)	$(Ce,La,Nd,Y)_2Fe^{2+}Be_2Si_2O_{10}$
Gadolinite-(Y)	$Y_2Fe^{2+}Be_2Si_2O_{10}$
Götzenite	$(Ca,Na,REE)_3(Ti,Al)Si_2O_7(F,OH)_2$
Hellandite-(Y)	$(Ca,Y)_6(AlFe^{3+})Si_4B_4O_{20}(OH)_4$
Hingganite-(Ce)	$CeBeSiO_4OH$
Hingganite-(Y)	$(Y,Yb,Er)BeSiO_4OH$
Hingganite-(Yb)	$(Yb,Y)BeSiO_4OH$
*Hydrocerite	$(Ce,Th,Ca)(Al,Fe,Ti,Nb)(Si,P)_2O_7 \cdot 5H_2O$
Iimoriite-(Y)	$Y_2SiO_4CO_3$
Ilimaussite-(Ce)	$Ba_2Na_4CeFe^{3+}Nb_2Si_8O_{28} \cdot 5H_2O$
*Ilmajokite	$(Na,Ca,Ba)_2TiSi_3O_5(OH)_{10} \cdot nH_2O$
Iraqite-(La)	$K(La,Ce,Th)_2(Ca,Na)_4(Si,Al)_{16}O_{40}$
Joaquinite-(Ce)	$Ba_2NaCe_2Fe^{2+}(Ti,Nb)_2Si_8O_{26}(OH,F) \cdot H_2O$
*Johnstrupite	≡ mosandrite
Kainosite-(Y)	$Ca_2(Y,Ce)_2Si_4O_{12}CO_3 \cdot H_2O$
Karnasurtite-(Ce)	$(Ce,La,Th)(Ti,Nb)(Al,Fe^{3+})(Si,P)_2O_7(OH)_4 \cdot 3H_2O$

SILICATES

Keivyite-(Y)	$(Y,Yb)_2Si_2O_7$
Keivyite-(Yb)	$(Yb,Y)_2Si_2O_7$
Kuliokite-(Y)	$Y_4Al(SiO_4)_2(OH)_2F_5$
*Manganosteenstrupine	$(La,Th,Ca)MnSiO_3(OH)_3 \cdot 2H_2O$
Melanocerite-(Ce)	$(Ce,Ca)_5(Si,B)_3O_{12}(OH,F) \cdot nH_2O$
Minasgeraisite-(Y)	$Y_2CaBe_2Si_2O_{10}$
Miserite	$K(Ca,Ce)_4Si_5O_{13}(OH)_3$
Monteregianite-(Y)	$(Na,K)_6(Y,Ca)_2Si_{16}O_{38} \cdot 10H_2O$
Mosandrite	$(Ca,Na,Ce)_{12}(Ti,Zr)_2Si_7O_{31}H_6F_4$
*Muromontite	Near $Be_2FeY_2Si_3O_{12}$
Nacareniobsite-(Ce)	$NbNa_3Ca_3(Ce,La,Nd)(Si_2O_7)_2OF_3$
Nordite-(Ce)	$(Ce,La)(Sr,Ca)Na_2(Na,Mn)(Zn,Mg)Si_6O_{17}$
Nordite-(La)	$(La,Ce)(Sr,Ca)Na_2(Na,Mn)(Zn,Mg)Si_6O_{17}$
Okanoganite-(Y)	$(Na,Ca)_3(Y,Ce)_{12}Si_6B_2O_{27}F_{14}$
*Orthite	$\equiv$ allanite-(Ce)
Orthojoaquinite-(Ce)	$Ba_2NaCe_2Fe^{2+}Ti_2Si_8O_{26}(O,OH) \cdot H_2O$
Perrierite	$(Ca,Ce,Th)_4(Mg,Fe^{2+})_2(Ti,Fe^{3+})_3Si_4O_{22}$
*Rinkite	$\equiv$ mosandrite
*Rinkolite	$\equiv$ mosandrite
Rowlandite-(Y)	$(Y,Fe,Ce)_3(SiO_4)_2(F,OH)$
Saryarkite-(Y)	$Ca(Y,Th)Al_5(SiO_4)_2(PO_4,SO_4)_2(OH)_7 \cdot 6H_2O$
Sazhinite-(Ce)	$Na_3CeSi_6O_{15} \cdot 6H_2O$
Semenovite	$(Ca,Ce,La,Na)_{12}(Si,Be)_{20}O_{40}(O,OH,F)_8 \cdot H_2O$
*Shentulite	Near $(Th,Fe < Ca,Ce)[(Si,P,As)O_4,CO_3OH]$
Steenstrupine-(Ce)	$Na_{14}Ce_6Mn^{2+}Mn^{3+}Fe^{2+}{}_2(Zr,Th)(Si_6O_{18})_2$ $(PO_4)_7 \cdot 3H_2O$
Stillwellite-(Ce)	$(Ce,La,Ca)BSiO_5$
Strontiochevkinite	$(Sr,La,Ce,Ca)_4(Fe^{2+},Fe^{3+})_2(Ti,Zr)_4Si_4O_{22}$
Tadzhikite-(Y)	$Ca_3(Y,Ce)_2(Ti,Al,Fe)B_4Si_4O_{22}$
Thalenite-(Y)	$Y_3Si_3O_{10}OH$
Thortveitite	$(Sc,Y)_2Si_2O_7$
Tombarthite-(Y)	$Y_4(Si,H_4)_4O_{12-x}(OH)_{4+2x}$
Törnebohmite-(Ce)	$(Ce,La)_2Al(SiO_4)_2OH$
Törnebohmite-(La)	$(La,Ce)_2Al(SiO_4)_2OH$
Tranquillityite	$Fe^{2+}{}_8(Zr,Y)_2Ti_3Si_3O_{24}$
Trimounsite-(Y)	$Y_2Ti_2SiO_9$
Tritomite-(Ce)	$(Ce,La,Y,Th)_5(Si,B)_3(O,OH,F)_{13}$
Tritomite-(Y)	$(Y,Ca,La,Fe^{2+})_5(Si,B,Al)_3(O,OH,F)_{13}$
Tundrite-(Ce)	$Na_3(Ce,La)_4(Ti,Nb)_2(SiO_4)_2(CO_3)_3O_4(OH) \cdot 2H_2O$
Tundrite-(Nd)	$Na_3(Nd,La)_4(Ti,Nb)_2(SiO_4)_2(CO_3)_3O_4(OH) \cdot 2H_2O$
Vyuntspakhite-(Y)	$Y_4Al_3Si_5O_{18}(OH)_5$
Wöhlerite	$Na(Ca,REE)_2(Zr,Nb)Si_2O_7(O,OH,F)_2$

GLOSSARY OF RARE EARTH MINERALS

Yftisite-(Y) $(Y,Dy,Er)_4(Ti,Sn)O(SiO_4)_2(F,OH)_6$
Yttrialite-(Y) $(Y,Th)_2Si_2O_7$

A.6 Phosphates

Belovite-(Ce) $(Sr_3,Na,Ce)_5(PO_4)_3OH$
Brockite $(Ca,Th,Ce)(PO_4)\cdot H_2O$
Cheralite $(Ca,Ce,Th)(P,Si)O_4$
Churchite-(Y) $YPO_4\cdot 2H_2O$
*Fenhuanglite $(Na,Ca,Ce,Th)_5(P,Si)_3O_{12}(OH,F)$
Florencite-(Ce) $CeAl_3(PO_4)_2(OH)_6$
Florencite-(La) $(La,Ce)Al_3(PO_4)_2(OH)_6$
Florencite-(Nd) $(Nd,Ce)Al_3(PO_4)_2(OH)_6$
Fluorapatite $(Ca,REE,Na)_5(PO_4)_3(F,OH)$
Francoisite-(Nd) $(Nd,Ce,Sm)(UO_2)_3O(OH)(PO_4)_2\cdot 6H_2O$
*Koivinite $YAl_5(PO_4)_4(OH)_4\cdot 2H_2O$
Laplandite-(Ce) $Na_4CeTiPO_4Si_7O_{18}\cdot 5H_2O$
Monazite-(Ce) $(Ce,La,Nd,Th)PO_4$
Monazite-(La) $(La,Ce,Nd)PO_4$
Monazite-(Nd) $(Nd,La,Ce)PO_4$
*Ningyoite $(U,Ca,Ce)(PO_4)\cdot nH_2O$
Petersite-(Y) $(Ca,Fe^{2+},Y,Ce)Cu_6(PO_4)_3(OH)_6\cdot 3H_2O$
Phosinaite $Na_3(Ca,Ce)SiPO_7\cdot H_2O$
*Phosphocerite $(La,Ce)PO_4$ (?)
Rhabdophane-(Ce) $(Ce,La)PO_4\cdot H_2O$
Rhabdophane-(La) $(La,Ce)PO_4\cdot H_2O$
Rhabdophane-(Nd) $(Nd,Ce,La)PO_4\cdot H_2O$
*Smirnovkite $(Th,Ce,Ca)(P,Si,Al)(O,F,OH)_5$
Vitusite-(Ce) $Na_3(Ce,La,Nd)(PO_4)_2$
*Vudyavrite   near $Ce_2Ti_3O_9\cdot nSiO_2\cdot mH_2O$
Xenotime-(Y) YPO_4

A.7 Arsenates, sulphates and vanadates

*Agardite-(Ce)   see Agardite-(La)
Agardite-(La) $(La,Ca)Cu_6(AsO_4)_3(OH)_6\cdot 3H_2O$
Agardite-(Y) $(Y,Ca)Cu_6(AsO_4)_3(OH)_6\cdot 3H_2O$
Arsenoflorencite-(Ce) $(Ce,La)Al_3(AsO_4,PO_4)_2(OH)_6$
Arsenoflorencite-(La) $LaAl_3(AsO_4)_2(OH)_6$
Arsenoflorencite-(Nd) $NdAl_3(AsO_4)_2(OH)_6$
Chernovite-(Y) $YAsO_4$
Chukhrovite-(Ce) $Ca_3(Ce,Y)Al_2(SO_4)F_{13}\cdot 10H_2O$
Chukhrovite-(Y) $Ca_3(Y,Ce)Al_2(SO_4)F_{13}\cdot 10H_2O$
Gasparite-(Ce) $(Ce,REE)AsO_4$

Goudeyite	$Cu_6(Al,Y)(AsO_4)_3(OH)_6 \cdot 3H_2O$
Kemmlitzite	$(Sr,Ce)Al_3AsO_4SO_4(OH)_6$
†Mineevite-(Y)	$Na_{25}Ba(Y,Gd,Dy)_2(CO_3)_{11}(HCO_3)_4(SO_4)_2F_2Cl$
*Paranite-(Y)	$Ca_y YAsO_4(WO_4)_2$
Retzian-(Ce)	$Mn_2CeAsO_4(OH)_4$
Retzian-(La)	$(Mn,Mg)_2(La,Ce,Nd)AsO_4(OH)_4$
Retzian-(Nd)	$Mn_2(Nd,Ce,La)AsO_4(OH)_4$
Wakefieldite-(Ce)	$(Ce,Pb^{2+},Pb^{4+})VO_4$
Wakefieldite-(Y)	YVO_4

A.8 Uranyl-carbonates and uranyl-silicates

Astrocyanite-(Ce)	$Cu_2(Ce,Nd)_2UO_2(CO_3)_5(OH)_2 \cdot 1.5H_2O$
Bijvoetite-(Y)	$(Y,Dy)_2(UO_2)_4(CO_3)_4(OH)_6 \cdot 11H_2O$
Kamotoite-(Y)	$4UO_3 \cdot Y_2O_3 \cdot 3CO_2 \cdot 14.5H_2O$
Lepersonnite-(Gd)	$CaO(Gd,Y)_2O_3 \cdot 24UO_3 \cdot 8CO_2 \cdot 4SiO_2 \cdot 60H_2O$
Shabaite-(Nd)	$Ca(Nd,Sm,Y)_2UO_2(CO_3)_4(OH)_2 \cdot 6H_2O$

* Refers to synonyms, variants or not well-characterized mineral species (see Clark 1993).
† In Jambor, Puziewicz and Robert (1994).

This compilation is based on several previously published lists of rare earth minerals, those from Clark (1984), Burt (1989), Cesbron (1989) and augmented with data from Miyawaki and Nakai (1987, and annual supplements), Fleischer and Mandarino (1991) and Clark (1993), and from new mineral lists published in *American Mineralogist*. Formulae are mostly from Clark (1993), using where possible the Bayliss and Levinson (1988) rare earth element nomenclature scheme. The rare earth elements are defined here using the IUPAC recommendations (Leigh, 1990).

References

Bayliss, P. and Levinson, A. A. (1988) A system of nomenclature for rare earth mineral species: revision and extension. *Am. Mineral.* **73**, 422–3.

Burt, D. M. (1989) Compositional and phase relations among rare earth element minerals, in *Geochemistry and Mineralogy of Rare Earth Elements* (eds B. R. Lipin and G. A. McKay), Mineralogical Society of America, *Reviews in Mineralogy* **21**, 259–307.

Clark, A. M. (1984) Mineralogy of the rare earth elements, in *Rare Earth Element Geochemistry* (ed. P. Henderson), Elsevier, Amsterdam, pp. 33–61.

Clark, A. M. (1993) *Hey's Mineral Index*, Chapman & Hall, London, 851 pp.

Cesbron, F. P. (1989) Mineralogy of the rare-earth elements, in *Lanthanides, Tantalum and Niobium* (eds P. Möller, P. Černý and F. Saupé), Berlin, pp. 3–26. Springer-Verlag.

Fleischer, M. and Mandarino, J. A. (1995) *Glossary of Mineral Species 1995*, Mineralogical Record, Tuscon, 280 pp.

Jambor, J. L., Puziewicz, J. and Robert, A. C. (1994) New mineral names. *Am. Mineral.*, **79**, 763-7.

Leigh, G. J. (1990) *Nomenclature of Inorganic Chemistry, Recommendations 1990*, IUPAC, Blackwell Scientific Publications, Oxford.

Miyawaki, R. and Nakai, I. (1987) *Crystal Structures of Rare-earth Minerals*, The Rare Earth Society of Japan, Tokyo.

Index

Page numbes in *italics* refer to tables, in **bold** refer to figures. Some textual material may appear on the same page.

Absorption spectroscopy 328
Adamello veins **132**, **133**, 140
Adiounedj (Mali) 217, **219**
Adiouneji (Mali) *195*
Aegirine
 in China 294, 295, 297, 300, 302, 303
 in Oslo region 151
 in Thor Lake deposit 171
Aegirine-augite 295, 297
Aegirine-nepheline syenites 302
Aegirine-syenite 286
Aeschynite
 in carbonatite *195*
 chemical zonation 110
 in China 282, 294
 hydrothermal systems *107*, 131
 in Kola Peninsula *313*, 315
Aeschynite-(Ce) 23
Aeschynite-(Y) 25, 30, **31**, 33
Afrikanda 63, 64
Agardite 34
Agardite-(Ce) 25
Agrellite 23, 25
Albite 171, 282
 in China 286, 295, 297, 300, 304
 Ga-bearing 171
 in Thor Lake deposit 171
Albitites 317
Albitization 316
Aldanite 303
Alkali amphibole 297
Alkali pegmatites 286
Alkaline complexes, laterites formed on 227–55
Allanite
 alteration 125
 bastnäsitization 125
 in carbonatites *196*, 220
 chemical zonation 110
 in China 282, 286, 295, 297, 298, 304
 hydrothermal systems *107*, 112
 in Kola Peninsula *313*, 315, 316
 partitioning 112, **113**
Allanite-(Ce) 126
 crystal structure 23
 in Kola Peninsula *313*
 in Oslo region *152*, 162
Allanite-(La) 126
Alnoites 59
 perovskites from 67
Alteration
 apatite 231
 during laterization 231
 hydrothermal RE minerals 121, 123–7
Amazonite pegmatites 316, 317, 323–4, **324**
Amphibolites 315
Analysis
 bulk techniques 329–35
 compositional imaging 329
 gravimetric methods 66
 historical background 327–8
 in situ microchemical techniques 328
 isotope 184–6, *187*, *188*
 laterites 235
 mechanical separation 329–30
 microanalytical techniques 335–45
 perovskites 59–60
 whole-rock 133–4
 see also Chemical analysis; Spectroscopy *and individual methods*
Analytical transmission electron microscopy (ATEM) 335, 342–3
Anatase 287, 297, 298
Ancylite
 in carbonatites *195*, 198, 220

INDEX

Ancylite *contd*
  hydrothermal origin 106
  in Kola Peninsula *314*, 317
Ancylite-(Ce)
  in carbonatite *195*
  isomorphous substitution 34
  in Oslo region *152*, 157
Anhydrite 106
Ankerite 172, 200
Anorthoclase 286
Apatite 10, 80
  alteration during laterization 231
  in carbonatites 196, 198, 209–10, 211, *212–13*, **215**, 216
  in China 282, 286, 293, 295, 297, 298, 302, 304
  crystallization 121–2
  hydrothermal replacement 220
  hydrothermal systems 106, *107*, 112, 121
  incipient mineralization in carbonatites 209–10
  isomorphous substitution 33
  laterites 228
  in Oslo region *153*, 155
  REE carriers 240, 242, **243**, **244**, 245, **246**
  replacement of 217, 220
  solubility 119, 122
Aqueous systems
  complexing and hydrolysis in 12, 14–15
  geochemical behaviour 12, 14–16
  see also Hydrothermal systems
Ar geochronology 179–81
Araxá (Brazil) *195*, 220
  geology 232–4
  laterites 227–55
Arfvedsonite 286, 297
Ashcroftine-(Y) 25
Astrophyllite 151, 300
Atomic absorption spectroscopy 331
Authapuscow aulacogen 167
Authigenic minerals 266–70
  chemical properties 259–64
  in karstic nickel deposits 257–79
  X-ray powder diffraction study 264–6

β-autoradiography 345

Backscattered electron images (BEIs) 59, 110, **111**, 200, **201**, 329
Baerzhe deposit (China) 300–1
Baltic Shield 316
Baotite 294
Barium pentatitanate 62
Barkevikite 152
Baryte
  in carbonatites *196*, 198, 200, 209, 216
  in China 282, 293, 295, 297
  hydrothermal systems 106
Barytolamprophyllite 62
Bastnäsite
  alteration 124, 125
  authigenic 259, *261*, 264, *265*, **266**, **267**, 269
  in carbonatites *195*, *196*, 198, 209, **210**, 220
  in China 282, 294, 295, 297–8
  in hydrothermal systems 105, *107*
  in karstic deposits 275
  in Kola Peninsula *314*, 315, 316, 317
  mineral host 80–1
  phase equilibrium studies 84–7
  phases **82**, **83**
  stability range 82–4
  synthesis 81–4
  in Thor Lake deposit 172, 173, **179**
  X-ray powder diffraction study 264, *265*, **266**
Bastnäsite-(Ce)
  in carbonatite *196*, 200, 201, 216
  crystal structure 22, 30, **32**
  in karstic deposits 266, **267**, 269, 275
  in Oslo region *152*, 156, 157
Bastnäsite-(La) 31, 268, 275
Bastnäsite-(Y) 30
Bauxites
  in China 287
  horizon 230
Bayan Obo (China) 105, 293–5
  bastnäsite mines 227
  carbonatites in 193, 194, *195*, 220
  deposits 281, *283*
  distribution of deposit 288, **290**, *291*
  economic deposits 11
  Fe-Nb-REE deposit 293–5
  geochemistry 288–9

INDEX

geological map **293**
hydrothermal origin of deposits *109*
isotope geochemistry 289, 292
Bear Lodge (Wyoming) *195*, 217
Bearpaw Mountains 220
Beforsite 198, 209
Belovite *314*
Belovite-(Ce) 25
Bergell Intrusion 110, **112**, **113**
Beryl 312
Betafite 297, 303, *313*
Biotite
   in China 286, 294, 295, 297, 302, 303
   in Thor Lake deposit 171
Blatchford Lake igneous complex 167
   geological setting 168–9
   magmatic events 169
Brannerite
   hydrothermal origin 106
   isomorphous substitution 32
Breccia 171
Brindleyite **271**, 275
Britholite
   alteration 125
   in China 296, 298, 303
   hydrothermal systems *107*
   in Kola Peninsula *313*, 315, 316
   in Oslo region *152*, 158–9
Britholite-(Ce) 34, *152*, 158–9
Britholite-(Y) 315
Brockite *107*
Brownmillerite 49
Buffalo Mine (South Africa) *109*
Burbankite
   in carbonatite *195*
   coordination polyhedra 25
   in hydrothermal environment 105, 106
   in Kola Peninsula *314*, 317
   replacement 220–1

Calcian ancylite 298
Calcian niobian loparite 64
Calcio-ancylite-(Ce), in Oslo region *152*, 157
Calciogadolinite 34
Calcite
   in carbonatites *195*, 198, 220
   in China 293, 294, 295, 297, 300

hydrothermal systems 106
   REE carrier 242, **243**
Calkinsite cappelenite 106
Calkinsite-(Ce) 29
Cancrinite, in China 302, 303
Capitan Pluton (New Mexico) 112
Cappelenite 151, *152*, 162
Cappelenite-(Y) 35, **36**, *152*, 162
Carbocernaite 220–1, *314*, 317
   in carbonatite *195*, *196*
   in China 298
   coordination polyhedra 25
   in hydrothermal environment 106
   isomorphous substitution 34
Carbon and oxygen isotope analysis, Thor Lake deposits 187
Carbonate-rich melts
   formation in mantle 78–9
   metasomatism 79–80
Carbonates, in Thor Lake deposit 171
Carbonatites 10
   ankerite 299
   apatite replacement 217, 220
   apatite–magnetite variety 80–1
   Bayan Obo (China) 193, 194, *195*, 220
   biotite–calcite 299
   burbankite replacement 220–1
   calcite 299
   in China 193, 194, *195*, 220, 286, 294
   comparisons 216–21
      with Mountain Pass 216–17
   crystallization from magma 198
   early 198
   enrichment in 194
   extrusive 196, 197
   florencite–goyazite series 200, **201**, 204, 205, *206*, 208
   formation of minerals in 197–8
   glasses 92, **93**
   host deposit 77, 80–1, 105, 193–4
   hydrothermal mineralization 197–8
   hydrothermal/carbothermal deposition 87
   incipient mineralization in 209–10
   intrusive 196–7
   isotope investigations 198
Kaiserstuhl 70
Kangankunde (Malawi) 193, *195*,

Carbonatites *contd*
    198–216
        apatite replacement 217, 220
        burbankite replacement 220–1
        comparisons 216–21
            with Mountain Pass 216–17
        deposit 193
        early 198
        florencite–goyazite series 200, **201**, 204, 205, *206*, 208
        geological map **199**
        incipient mineralization in 209–10
        isotope investigations 198
        late 198
        monazite *202–3*, 204–5
        quartz rocks 200–9
            outside carbonatite 211–16
        RE mineral assemblage 200, 201–9
        replacements 217, 220–1
        weathering 208–9
    Kovdor 68
    late 194, 196–7, 198
    monazite *202–3*, 204–5
    Mountain Pass (California) 193, *195*, 198, 216–17
        glasses 92, **93**
        hydrothermal/carbothermal deposition 87
        phase relationships 89–92
        primary magmatic crystallization 87
        synthetic carbonatite magma 87–93
    occurrences 194, *195–6*, 197
    perovskite minerals in 41, 67, *68*
    phase equilibrium studies in magmas 82–7
    phase relationships 89–92
    Polino 68
    primary magmatic crystallization 87
    quartz rocks 200–9
        outside carbonatite 211–16
    rare earth mineral variety 80, 81
    RE elements in 80–1, 193–4
    RE mineral assemblage 200, 201–9
    RE mineralization in 81
    RE minerals in 193–225
    replacements 217, 220–1
    silicate–carbonate liquid immiscibility 93–5
    synthetic carbonatite magma 87–93
    'typical' intrusive sequence 196–7
    weathering 198, 208–9
Carriers, in laterites 240–9
Carro Manomó (Bolivia) *196*
Caryocerite 152
Cassiterite 287
Catalão (Brazil)
    geology 232–4
    laterites 227–55
Catapleiite 151, *153*, 163, 303
Cathodoluminescence (CL) 6, **111**, 345
Caysichite *313*, *320*
Ce-group 21
Ceramics 41
Cerianite
    in carbonatite *196*
    in laterites 247
    precipitation 120
Cerianite-(Ce) 23
Cerite
    alteration 125
    in hydrothermal environment 105, *107*
    in Kola Peninsula *313*, 317
Cerite-(Ce) 23, 25, *152*, 158
Cerussite 300
Chalcopyrite 139
Chemical analysis
    historical background 327–8
    in situ microchemical techniques 328
    'wet' 328
Chemical behaviour 1–6
Chemical formulae 349–55
Chemical symbols 2
Chemical zonation patterns 110–12, 129, **130**
Chernovite 106
Chevkinite 282, *313*, 315, 316
Chevkinite-(Ce) *152*, 161–2
Chilwa Island (Malawi) 197
China
    age of formation of deposits 288
    alkali complexes 286
    alkali granites 282, 286, 300–1
    alkali pegmatites 286

Baerzhe deposit 300–1
bauxites 287
Bayan Obo deposit, *see* Bayan Obo (China)
  carbonatitic rocks 282
  deposits in 281–310
    individual deposits e.g. Bayan Obo (China)
  distribution of deposits 288
  distribution patterns 288–9
  geochemistry 288–92
  Guangshui deposit 304, **305**
  isotope geochemistry 289, 292
  lateritic weathering crusts 287
  Maoniuping deposit 281, 282, 295–7
  metamorphic rock 286
    Zr–Y–heavy REE mineralization 304–5
  Miaoya deposit 299–300
  phosphorites 286–7
  placers 287–8
  quartz syenites 282
  Saima deposit 301–3
  types of deposit 282, *283–5*, 286–8
  Weishan deposit 281, 282, *284*
    bastnäsite-carbonate vein deposit 297–8
Chipman lake 221
Chishan ('101') deposit (China), *see* Weishan deposits (China)
Chlorine, in Thor Lake deposit 178
Chlorite
  in China 287, 297, 300
  in Thor Lake deposit 171
Chondrite normalizing 9–10
Chondritic meteorites *9*
Churchite *314*, *322*
  in carbonatites *196*, 198
Clays 10
Colorimetry 328
Columbite 171
  in China 282, 294, 297, 298, 300
  in Kola Peninsula 312
  in Thor Lake deposit 171
Complex perovskites 48, **49**
Complexing 12, 14–15
Contaminated syenite 299–300
Coordination polyhedra 23, **24**, 25–6

Cordylite *195*, *314*, 317
Corrosion, *see* Alteration
Crandallite *196*, 263
  in carbonatites 198
  in hydrothermal environment 106
  laterites 228
Crichtonite 79
Crystal chemical aspects 21–40
  Ce-group 21
  Y-group 21
Crystal structures 22–3
  coordination polyhedra 23, **24**, 25–6
  ionic radii and 26, 29
  isomorphous substitutions 22, 31–4
    coupled substitution at independent sites 33–4
    coupled substitution within rare earth site 33
    gadolinite–datolite group 34–7
    phase relationships **35**
    substitution accompanying vacancies 33
    substitution accompanying valence variation 34
  structural difference between Ce- and Y-group 26–32
Crystallization
  fluorapatite 139
  gangue minerals 121–3
  zirconolite **133**

Daqingshanite *196*, 294
Datolite 34, **35**, **36**
Daughter minerals 106, 121
Davidite 106
Davidite-(La) 25, 33
Deposition
  during laterization 231
  hydrothermal systems 137–9
  hydrothermal/carbothermal deposition 87
Diaspore 287
Didymia 328
Diopside 286, 297
Dissolution, *see* Alteration
Dititanates 51, **52**
Dollaseite-(Ce) 23

Dolomite
  burbankite in 220
  in carbonatites 198
  in China 293, 297
  formation in peridotite 78
Dolomite–lherzolite 78
Donnayite *314*
Dysanalyte 56, 57, 66

East Sayan (Siberia) 220
Economic deposits 11
Electron backscattered images (BEIs) 59, 110, **111**, 200, **201**, 329
Electron probe microanalysis (EPMA) 328, 335, 336–42
  bastnästite 209
  perovskites 59
  wavelength- and energy-dispersive techniques 60
Electronic configurations 2, *3*
Electronic spectra 6
Electrum 300
Element abundancies 6–11
Element partitioning 3, 5, 11, 12, *13*, **14**, 112
Elemental X-ray maps 329
Elpidite *153*
Emission spectroscopy 328
Enrichment 77
  in carbonatites 194
  in metasomatized rocks 79–80
  in Thor Lake deposits 172–3
Epidote 10, 287, 295, 297, 304
Erbia 328
Eudialyte 56, *313*, 316, 317
  in China 302, 303
  in Oslo region *153*, 155, 163
Eudidymite 151
Euxenite 312, *313*, 314, 315
Euxenite-(Y) 34
Ewaldite *314*
Extended X-ray absorption fine structure (EXAFS) spectroscopy 6

Feldspar 171, 293, 294, 297, 302
Fen (Norway) *195*, 221
Fenites, Sarambi and Salitre 62
Fergusonite
  in China 282, 286, 304
  in hydrothermal environment 106
  in Kola Peninsula *313*, 314, 315, 316
  in Oslo region *152*, 156
Fergusonite-β-(Y) 25
Fergusonite-(Y) *152*, 156
Ferrothorite 300
Fersmite 34, 51, 106, *313*
Florencite
  authigenic 263
  in carbonatites *196*, 198, 200, 216
  in hydrothermal environment 106
Florencite–goyazite, in carbonatite *195*, 200, **201**, 204, 205, *206*, 208
Fluid inclusion analysis 181–6
Fluid mixing 120
Fluocarbonates, transformations within group 124, 172, **173**, 174, **176–7**
Fluocerite
  alteration 124, 126
  hydrothermal systems *107*
  in Kola Peninsula *313*
Fluocerite-(Ce) 23
Fluorapatite 133, 139–40
Fluorcarbonates, *see* Fluocarbonates
Fluorite
  alteration 125
  in carbonatites *195*, 198, 216
  in China 282, 293, 295, 297, 300
  crystallization 121–2
  hydrothermal systems 106, *107*, 112, 121, 122–3
  in Kola Peninsula *313*
  in Thor Lake deposit 171
Fluorite-(Ce) *313*
Fluorite-(Y) *313*
Fluoritization 318
Formanite *313*
Fourier-transform infrared spectroscopy (FTIR) 345

Gabbro-labradorites 312, 315
Gadolinite
  in China 282, 286, 304
  hydrothermal systems *107*
  in Kola Peninsula *313*, 315, 316
  in Oslo region *152*, 160
  in Thor Lake deposit 171

Gadolinite-(Ce) 21, *152*, 160, *313*
Gadolinite-(Y) 21, *313*
   coordination polyhedra 25
   crystal structure 30, **31**
   isomorphous substitution 34, **35**, **36**, 37
Gagarinite-(Y)
   coordination polyhedra 25
   isomorphous substitution 33
Gakara (Burundi) *109*
Galena 297, 300
Gallinas Mountains (USA) *109*
Gamma-ray spectroscopy 328
Garnet 10
   chemical zonation 110
   in China 287, 304
   hydrothermal systems 106
Gasparite 105, 106, 124
Gatineau (Quebec) *195*
Gem Park (Colorado) *195*, 217
Genthelvite 300
Geochemistry
   aqueous systems 12, 14–16
   Bayan Obo (China) 288–9, 292
   laterites 234–42
      analytical methods 235
      major elements 235–7
      REE contents 237–40
   magmatic systems 11–12
   transportation 116–18
Geochronology 16
   Ar 179–81
Gibbsite 287
Glasses 92, **93**
Glimmerites 233
Glossary 349–55
Goethite 287, 300
Goldschmidt tolerance factor 47
Gorceixite 198, 240, 247
Götzenite *153*
Goyazite
   authigenic 263
   in carbonatites 198, 200
Granite
   aegirine alkalic 282
   alkali 282, 286
   pegmatites 312
   peralkaline 315
Graphite 139

Gravimetric analysis, perovskites 59, 66
Great Slave Lake 167
Guangshui deposit (China) 304, **305**

Haast River (New Zealand) 205
Halides, complexing groups 15
Halloysite 287
Hambergite 151
Harzburgite 78
Hastingsitic hornblende 152
Hatchettolite 317
Hellandite 33
Hellandite-(Y) 35, **36**
Hematite
   in carbonatites 198
   in China 298, 300
Hemlo gold deposits (Canada) 127
Hibonite 34
Hingganite
   in China 282, 300
   in Kola Peninsula *313*, 318, *320–3*, **322**
Hingganite-(Y) *313*, 318
   isomorphous substitution 34, **35**, **36**
   in Oslo region *152*, 160–1
Hingganite-(Yb) *313*, 318, *320*, *320–3*, **320**, **322**
Hiortdahlite 151, *153*
Hogan Group (China) 304
Hollandite 62
Holmquistite 312
Homilite
   isomorphous substitution 34, **35**, **36**
   in Oslo region 151, *153*
Huanghoite *314*, 317
   in China 294
   in hydrothermal environment 105, 106
Huanghoite-(Ce) 25
Hyalophane 286
Hydrothermal systems 105–49
   alteration 121, 123–7
   case study 127–39
      characterization of minerals 129–33
      deposition 137–40
      transportation 136–7
      vein formation conditions 135–6
      whole-rock chemical composition of veins 133–4
   chemical zonation patterns 110–12,

Hydrothermal systems *contd*
  129, **130**
  deposition of REE 87, 137–40
  fluid mixing 120
  formation of RE minerals 118–23
    temperature and pressure changes
      119–20
  mineral deposits *109*, 110
  mineralization 118–23, 197–8
  paragenetic relationships 118
  petrographic features 127–9
  precipitation 118–23
    crystallization of gangue minerals
      121–3
    fluid mixing 120
    interaction of fluids with wall rock
      120–1
    secondary RE minerals 124
    temperature and pressure changes
      119–20
  RE minerals originating in 105, 106,
    107–8, *109*, 110
  REE in hydrothermal fluids 113–16
    ancient systems 115–16, 117
    concentrations *114*
  speciation of REE 116–17
  submarine vents *114*, 117, 119, 120
  transportation 116–18, 136–7
  weathering 121
  zoning 110–12, 129, **130**
Hydrothermal/carbothermal deposition 87
Hydroxide complexes 15
Hydroxylbastnäsite *314*
  authigenic 259, **260**, *262*, 265, 274, 276
  deposits 268
  in karstic deposits 274, 276
  REE carriers in laterites 268–70
Hydroxylbastnäsite-(La) 84–7, 276
Hydroxylbastnäsite-(Nd) 276
Hydroxylherderite 34, **35**, 36

Igdloite 56
Iimoriite *313*, *321*
Ilimaussaq (Greenland) 153
Ilimaussite 106, *313*
Ilmajokite *313*
Ilmenite 59, 286, 287, 300, 302
Ilmenite–rutile reaction 136

Instrumental neutron activation analysis
  (INAA) 235, 331–3
Ion chromatography 331, 335
Ion probe analysis 328
Ionic radius 2, 3
  crystal structures and 26, 29
  trivalent rare earth ions 21, **22**
Irinite 56, 57, 66
Iron hydroxides, in laterites 240, 242, 247, 249
Iron oxides, limonitic 200
Isokite *196*
Isomorphous substitutions 22, 31–4
  coupled substitution
    at independent sites 33–4
    within rare earth site 33
  gadolinite–datolite group 34–7
  phase relationships **35**
  substitution accompanying vacancies 33
  substitution accompanying valence variation 34
Isotope analysis 184–6, *187*, *188*
Isotope dilution mass spectrometry
  (IDMA) 331, 334
Isotopes, use in petrogenesis and geochronology 16

Joaquinite 106
Johnstrupite 152

K-feldspar 171
Kainosite
  in hydrothermal environment 106
  in Kola Peninsula *313*, *321*
  in Oslo region 163
Kamphaugite *314*
Kangankunde (Malawi) 193, *195*
  carbonatites 198–216
    apatite replacement 217, 220
    burbankite replacement 220–1
    comparisons 216–21
      with Mountain Pass 216–17
    deposit 193
    early 198
    florencite–goyazite series 200, **201**, 204, 205, *206*, 208
    geological map **199**
    incipient mineralization in apatite

INDEX

dolomite carbonatites 209–10
  isotope investigations 198
  late 198
  monazite *202–3*, 204–5
  quartz rocks 200–9
    outside carbonatite 211–16
  RE mineral assemblage 200, 201–9
  replacements 217, 220–1
  weathering 208–9
  formation of RE minerals in carbonatite 198–9
  geological map of complex **199**
Kaolinite 287
Karnasurtite *313*
Karonge/Gakara (Burundi) *109*, 124
Karst-bauxites/karstic nickel deposits
  age 259
  authigenic rare earth minerals in 257–79
  bauxite deposits in China 287
  chemical properties of minerals 259–64
  distribution of minerals 258–9
  hydroxylbastnäsite, authigenic 259, **260**, *262*, 265, 274, 276
  leaching 258, 273
  mobility 274
  occurrence of minerals 258–9
  REE distribution in 268–70, 271
  synchysite-(Nd) 266, **267**
  X-ray powder diffraction study 264–6
Katete (Zimbabwe) 205
Keivyite-(Y) *313*, 318–23
Keivyite-(Yb) *313*, 318–23, *320–1*, **322**, 323
Kersantite 297
Khibina complex (Kola Peninsula, Russia) 66, *195*, 220
Kimberlite xenoliths, MARID suite 79
Kimberlites 10, 59
  perovskite minerals in 41
  perovskites 63, 67, *68*
Kimuraite-(Y) 26, 29
Kizilcaören (Turkey) *195*, 221
Knopite 56
Kola Peninsula (Russia) 153, 311–26
  carbonatites *195*
  early proterozoic–Karelian stage 315–16
  economic mineralization 323–4

late Archaean–Kola stage 312, 314
  mineralization 316–17, 323
    economic 323–4
    occurrences 311, **312**
    minerals found *313–14*
  palaeozoic–Caledonian–Hercynian stage 316–17
  stages of ore genesis 311–17
Kovdor carbonatite 68
Kuliokite *313*, *320–31*, **322**

Lamproites
  composition 60, *61*
  olivine 59
Lamprophyllite 302, 303
Lamprophyres 10
Lanthana 328
Lanthanides 1
Lanthanite 105, 106
Lanthanite-(La)
  coordination polyhedra 25
  crystal structure 22, 26, **28**, 29
Laplandite *313*
Larvikite 153, 155
Laser ablation microprobe (LAMP) ICPMS 328, 335, 344–5
Laterites
  analytical methods 235
  Araxá and Catalão geology 232–4
  bauxite horizon 230
  formation 16
  genetic considerations 270–7
  geochemistry 234–42
    analytical methods 235
    major elements 235–7
    REE contents 237–40
  laterization process 229
  phosphates in 242, 245, 247
  profile 228, **229**, 230–1
  REE carriers 240–9
    apatite 240, 242, **243**, **244**, 245, **246**
    bastnäsite-(La) 268
    calcite 242, **243**
    gorceixite 240, 247
    hydroxylbastnäsite 268–70
    iron hydroxides 240, 242, 247, 249
    monazite-(La) 270
    monazite-(Nd) 270

Laterites contd
    phosphates 242, 245, 247
    REE contents
        Araxá 240, **241**
        Catalão 237–40
    residual soil horizon 229–30
    saprolite formation 235, **236**
    saprolite horizon 230, 234, 239
    weathering 228, **229**, 231, 287
Lateritic bauxite, see Laterites
Laterization
    process 229
    REE process during 231–2
Latrappite *50*, 51, 52, 66–7
    analysis 60
Laumontite 106
Lavenite 151
Leaching 257–8, 273
Leucite Hills (Wyoming) 60–1
Leucophanite 151, *153*
Levinson criteria 21
Lherzolite 78, 80
Limonitic iron oxides 200
Liquid miscibility 93–5
Little Murun complex (Siberia) 61, 68
Lokkaite-(Y) 29
Loparite 49, *50*
    analysis 60
    calcian niobian 64
    in China 303
    composition *64*
    in Kola Peninsula *313*, 316, 317
    niobian 56, 64
    strontian *63*
    synthetic 53
Loparite perovskites 41
Loparite-(Ce) *152*, 155
Lovozero (Kola Peninsula, Russia) 153
Lucas Heights (Australia) 227
Lueshite 49, *50*, 51, 53, 57
    analysis 60
    composition *64*
    in Kola Peninsula *313*, 317
    perovskites from 68
Lueshite–loparite–perovskite system 63–4
Lueshite–loparite–tausonite system 65–6
Lujavrite 286

Macedonite 50, 67
Mckelveyite *314*
Magma
    geochemical behaviour of systems 11–12
    glasses 92, **93**
    silicate–carbonate liquid immiscibility 93–5
Magnesio–arfvedsonite 293
Magnesium ilmenite 136
Magnetite
    in China 282, 295, 297, 298, 300, 304
    in Thor Lake deposit 171
Mantle
    bastnäsite synthesis and stability range 81–4
    carbonate-rich melt formation 78–9
    glasses 92, **93**
    miscibility 93–5
    phase equilibrium studies in magmas 82–7
    REE in metasomatized rocks 79–80
    silicate–carbonate liquid immiscibility 93–5
    see also Mountain Pass carbonatite
Maoniuping deposit (China) 281, 282, 295–7
    bastnäsite–carbonate–baryte vein deposit 295–7
Marmara bauxite deposit (Greece) 263, 268, 270
Marmeiko deposit 265, 273
Mary Kathleen (Australia) *109*, 110
Melanite 302
Melanocerite 151, 152, 159, 160
Melanocerite-(Ce) *152*, 159, 160
Melilitites 59
    perovskites in 41, 63
Meliphanite 151, *153*
Metaloparite 57
Metasomatism 79–80
Metasomatites 315
    apobasic 316
Meteorites
    chondritic 9
    perovskite minerals in 41
Miaoya deposit 299–300
Miaskites 286, 302, 303

Microcline 297, 303, 304, 316
Microperthite 295
Mineralization
   apatite dolomite carbonatites 209–10
   in carbonatite 81
   in China 304–5
   hydrothermal 118–23, 197–8
   Kangankunde (Malawi) 209–10
   in Kola Peninsula 311–12, 316–17
      economic 323–4
   supergene 121
   Thor Lake (Canada) deposits 167–92
      Blatchford Lake igneous complex 167,168–9
      chemical analysis of minerals *174, 176–9*
      geological setting 168–9
      location of deposits 167
      magmatic events 169
      minerals present 171–3
      North T-zone deposit **170**, 171–9
      REE–Y–Be deposit **170**, 171–9
      T-zone deposit 167, 170–1
      timing of mineralization 169
      wall zone 171
Miscibility gap **94**, 95
Molybdenite 297
Monazite 80
   alteration 124, 125
   authigenic 259, *263*
   in carbonatites *195*, *196*, 198, **210**, 220
   in China 282, 286, 287, 294, 295, 297, 298, 304
   hydrothermal systems *107*
   in karstic deposits 275
   in Kola Peninsula *314*, 315, 317
   placer deposits 11
   precipitation 119, 120
   solubility 119, 120
   in Thor Lake deposit 171
Monazite-(Ce) 21, 127
   bright green 199, 204–5
   in carbonatite *195, 196*, 200, **201**, *202–3*, 216
   crystal structure 23, 29, **30**
   in karstic deposits 275
   in Oslo region *152*, 158
Monazite-(La) 21

in laterites 270
Monazite-(Nd) 127
   authigenic 263
   in carbonatite *195*
   in karstic deposits 275
   in laterites 270
Monsandrite 151
Mont St Hilaire (Canada) 153
Montenegro bauxite deposits 258, 259, 260
Monteregianite-(Y) 23
Montmorillonite 287
Monxonite 286
Mosandrite *313*, 317
   crystal structure 23
   discovery 328
   in Oslo region 152, 161
Mosandrite-(Ce) *152*
Mountain Pass (California)
   bastnäsite mines 227
   carbonatites 11, 77, 82, 193, *195*, 198, 216–17
   glasses 92, **93**
   hydrothermal/carbothermal deposition 87
   phase relationships 89–92
   primary magmatic crystallization 87
   synthetic carbonatite magma 87–93
Mt Weld (Australia) 227
   carbonatite in *196*
   laterites 231–2
Muscovite 287, 297, 304

Na-tremolite 293
Nagyharsány bauxite deposit (Hungary) 257, 268, 275
Natrolite 302, 303
Natroniobate 57, *313,* 317
Nazda deposit 268
Nb-rutile 282, 294
Nenadkevite 303
Neodymium-goyazite 263, *264,* **272**
Nepheline 302, 303
Nepheline syenite 63, 64, 286, 316
   perovskite minerals in 41
Neutron activation-induced β-autoradiography 345
Nisi deposit 265, 273

INDEX

Nkombwa Hill (Zambia) *196*, 204
Nordenskiöldine 151
Nordite-(Ce) *313*
Nordite-(La) 35, **36**, *313*
Nordmarkite 282, 295
North American shales composite *9*

Ocean systems 16
Oddo–Harkins effect 7, **8**
Oka (Quebec) 70
Oligoclase–albite 297, 300
Olivine lamproites 59
Olympic Dam (Australia) *109*
Orangeites, perovskites from 60, *61*
Orthite 286
Oslo region
    geological setting 153–5
    history of mineral discovery 151–2
    rare earth minerals 155–63
    see also individual minerals
Oxidation states 2, 3, *4*, 5
    iron in perovskites 58

Paraniite 105, 106
Parisite 80, 82
    alteration 125
    in carbonatites *195*, *196*, 198, 216
    in China 295, 298
    hydrothermal systems *107*
    in Oslo region *152*, 156, 157
    in Thor Lake deposit 173, 178, **179**
Parisite-(Ce) *152*, 156, 157, *195*, *196*, 216
Pegmatites
    alkali 286
    amazonite 67, 316, 317, **324**
    granitic and syenitic 10
    Kola Peninsula 312
    Oslo region 151–66
    rand-pegmatites 316, 317, 324, **324**
    syenite 151–66
Peralkaline systems
    granites 315
    Thor Lake 167–92
Peridotites 78–9, 233
Perovskites 41–76, 80
    in alkaline rocks 59–68
    analysis 59–60, 66
    cation ordering 48

    cerian *64*
    in China 298
    complex 48, **49**
    composition **54–7**
        analysis 59–60
        end-members 49, *50*, 57–8, 66
        evolution 64
        latrappite 66–7
        lueshite–loparite–perovskite system 63–4
        lueshite–loparite–tausonite system 65–6
        perovskite–loparite–tausonite system 60–3
    end-member composition 57–8, 66
    potential 49, *50*
    gravimetric analysis 59, 66
    high temperature superconducting structure **43**, 44
    latrappite 66–7
    loparite 41
    Mössbauer studies 58
    naturally occurring 41–2, 48–53
        compositional variation 48–52
        nomenclature 53–7
        potential end-member composition 49, *50*
        structure 52–3
    non-stoichiometric 44
    orangeites 60, *61*
    oxidation state of iron in 58
    paragenesis 59
    Pb-bearing 68
    REE distribution patterns **69**, 70
    Ruddlesden–Popper phases 42, **43**
    space 47, **48**, 49, **50**
    structural formulae 57–8
    structure
        distorted 44–7
        ideal 42, **43**, 44
        naturally occurring perovskites 52–3
    tolerance factors 47
    transformation twins 44
    vector representation 58
    Zr-rich 68
    see also individual minerals
Perrierite *313*
Perrierite-(Ce) *152*, 161–2

Perthite 300
Phenakite 171, 172
Phlogopite 286, 293, 297
Phonolite 286, 302
Phosinaite *314*
Phosphates
    in carbonatite *196*
    in laterites 242, 245, 247
Phosphorescence 6
Phosphorites 286–7
Pitchblende 303
Placers 11, 287–8
Plumbogummite 228
Plumbopyrochlore *313*
Pollucite 312
Polylithionite
    Ar geochronology 179–81
    in Thor Lake deposit 171, 173, 179–81
Polymignite 151
Precipitation 118–23
    crystallization of gangue minerals 121–3
    fluid mixing 120
    hydrothermal systems 118–23
        crystallization of gangue minerals 121–3
        fluid mixing 120
        interaction of fluids with wall rock 120–1
        temperature and pressure changes 119–20
    interaction of fluids with wall rock 120–1
    pyrrhotite 139
    secondary RE minerals 124
    temperature and pressure changes 119–20
Prehnite, hydrothermal systems 106
Proton probe analysis 328
Pyrite
    in China 294, 297, 298
    in Thor Lake deposit 172
Pyrochlore 80, 228
    in China 282, 294, 295, 297, 298, 300
    in hydrothermal environment 106
    in Kola Peninsula 312, *313*, 315, 316, 317
    in Oslo region 151, *153*

Pyrochlore-(Ce) *313*
Pyroxenites 59, 63, 233
Pyrrhotite 135
    precipitation 139

Quartz apatite rocks 211–16
Quartz fluorite 216
Quartz syenites 282
    porphyry 298

Rand-pegmatites 316, 317, 318, **324**
Rare earth elements mobility 12, 14–16
    during laterization 231
    in karstic deposits 274
Redox equilibria 15
Remobilization 124
Residual soil horizon 229–30
Rhabdophane 228
    in carbonatites *196*, 198
    in hydrothermal environment 106
    in Kola Peninsula *314*, 317
Rhyolite 295
Riebeckite 294, 297, 300, 302
Riebeckite alkalic granite 282
Rincolite 286, 302, 303
Rodeo de Los Molles (Argentina) *109*
Röntgenite 80, 82, 106
Rosenbuschite 151, *153*, 163
Rowlandite *313*
Ruddlesden–Popper phases 42, **43**
Rutile 230, 287, 297, 298

Sahamalite
    alteration 125
    in hydrothermal environment 105, 106
Sahamalite-(Ce)
    in carbonatite *195*, 216
    coordination polyhedra 25
Saima deposit (China) 301–3
St Honoré (Quebec) *196*
Salitre fenites 62
Samarskite *313*
San Giovani Rotondo deposit (Italy) 257
Saprolite
    formation 235, **236**
    horizon 230, 234, 239
Sarambi fenites 62
Sarnu Rajasthan (India) *196*, 221

Sazhinite *313*
Sazhinite-(Ce) 24
Scheelite 112
Schryburt Lake 63, 64
Semenovite-(Ce)
   crystal structure 34–5
   isomorphous substitution 33, 34–5, **36**
Senaite *153*
Shaxiongdong carbonatite complex 221
Shcherbakovite 62
Shonkinite 220
Siderite 172
Silcretes 230–1, 234, 239, 247, **248**
Silexites 230–1, 315, 323
Silicates *196*
Silicification 316
Smithsonite 300
Soil horizon
   residual 229–30
   truncation 231
Solubility 14
   apatite 119, 122
   monazite 119, 120
Speciation, hydrothermal fluids 116–17
Spectral properties 6
Spectroscopy
   absorption 328
   backscattered electron imagery (BEIs) 59, 110, **111**, 200, **201**, 329
   emission 328
   extended X-ray absorption fine structure (EXAFS) 6
   Fourier-transform infrared (FTIR) 345
   gamma-ray 328
   inductively coupled plasma emission (ICPES) 328, 331, 333–4
   inductively coupled plasma mass (ICPMS) 328, 331, 334
   isotope dilution mass (IDMA) 331, 334
   Môssbauer 6, 58
   proton-induced X-ray emission (PIXE) 335, 343
   scanning electron 329
      energy dispersive spectrometer 173
   secondary ion mass spectrometry (SIMS) 335, 344
   techniques 328
   X-ray 328

Sphalerite 172, 300
Sphene, *see* Titanite
Spodumene 312
Steenstrupine *314*
Stillwellite 105, *108*
Stillwellite-(Ce) *152*, 163
Stilpnomelane 300
Stitovo deposit 275
Strontianite
   in carbonatite *195*, *196*, 198, 200, **201**, 216, 220, 221
   in China 303
Strontiopyrochlore *313*
Submarine vents *114*, 117, 119, 120
Substitutions, isomorphous 22, 31–4
   coupled substitution
      at independent sites 33–4
      within rare earth site 33
   gadolinite–datolite group 34–7
   phase relationships **35**
   substitution accompanying vacancies 33
   substitution accompanying valence variation 34
Sulphides in carbonatites 198
Sulphur isotope analysis 187, *188*
Supergene mineralizations 121
Syenites
   aegirine 286
   aegirine–nepheline 302
   'contaminated' 299–300
   nepheline 41, 63, 64, 316
      in China 282, 286
      pegmatites 151–66
   ultrapotassic 59, 61, 68
Synchrotron X-ray fluorescence (SXRF) 328, 335, 343–4
Synchysite 80, 82
   alteration 124
   authigenic 259, *261*, 266, **267**, 274
   in carbonatites *195*, *196*, 198, 216
   in China 300
   hydrothermal systems *108*
   in Kola Peninsula *314*, 317
   in Thor Lake deposit 173
Synchysite-(Ce) *196*, 216, *314*
Synchysite-(Nd) 266, **267**, 274
   cerium content 274

Synchysite-(Y) *314*

Tadzhikite-(Ce) 35, **36**
Tadzhikite-(Y) *152*, 161
Tantalite 312
Tausonite 49, *50*, 54
  calcian **54**, 61, *63*
  synthetic 53
Tengerite *314*
Tengerite-(Y) 26, **28**, 29
Terbia 328
Thalenite *314*, 315, *321*, **321**, **322**
Thalenite-(Y) 23
Thor Lake (Canada) deposits
  Ar geochronology 179–81
  carbon and oxygen isotope analysis *187*
  enrichment 172–3
  fluid inclusion analysis 181–6
  grain sizes 173, **175**
  heating and freezing experiments 182
  homogenization 182, **183**, 184
  hydrogen and oxygen isotope analysis 184–6
  intermediate zones 172
  mineralization 167–92
    Blatchford Lake igneous complex 167, 168–9
    chemical analysis of minerals *174*, *176–9*
    geological setting 168–9
    location of deposits 167
    magmatic events 169
    minerals present 171–3
    North T-zone deposit **170**, 171–9
    REE–Y–Be deposit **170**, 171–9
    T-zone deposit 167, 170–1
    timing of mineralization 169
    wall zone 171
  quartz zone 171–2
  sulphur isotope analysis 187, *188*
Thorite
  in China 282, 295, 298
  formation 125
  in hydrothermal environment 106
  in Oslo region 151, *153*
  in Thor Lake deposit 171, 173
Titanite 10, 286, 287, 295–7, 302
  alteration 125

chemical zonation 110, 111
hydrothermal systems *108*, 131, 133
in Oslo region *153*
Tonalite 134
Törnebohmite 105, 106
Tourmaline
  in China 286, 287
  hydrothermal systems 106
Trachyte 286, 302
Transformation, *see* Alteration
Transportation
  aqueous geochemistry 116–18
  hydrothermal systems 136–7
Tremolite 297
Tritomite 151, *152*, 159
Tritomite-(Ce) *152*
Tundrite-(Ce) *314*
Tundulu carbonatite complex (Malawi) 124, *196*
Tveitite-(Y), coordination polyhedra 25

Uranmicrolite *313*
Uranothorianite 303
Uranpyrochlore
  in China 303
  in Kola Peninsula *313*, 317
Uyaynah (UAE) *196*, 197

Vermiculite 287
Vesuvianite scheelite 106
Vigezzite *313*
Vitusite *314*
Vitusite-(Ce) 25
Vlasenica deposit (Bosnia) 268, 275
Voloshin, A.V. 311–26
Vyuntspakhkite-(Y) *314*, *320–2*, **322**

Wakefieldite-(Ce) 29–30
Wasaki (Kenya) 197
Weathering 121, 198
  carbonatites 198, 208–9
  laterites 228, **229**, 231, 287
Weibyeite 157
Weishan deposits (China) 281, 282, *284*
  bastnäsite–carbonate vein deposit 297–8
Wigu Hill (Tanzania) *196*, 217, **218**
Wöhlerite 151, 153, *153*, 327

Wolframite 106, 112
Wulfenite 298

X-ray fluorescence (XRF) 59, 328, 331, 334–5
X-ray powder diffraction study 264–6
Xenotime 227
  in carbonatite *214*
  in China 286, 287, 296–7, 304
  hydrothermal systems *108*
  in Kola Peninsula 312, *314*, *321*, **322**
  in Thor Lake deposit 171, 172
Xenotime-(Y) *152*, 158
  coordination polyhedra 25, 29
  crystal structure 23, **29**, 30
  in Oslo region *152*, 158

Y-group 21
Yellowstone hydrothermal system 113, 114
Yttrialite *314*, 315, 316
Yttrialite-(Y) 23

Yttrofluorite 318
Yttrotungstite-(Y) 24

Zhonghuacerite 294, *314*
Zircon 10
  alteration 125
  in China 282, 286, 287, 295, 297, 300, 302, 304
  Ga-bearing 171
  hydrothermal systems *108*
  in Kola Peninsula 317
  in saprolite environment 230
  in Thor Lake deposit 171
Zirconolite
  alteration 125, **126**, 131
  chemical zonation 110, **112**
  compositional evolution **132**
  concentration during crystallization **133**
  hydrothermal systems *108*, 129–31
  in Oslo region 151, *153*, 155
Zirkelite *313*